The Microbiology of Poultry Meat Products

Food Science and Technology

A Series of Monographs

Series Editor

Bernard S. Schweigert
University of California, Davis

Advisory Board

S. Arai
University of Tokyo, Japan

C. O. Chichester
Nutrition Foundation,
Washington, D.C.

J. H. B. Christian
CSIRO, Australia

Larry Merson
University of California, Davis

Emil Mrak
University of California, Davis

Harry Nursten
University of Reading, England

Louis B. Rockland
Chapman College, Orange,
California

Kent K. Stewart
Virginia Polytechnic Institute
and State University
Blacksburg

The complete listing of books in this series is available from the publisher on request.

The Microbiology of Poultry Meat Products

Edited by

F. E. CUNNINGHAM
Department of Animal Sciences
Kansas State University
Manhattan, Kansas

N. A. COX
U.S. Department of Agriculture
Agricultural Research Service
Richard B. Russell Agricultural Research Center
Athens, Georgia

1987

ACADEMIC PRESS, INC.
Harcourt Brace Jovanovich, Publishers

Orlando San Diego New York Austin
Boston London Sydney Tokyo Toronto

COPYRIGHT © 1987 BY ACADEMIC PRESS, INC.
ALL RIGHTS RESERVED.
NO PART OF THIS PUBLICATION MAY BE REPRODUCED OR
TRANSMITTED IN ANY FORM OR BY ANY MEANS, ELECTRONIC
OR MECHANICAL, INCLUDING PHOTOCOPY, RECORDING, OR
ANY INFORMATION STORAGE AND RETRIEVAL SYSTEM, WITHOUT
PERMISSION IN WRITING FROM THE PUBLISHER.

ACADEMIC PRESS, INC.
Orlando, Florida 32887

United Kingdom Edition published by
ACADEMIC PRESS INC. (LONDON) LTD.
24-28 Oval Road, London NW1 7DX

Library of Congress Cataloging in Publication Data

The Microbiology of poultry meat products.

(Food science and technology)
Includes index.
1. Poultry—Microbiology.   2. Poultry plants.
I. Cunningham, Frank E.   II. Cox, N. A.   III. Series.
QR117.M53   1986   664'.93   86-14004
ISBN 0-12-199880-0 (alk. paper)

PRINTED IN THE UNITED STATES OF AMERICA

86 87 88 89   9 8 7 6 5 4 3 2 1

Contents

Preface		ix
1.	**Introduction** KENNETH N. MAY	
	Text	1
	Reference	3
2.	**Types of Microorganisms** DANIEL Y. C. FUNG	
	I. Microbial World	5
	II. Microbes Important in Poultry Microbiology	6
	III. Bacteria	7
	IV. Yeasts and Molds	15
	V. Microbial Growth and Death	17
	VI. Effects of Physical and Chemical Agents on Microbes	20
	VII. Conclusions	25
	References	26
3.	**Types of Microorganisms Associated with Poultry Carcasses** F. E. CUNNINGHAM	
	Incidence of Microorganisms	29
	References	38
4.	**Standard and Experimental Methods of Identification and Enumeration** JOHN R. CHIPLEY	
	I. Introduction	43
	II. Detection and Enumeration	44

	III. Sampling Techniques	79
	IV. Evaluation of Indicator Microorganisms as a Measure of Food Quality	87
	Appendix: Detection, Isolation, and Enumeration Methodologies—Reviews of Recent Studies	97
	References	138

5. Contamination of Poultry during Processing
J. S. BAILEY, J. E. THOMSON, and N. A. COX

I.	Introduction	193
II.	Bacteria Associated with Live Poultry	193
III.	Scalding	195
IV.	Bacteriology of Other Scalding Methods	197
V.	Picking	198
VI.	Evisceration	198
VII.	Chilling	200
VIII.	Bacteria Associated with Giblets	204
IX.	Summary	205
	References	206

6. U.S. Department of Agriculture Standards for Processed Poultry and Poultry Products
GEORGE J. MOUNTNEY

Text	213
References	221

7. Packaging of Processed Poultry
LARRY E. DAWSON

I.	Introduction	223
II.	Specific Package Function	224
III.	Package Materials and Their Development	226
IV.	Packaging Films for Poultry	227
V.	Packaging as It Affects Microbial Control on Poultry	228
VI.	Summary	231
	References	232

8. Radiation Preservation of Poultry Meat
RICHARD E. FAW and TSING-YUAN CHANG MEI

I.	Introduction	235
II.	A Brief History of Radiation Food Preservation	235
III.	Radiation Technology	236
IV.	Classification of Radiation Treatments	245
V.	Food Product Potentialities for Radiation Treatment	246
VI.	International Cooperation in Food Irradiation	247
VII.	U.S. Regulation of Radiation Food Processing	249
VIII.	Incentives for Radiation Preservation of Poultry Meat	250

IX. Medium-Dose Treatment of Poultry Meat	251
X. High-Dose Treatment of Poultry Meat	258
XI. Economics of Poultry Irradiation	268
References	271

9. Methods of Preservation of Poultry Products
F. E. CUNNINGHAM

I. Introduction	275
II. Heat Treatments	275
III. Use of Antibiotics	277
IV. Use of Sanitizers	277
V. Use of Food Acids	280
VI. Use of Controlled Atmosphere	281
VII. Use of Salts	282
VIII. Disinfectants	285
IX. Coatings	286
X. Polyphosphates	286
XI. Miscellaneous	287
References	288

10. Pathogens Associated with Processed Poultry
N. A. COX and J. S. BAILEY

I. Introduction	293
II. *Campylobacter*	294
III. *Clostridium*	299
IV. *Salmonella*	301
V. *Staphylococcus*	305
VI. *Yersinia*	307
References	310

11. Further-Processed Poultry Meat Products
G. W. FRONING

I. Introduction	317
II. Cutting	319
III. Deboning	319
IV. Precooked Poultry Products	324
V. Cured Poultry Products	329
References	330

12. Microbiology of Frozen Poultry Products
E. A. SAUTER

I. Introduction and History of Frozen Poultry Products	333
II. Microbiology of Frozen Poultry	333
III. Thawing of Poultry Products	336
IV. Microbial Buildup after Thawing	337
References	338

13. Microbiology of Cooked and Canned Poultry Products
W. J. STADELMAN

    I. Introduction   341
    II. Equipment   343
    III. Microorganisms and Canned Poultry Products   344
    References   345

14. Microbiology of Salted and Smoked Poultry Products
D. M. JANKY, J. L. OBLINGER, and J. A. KOBURGER

    I. Introduction   347
    II. Salted and Smoked Products   348
    III. Salting   349
    IV. Smoking   350
    V. Storage   351
    VI. Conclusions   353
    References   354

Index   355

Preface

Marketing poultry meat products is a major part of the food industry: Americans spend $12-13 billion annually for chicken products, according to USDA data. In 1985 consumer expenditures for chicken gained about 2% compared with the year before. Together with expenditures for turkey products of over $3 billion, consumers now spend between $15 and $16 billion annually for poultry meat products.

The continued growth and prosperity of the poultry meat products industry will depend, in large measure, on supplying the consumer with a safe and wholesome product. The intent of this book is to present current scientific knowledge on poultry meat and its products and to show the various disciplines required in the determination of poultry meat microbiology.

The first part of this book is devoted to the various types of microorganisms affecting poultry meat and their classification and identification. The second part addresses microbial considerations during processing and topics such as USDA regulations; contamination during processing; refrigerated, frozen, and canned storage problems; and proven methods of poultry preservation.

In common with all other major industries, future developments will depend on advances made in science and technology. Workers in every branch of the poultry meat industry will need the best possible training. It is with that belief that the book is primarily designed as a reference source for industry and quality assurance personnel, and for use in undergraduate courses in food science or microbiology. Whatever its final use, it should prove to be an invaluable education aid for the poultry industry.

F. E. Cunningham
N. A. Cox

CHAPTER 1

Introduction

Kenneth N. May

Holly Farms Poultry Industries, Inc.
Wilkesboro, North Carolina 28697

Most consumers of poultry meat products know little, if anything, about how such products are produced, processed, and delivered to retail and institutional outlets. Many consumers would be surprised, and perhaps horrified, to learn that fresh poultry meat has a diverse natural microflora including bacteria, yeast, molds, and viruses. Some of these organisms may be pathogenic to humans under certain circumstances.

Livestock and poultry are not grown in a sterile environment. Contrary to much publicity and a misconception in many consumer's minds, ducks, turkeys, and broiler chickens are not grown in cages; they are produced on litter such as wood shavings over dirt floors. Thus, live birds arriving at a processing plant have been subjected to a variety of microorganisms from dust, soil, feces, insects, humans, and, perhaps, rodents and reptiles. If any of the birds have come in contact with pathogenic microorganisms, then many others will likely have been contaminated through contact. This points up one reality—the processor cannot produce pathogen-free raw products if incoming birds carry pathogens.

From a practical standpoint, poultry processors must concentrate on reduction of total numbers of microorganisms to assure adequate shelf life during distribution and retail display and in the consumer's home. Fortunately, methods of reduction of total numbers of microorganisms also reduce pathogens in most instances.

May (1974) stated that "there exists in all plants numerous points of potential cross contamination. . . . In fact, anything that touches a single bird might cause contamination and anything that touches more than one bird might cause cross-contamination." Economics dictate that poultry be processed on a production line using automated mechanical equipment at every step where such equipment is available. In such systems, it becomes impossible to isolate individual carcasses from other carcasses or from equipment, employees' hands, process waste, packaging material, and other essential components of the system.

Birds commonly arrive at plants in coops or cage systems in which they are confined in a very limited space. The birds are removed from coops and are manually hung by their feet on overhead conveyors. The conveyor moves the birds through an electrical stunner, a killing machine, a scalder, and pickers. Cross-contamination possibilities exist from air, dust, bird to bird, scald water, equipment contact, condensate, and employee contact.

Most plants remove carcasses from the picking line by conveying them through a hock cutter which cuts the hock, thus releasing carcasses onto a chute and transfer belt. Carcasses are rehung on the evisceration lines by hand. Cross-contamination may result from equipment, employee contact, and process water contact, as well as from air and rail dust.

Carcasses must be opened mechanically, by hand, or by a combination of these. The viscera must be drawn and exposed for inspection. After inspection, edible giblets are removed and cleaned. Cross-contamination may result from bird-to-bird contact or contact with employees and inspectors, equipment (such as opening machines, oil-sac removers, drawing machines, lung removers, and head removers), or tools (such as knives, scissors, lung rakes, and shears). Process water, flumes, chutes, bins, pumps, and other necessary items in this department may also cause cross-contamination.

Cross-contamination may occur in final wash or chilling from process water, ice, and bird-to-bird contact. Employee tools (such as knives and saws) and packaging materials can also cause cross-contamination. The chilling system, often condemned by the news media, actually is an excellent decontaminating process.

Fortunately, pathogens usually find no place to persist and multiply from day to day in poultry plants. This is true, because of generally good sanitary design of equipment and cleanup requirements promoted by the U.S. Department of Agriculture.

If incoming birds are contaminated with pathogens, then some dissemination among carcasses is impossible to prevent. Thus, carcasses from the same plant will be negative for specific pathogens at times and positive at others.

The poultry meat products industry is continually working on practical methods to reduce total counts of microorganisms on products. Some efforts which appear useful at present are: more effective final washers, increased automation of hand operations, better inspection systems, plant prepackaging, and employee education programs.

This volume is the first single-source compilation of research in this segment of the food industry. Various chapters have been written by known authorities in the field. This book should be useful to students, microbiologists, food technologists, and any producer, distributor, or retailer of poultry meat products. It is hoped that use of the book will lead to poultry meat products of better microbial quality.

REFERENCE

May, K. N. 1974. Changes in microbial numbers during final washing and chilling of commercially slaughtered boilers. Poultry Sci. 53: 1281.

CHAPTER 2

Types of Microorganisms

Daniel Y. C. Fung

Food Science Graduate Program
Kansas State University
Manhattan, Kansas 66506

I. MICROBIAL WORLD

Microorganisms are minute living entities too small to be seen by the naked eye. They are ubiquitous in our environment and under proper growth conditions can affect our daily lives. All living things less than 0.1 mm in diameter fall into the world of microbes. The refinement of the lens by Antony van Leeuwenhock in the 1670s allowed him to discover the existence of these microbes. The microbial world includes viruses, bacteria, yeasts, molds, protozoa, and algae. Microorganisms are beneficial to humans through their roles in the various geochemical cycles such as the phosphorous, carbon, oxygen, nitrogen, and sulfur cycles; without microbes the earth would not be livable for humans. They are also important in various fermented foods such as wine, cheese, beer, vinegar, bread, and soybean products and in the production of industrially important acids, solvents, antibiotics, steroids, enzymes, etc. They can even be eaten as foods such as mushroom, yeast, and single-cell protein. However, they can also spoil our food supplies and cause devastating diseases in animals and humans.

From the standpoint of the microorganisms, however, they are simply trying to fulfill their biological activities of growth and perpetuation in the form of sexual and asexual reproduction. They need water, carbohydrate, protein, fat, minerals, vitamins, and the right combination of gases, temperature, pH, and other conditions in order to grow and to multiply. Therefore, there are no "good" or "bad" microorganisms in nature; it is according to how they affect us that we consider them harmful or beneficial (Fung, 1983).

II. MICROBES IMPORTANT IN POULTRY MICROBIOLOGY

A. Protista

Plants are nonmotile organisms with a rigid cell wall, containing chlorophyll, and have a well-defined nucleus within individual cells; starch is utilized as a food reserve. Animals are motile organisms with flexible cell membranes, no chlorophyll, and a well-defined nucleus within individual cells; glycogen or fat is utilized as a food reserve. There is another group of microscopic living organisms not conveniently classified as plant or animal, called the Protista. The "higher protists" are organisms with a defined nuclear membrane surrounding the hereditary materials, deoxyribonucleic acid (DNA) in the cell (eukaryotes). Examples of this are protozoa, algae, and fungi (yeasts and molds). The "lower protists," such as bacteria and blue-green algae are organisms that do not have a nuclear membrane surrounding DNA in the cell and are called prokaryotes.

B. Definitions of Bacteria, Yeasts, Molds, and Viruses

Bacteria are unicellular organisms ranging in size from 0.1 to 2.0 μm and occurring in cylindrical (rod), spherical (coccus), or curved (spiral) shapes. They divide asexually by binary fission. No sexual stage occurs, but genetic materials can be transferred from one cell to another.

Molds are multicellular filamentous fungi, highly structured and organized in morphology, reproducing by sexual and asexual stages in the form of sexual and asexual spores.

Yeasts are unicellular fungi occurring singularly with round or oval shape. Asexual reproduction is by budding, and sexual reproduction is by sexual spores.

Viruses are submicroscopic entities which cannot reproduce without a living host. They occur with a protein coat surrounding a coil of DNA or ribonucleic acid (RNA). The shapes and sizes vary greatly between groups of viruses.

III. BACTERIA

There are thousands of genera and species of bacteria in nature. Many are not related to poultry products. For a detailed description and classification of bacteria the new series of "Bergey's Manual of Systematic Bacteriology" (Holt, 1984) should be consulted.

Typically a bacterium has a cell wall, a cell membrane, and cytoplasmic materials inside the cell membrane. The cytoplasm contains all the metabolic, anabolic, and reproductive activities of the cell. The DNA in the cell is not confined within a nuclear membrane. In addition, cells may possess a capsular material outside of the cell wall. This thick layer acts as a protective barrier for the bacterium against phagocytosis by higher organisms such as protozoa or white blood cells of animal origin. Some bacteria, mainly the rod-shaped cells, are motile by means of polar or peritrichous flagella. Some bacteria can exist in the form of a spore. Spores are highly resistant to environmental conditions such as heat, cold, radiation, and chemical agents. These spores act as a dormant form of life of the bacterium. Under suitable conditions the spore wall will be dissolved, and the cell can germinate from the spore and continue a regular life cycle (vegetative stage).

The cell wall materials of different bacteria are important in the separation and classification of bacteria into two large groups. One group, the gram-positives and another group, the gram-negatives are differentiated by a staining procedure developed by Christian Gram in 1884. The gram-positive cells retain the purple color dyes (crystal violet) in the first step of the staining procedure, whereas the gram-negative cells will not retain the primary staining when alcohol is used to wash the cells but do retain the red counterstain (safranin). This staining procedure is usually one of the first tests used to classify or to identify a bacterial culture isolated from the environment. Once the cell has been determined as gram-positive or gram-negative and the shape of the cell determined under the microscope as a rod or coccus, a bacteriologist can further divide the bacterial groups into gram-positive or negative rods or cocci. Further differentiation of bacterial groups can be made by studying a variety of morphological, physical, and chemical characteristics, thereby identifying them to genus and species. Some of the more important diagnostic manuals are by Bailey and Scott (1974), Lennette *et al.* (1980), and Gerhardt *et al.* (1981). Figure 1 illustrates some typical morphologies and characteristics of bacteria.

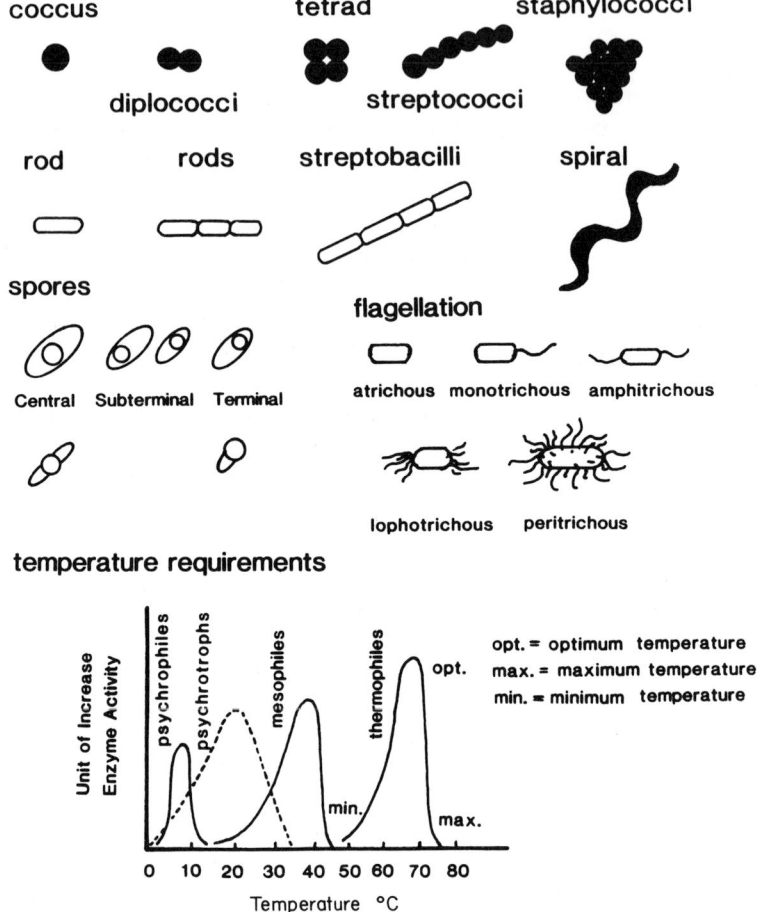

Fig. 1. Some typical morphologies and characteristics of bacteria.

The following are important groups of bacteria related to poultry microbiology.

A. The Lactic Acid Bacteria

1. General Characteristics and Importance

The lactic acid bacteria are gram-positive, non-spore-forming bacteria producing lactic acid as the major or sole product of fermentation. The homofermentative lactic acid bacteria produce two molecules of lactic acid from one molecule of glucose, with the generation of two adenosine triphosphate (ATP) molecules. The heterofermentative lactic acid bacteria produce one lactic acid, one ethanol, and one carbon dioxide molecule from one molecule of glucose, with the generation of one ATP molecule. These organisms lack iron porphyrin compounds, cytochrome C, and oxidative phosphorylation. They can all grow aerobically and anaerobically (facultative anaerobes), obtain energy from sugar, have limited biosynthetic ability, need a complex nutritional supply for growth, and do not produce an important enzyme called catalase, which can break down hydrogen peroxide to water and molecular oxygen. As a group they are important in spoilage of food by causing souring and discoloration. However, from a positive standpoint, they are very important in pickling, cheese making, fermented dairy products, and silage microbiology.

2. Major Divisions of Lactic Acid Bacteria

Streptococcus. The genus *Streptococcus* consists of cocci in chain formation and is homofermentative. It is divided into four groups: pyogens (pus forming), viridans, fecals, and lactics. The pathogenic streptococci are *S. hemolyticus, S. pyogens,* and *S. faecalis.* Many streptococci, such as *S. lactis, S. cremoris, S. thermophilus,* and *S. diacetilactis,* are very important in the dairy industry.

Pediococcus. The genus *Pediococcus* consists of cocci cells arranged in pairs and tetrads and is homofermentative. These bacteria produce large quantities of lactic acids and therefore are potential spoilage bacteria. They are also used as starter culture in the curing of meat. *Pediococcus cerevisiae* is the most important species in this genus as a starter culture.

Leuconostoc. The genus *Leuconostoc* consists of gram-positive cocci in chains and is heterofermentative. The cells produce a variety of compounds from sugar in addition to lactic acid and are used to produce flavor compounds in dairy products. Due to their ability to form gas, they can spoil food by gas production in products such as candy. The species *L. mesenteriodes* is able to form slime by the production of dextran in the presence of carbohydrate.

Lactobacillus. This is a heterogeneous group of organisms consisting of slender, gram-positive rods. One group is homofermentative (acid producers) and another group is heterofermentative (gas producers). They are important in dairy, meat, and silage fermentation but are undesirable in meat products due to the formation of large quantities of lactic acid. Important homofermentative organisms include *L. bulgaricus, L. acidophilus,* and *L. lactis.* Important heterofermentative organisms are *L. fermenti, L. buchneri,* and *L. brevis.*

B. Aerobic, Gram-Positive, Catalase-Positive Cocci

1. General Characteristics and Importance

This group of bacteria consists of gram-positive, cocci in pairs, short chains, and irregular clusters. They produce catalase and form acid from carbohydrate. Some are heat resistant and salt tolerant and produce a variety of color pigments in food and culture media. Some species grow well under refrigerated temperatures. Some are important human pathogens, while others are important in food spoilage.

2. Major Divisions of Gram-Positive, Catalase-Positive Cocci

Micrococcus. These are aerobic, gram-positive cocci occurring widely in nature, producing large quantities of catalase and oxidatively metabolize carbohydrate. Many species, such as *M. luteus* and *M. varians,* grow in refrigerated temperatures and are found in inadequately cleaned equipments.

Staphylococcus. This is an important genus of gram-positive cocci. The organisms are facultative anaerobic, occurring in irregular clusters. They are ubiquitous in nature and occur on human skin and contaminated food products. The organism produces a variety of extracellular enzymes and metabolites such as hemolysins, catalase, coagulase, thermoendonuclease, etc. The most important metabolite they produce is a group of heat-stable toxins called staphylococcal enterotoxins. These are lower-molecular-weight protein toxins (32,000) and can elicit violent vomiting, nausea, abdominal cramps, prostration, and diarrhea in susceptible human subjects. The bacteria are salt tolerant and can be found in meat, poultry, cream pies, potato salads, and many other foods. The most important species is *S. aureus.* This organism causes about one-third of all foodborne disease cases in the United States annually.

2. Types of Microorganisms

C. Spore-Forming Bacteria

1. General Characteristics and Importance

Spore-forming bacteria are gram-positive, spore-forming rods. The spore may occur in the center of the cell or at the near or extreme end of the vegetative cell. Motility of a vegetative cell is by peritrichous flagella. The spores are highly resistant to environmental conditions. The organisms produce a variety of odoriferous compounds from proteins and produce much gas from metabolism of carbohydrate. One group is characteristically anaerobic, and another group is aerobic. Both groups contain important human pathogens as well as important spoilage bacteria in food supplies, especially canned and vacuum-packaged goods.

2. Major Divisions of Spore-Forming Bacteria

Bacillus. This genus consists of a heterogeneous collection of aerobic, as well as facultative anaerobic, gram-positive spore-forming large rods, which are catalase positive. The majority of species of *Bacillus* are soil bacteria and can contaminate food through dust, air, soil, water, and animal carriers. They possess a wide variety of physiological activities including fermentation of sugar, peptonization of protein, hydrolysis of starch, and rennin coagulation of milk. They can be pathogenic, such as *B. anthracis.* Other species, such as *B. subtilis, B. cereus,* and *B. stearothermophilus,* are environmental contaminants. Some are thermophilic, and some are thermoduric.

Clostridium. These are gram-positive, anaerobic, spore-forming rods which are catalase negative and motile by peritrichous flagella. Some are highly anaerobic and can be killed in the presence of molecular oxygen, while others are aerotolerant. Clostridial species that are important in food spoilage include *C. butyricum, C. putrefaciens,* and *C. sporogenes.* From the standpoint of food hygiene, the species *C. botulinum* is most important, since this species produces a group of protein toxins called botulin (molecular weight of ca. 1 million), which can kill human subjects when it is ingested, and are known to be the most potent toxins produced by a biological system. Fortunately the toxin is sensitive to heat and can be inactivated by boiling for 10 min. The toxins are responsible for the disease called botulism. Another species, *C. perfringens,* although less toxic in nature, causes about one-third of the food poisoning cases in the United States. This organism produces a toxin in the intestines of human subjects and causes a mild diarrhea.

D. Gram-Positive, Irregularly Shaped Bacteria

1. General Characteristics and Importance

Gram-positive, irregularly shaped bacteria are environmental contaminants mainly involved in food spoilage, although some can cause severe disease.

2. Major Divisions of Gram-Positive, Irregularly Shaped Bacteria

Propionibacterium. This genus consists of small, anaerobic, pleomorphic rods (cells with irregular shapes), which are nonmotile. Some species produce pigments and can spoil foods (such as *P. jensenii*). From a practical standpoint the most important species is *P. shermanii*, which is responsible for the production of "eyes" in Swiss-type cheese. These organisms also impart desirable flavor in cheese fermentation.

Corynebacterium. This is a pleomorphic rod, arranged in a characteristic "Chinese letter" morphology under the microscope. The most important organism is *C. diphtheriae*, which is the causative agent of diphtheria. Some environmental contaminants with similar morphology are generally called coryneforms.

Microbacterium. This genus consists of small, aerobic, heat-resistant rods. They can withstand heat at 80°C for 10 min, which is very heat resistant for a vegetative cell. The most important species is *M. thermosphactum*, which may occur in vacuum-packaged meat products.

E. Gram-Negative, Polarly Flagellated Bacteria

1. General Characteristics and Importance

This group occurs in soil and water and contaminates food supplies. The flagella are present only at either pole of the rod cells. They are aerobic organisms, possessing proteolytic enzymes and a variety of pigments. Many are able to grow in refrigerated temperatures and are called psychrotrophic organisms. Although these psychrotrophs can grow and metabolize at 0–10°C, they will grow much faster at temperatures of 20°C. They cause spoilage of food and produce off-flavor and off-colors and sometimes form slimes.

2. Major Divisions of Gram-Negative, Polarly Flagellated Bacteria

Pseudomonas. This genus is the most important psychrotroph. Bacteria of this genus are extremely prolific in their metabolic activities and are able to degrade almost all organic compounds, including kerosene and gasoline. From the standpoint of food spoilage they produce slime, fluorescent compounds, and a variety of pigments on cold-stored foods such as poultry meat. They putrify food due to their prolific proteolytic activities. Important species include *P. fluorescens, P. aeruginosa, P. putida,* and *P. syringae.*

Acetobacter. This genus is important in the souring of foods by the production of acetic acid from alcohol. However, it is important for the commercial production of vinegar. The important species is *A. aceti.*

Photobacterium. Species of this genus, such as *P. phosphoreum,* can cause phosphorescence of meat and fish when incubated in suitable conditions.

Halobacterium. This genus can grow in high salt concentrations up to 30% salt. Certain species, e.g., *H. salinarum,* produce color compounds and spoil salty fish.

F. Gram-Negative, Short Rods

1. General Characteristics and Importance

This heterogeneous group of organisms consists of gram-negative small rods and, when motile, possesses peritrichous flagella. They are facultative anaerobic organisms and are found in water and soil and in human and animal populations. Some of these bacteria are very important pathogens, and others are important food spoilage organisms. All species metabolize glucose and produce acids. Metabolism of other carbohydrates varies greatly and is one of the bases for differentiation of these organisms. The most important family in this group is the Enterobacteriaceae, or organisms found in the intestines of humans or animals. According to the newest classification (Holt, 1984) there are 14 genera in this family (*Escherichia, Shigella, Salmonella, Citrobacter, Klebsiella, Enterobacter, Erwinia, Serratia, Hafnia, Edwardsiella, Proteus, Providencia, Morganella,* and *Yersinia*). Most genera in this family are important in poultry microbiology as potential human pathogens.

2. Major Divisions of Gram-Negative, Short Rods

Escherichia. This is the true fecal coliform. The type species is the well-known organism, *E. coli.* Although *E. coli* can cause disease, the most important role from a practical standpoint is that occurrence of this organism in food and poultry indicates the potential presence of other, more pathogenic, enteric organisms. Thus, the presence of this organism in cooked food is highly undesirable. These organisms produce acid and gas from lactose and are positive for indole and methyl red tests.

Shigella. This organism is very important in the Far East as an etiology for diarrhea. The organism *S. dysenteriae* is responsible for many water-related outbreaks in countries with poor sanitation. The organism is nonmotile.

Salmonella. *Salmonella* is a most important pathogenic bacterium in poultry products, since it is found in most poultry flocks. Some raw products of poultry origin harbor *Salmonella;* thus, all poultry products should be well cooked before consumption. The organism, when ingested in large quantities (10^6 *Salmonella* cells per gram of flesh) will cause a disease called salmonellosis, which includes nausea, vomiting, headache, chills, diarrhea, and fever. Mortality can occur in the very young, old, or infirmed. There are currently more than 2000 different serotypes of *Salmonella* reported. Each serotype is potentially pathogenic to humans, animals, or both. *Salmonella* is usually motile, does not ferment lactose, and can utilize citrate as a sole carbon source.

Citrobacter. This genus is closely related to *Salmonella* and can be differentiated from *Salmonella* by its inability to decarboxylate lysine. This organism has been found in humans, poultry, and mammals, and is also found in soil, water, sewage, and food.

Klebsiella. This organism is nonmotile and produces capsules. It occurs widely in nature and, in clinical cases, can cause pneumonia in humans by the species *K. pneumoniae.*

Enterobacter. This organism is also grouped in the coliform group. Many of the isolates can be found in water and soil. They can be distinguished from *E. coli* by their inability to produce indole and their negative response to the methyl-red test. They can also be differentiated from fecal coli by their inability to form large colonies at 45°C in solid media such as violet-red bile agar.

Erwinia. This genus is usually associated with plants and is a plant pathogen. The species *E. carotovora* can dissolve plant materials and can cause rotting of plants.

Serratia. This genus consists of some species, such as *S. marcescens,* which are easily recognized by their ability to produce a characteristic red color on agar. The organism occurs naturally in soil and water and may be an opportunistic human pathogen.

Hafnia. This genus is motile at room temperature but nonmotile at 35°C. Lactose is not fermented. The majority of the species can utilize citrate, acetate, and malonate as sole carbon sources. It occurs in the feces of mammals and birds and in sewage, soil, and water. The type species is *H. alvei.* This organism was formerly known as *Enterobacter hafnia.*

Edwardsiella. This genus is relatively inactive biochemically compared with other members of the Enterobacteriaceae, is usually isolated from cold-blooded animals, and is rare as a human pathogen. *Edwardsiella tarda* is the human pathogen of concern.

Proteus. This is an actively motile organism and can cause swarming on agar surfaces. *Proteus* metabolizes lactose slowly and produces an enzyme that can hydrolyze urea. This is an environmental organism and also occurs in a wide variety of animals and causes urinary tract infections in humans.

Providencia. This organism is motile but does not cause swarming on agar. It can be found in diarrhetic stools, urinary tract infections, and wounds, but it is not a major problem in poultry products.

Morganella. Species of this genus are also motile but do not cause swarming on agar. They are differentiated from *Providenica* by their inability to utilize citrate. They are not particularly important in poultry microbiology.

Yersinia. This genus includes the dreaded plague organism, *Y. pestis.* The important species in food microbiology is *Y. enterocolitica.* This is a human pathogen that is able to grow in vacuum-packaged food under cold storage and is considered to be one of the new "emerging food pathogens."

Some other bacteria are discussed in other chapters in this book.

IV. YEASTS AND MOLDS

Yeasts and molds belong to higher protista and are both included in the group called fungi. A mold can start from a spore or from a vegetative piece called thallus. As the cell elongates it forms a hypha. There are two types of hyphae. One type has no cross-wall (nonseptated or coenocytic), and another type has cross-walls (septated). As the hyphae grow longer, branches occur, and eventually they will form a mass of cellular materials called mycelium. From the mycelium two types of hyphae develop. One type is for nutrient absorption, and the other for reproduction. At the tip of the reproductive hyphae, specialized cellular arrangements occur and support the formation of asexual spores. Two distinctive types of asexual spores are found at the tip of the reproductive hyphae. One is the sporangiospores, which are housed in a saclike structure called the sporangium at the tip of the special hyphae called sporangiophore. Another type is the conidiospores, which are borne free but grow out of, and are attached to, the tip of the hyphae called the conidiophore. There are a variety of distinct morphologies and arrangements at the tip of the conidiophores which would allow a mycologist to identify the particular mold. The asexual spores can also occur in small groups called microconidia or in more organized units called macroconidia. The shapes of these macroconidia are also useful for the identification of the genus of molds.

When asexual spores are split out of the tip of a hypha they are called arthrospores. Some spores are formed in the middle of a piece of hypha and are called chlamydospores.

There are a variety of sexual spores. The most important ones in food microbiology are zygospores, ascospores, and basidiospores. Zygospores are formed when two different hyphae fuse and exchange genetic materials. Ascospores are formed by the fusion of two different sexual spores, and the daughter cells are housed in a saclike structure called the ascus. Basidiospores are formed at the tip of the basidium in a highly organized structure, such as a mushroom.

The fungi are divided into four subdivisions: Phycomycetes, Ascomycetes, Basidiomycetes, and Deuteromycetes.

Phycomycetes. The hyphae are nonseptated; asexual spores (sporangiospores) are housed in the sporangium; and the sexual spores are free zygotes. Important genera are *Mucor* and *Rhizopus*. Both genera are important as food spoilage organisms.

Ascomycetes. The hyphae are septated; asexual spores (conidiospores) are borne free; and sexual spores are housed in the ascus. Examples are *Aspergillus, Penicillium, Fusarium, Alternaria, Cephalosporium, Botrytis, Sporotrichum, Trichothecium,* and *Cladosporium.*

Included in this group are also the true yeasts *Saccaromyces, Candida,* and other yeasts.

Diagrams of commonly occurring yeast and mold are given in Fig. 2.

Basidiomycetes. The hyphae are septated; asexual spores are rarely observed; and sexual spores (basidiospores) occur at the tip of basidium. Common examples of this group are mushroom, toadstools, smuts, etc.

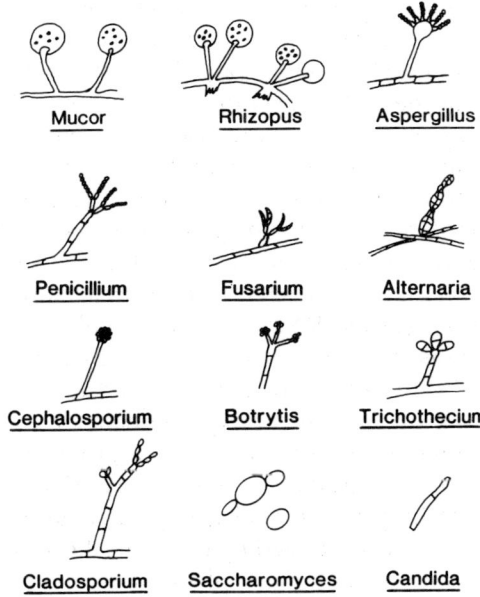

Fig. 2. Typical morphologies of some common yeasts and molds.

Deuteromycetes or *Fungi Imperfecti*. This is a catch-all grouping of all those fungi that show asexual, but not sexual, cycles. Once both sexual and asexual cycles are characterized, the culture will be assigned accordingly to one of the aforementioned subdivisions, most likely the Ascomycetes.

The role of yeasts and molds in poultry products is mainly in food spoilage. When poultry products are processed and kept at low temperatures for a long time there may be chances for spoilage from yeasts and molds. The only real food poisoning concern is the possible production of a carcinogenic toxin called aflatoxins by species of *Aspergillus* (*A. flavus* and *A. parasiticus*). Other yeasts and molds are chance contaminants.

V. MICROBIAL GROWTH AND DEATH

Growth of microorganisms can be viewed as growth in size of an individual organism or growth in numbers of a population. Since microorganisms are minute in size, growth is usually ascertained by monitoring numbers. Thus, in terms of studying growth and death of microorganisms, it is common to monitor population dynamics. This assumption is valid only for the truly unicellular organisms such as bacteria and yeasts. Occasionally, mold counts are also made, but the data should be interpreted with caution, since fragmentation of a fungal hypha or sporangium may result in hundreds of new fungal colonies.

Typically, a population of microorganisms in a suitable environment will first adjust to the environment of its existence. This is the lag phase of the growth curve. At this phase, nutrients are absorbed into the cells, and cellular materials (DNA, RNA, macromolecules, cytoplasmic materials, etc.) will increase per cell. No appreciable increase of cell numbers will be observed. As the cells start to multiply, they will follow the first-order reaction and grow in number exponentially. This is the log phase of growth. A population will not continue to grow forever, because of depletion of nutrients, accumulation of toxic materials, limiting of space for growth, and limiting of essential gases. A stage will be reached at which the number of cells increased by multiplication equals the number of cells dying. This is the stationary phase. Eventually, the number of dying cells exceeds the number of new cells being generated, and the population will experience a death phase. Figure 3 is a typical growth and death curve of a unicellular population.

The growth curve of microorganisms in foods is controlled by two major factors in a food system: the intrinsic factor and the extrinsic factors. Intrinsic factors include the nutritive value of foods, pH (acidity or alkalinity) of the food, water activities of the food (the availability of water for microbial growth), oxidation–reduction potential (the degree of aerobic and anaerobic conditions), presence of natural inhibitory substances in food, and the physical structure of the food. These factors are an inherent part of the food and will have direct influence on the type of organisms which can survive and grow in the particular food system. The extrinsic factors of food are those factors that we can manipulate. These include storage of food in various gaseous environments, high- and low-temperature storage, increase or decrease of relative humidity, prevention of physical stress of the food, and the length of time of storage. Combination of intrinsic and extrinsic factors of food will dictate the microbial growth in a particular food system (Jay, 1978).

Although growth and death of microbes are usually monitored by the classical method of growing the culture in nutrient agar after appropriate dilutions in buffer solutions and then counting the colonies after an adequate incubation time (i.e., the standard plate-count method), many other methods have been developed for monitoring microbial numbers

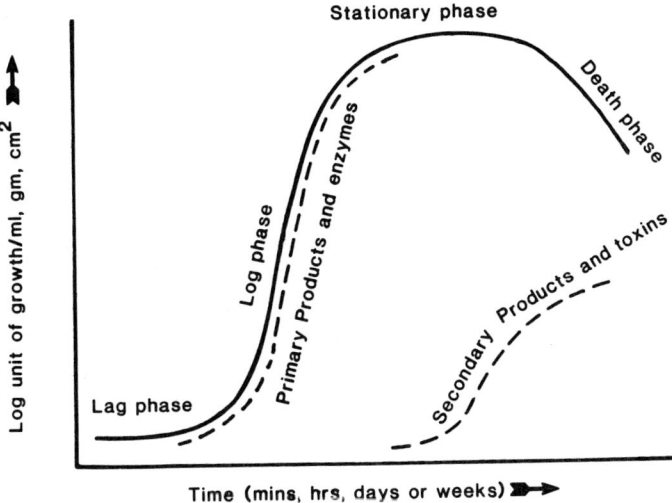

Fig. 3. Typical growth curve of microorganisms.

and activities. These include monitoring turbidity of the culture, increase of the total dry weight of a culture, increase in nitrogen content of a culture, increase in enzymatic activities, gas production, and DNA contents. All these are indirect evaluations of microbial growth and death. More recently several sophisticated procedures using sensitive instruments and computer-assisted data processing have been developed to monitor specific parameters of microbial activities. These include monitoring the amount of ATP in a population, the electric conductivity changes in food as a result of microbial activities, the evolution of radioactive carbon dioxide from a culture grown in a radioactive carbohydrate substrate, and minute changes of heat caused by microbial growth and metabolism (Fung, 1984).

All these newer methods are designed to rapidly estimate the total number of microbes in foods for the purpose of quality control and assurance. The aim is to have data on the microbial quality of food within a working day. All these methods, however, must relate back to the standard procedure of total viable cell counts. Therefore, for any method to be acceptable it must be able to demonstrate that the increase of activities or units of measurements is directly related to the increase of viable number microbes in a particular food system (Fung, 1984).

There is no specific guideline as to what is a "high" or "low" number of microbes in a particular food. The kind of foods involved, the condition under which they are processed or stored, and the particular kinds of organisms involved must be considered in making such a designation.

With those limitations in mind, a convenient guideline was developed by Fung et al. (1980) as follows: microbial load on meat surface is considered low when the count is log 0–2 colony-forming units (CFU) per square centimeter; intermediate, log 3–4 CFU/cm^2; high, log 5–6 CFU/cm^2; and very high, log 7 CFU/cm^2. Using this guideline samples with log 0–4 CFU/cm^2 would be considered as acceptable; samples with log 5–6 CFU/cm^2 would be considered questionable, and log 7 CFU/cm^2 would be considered unacceptable from a food spoilage standpoint. Counts above log 7.5 CFU/cm^2 will result in foods exhibiting odor and/or slime. This guideline does not include monitoring potential pathogens in foods such as *Salmonella*, *Staphylococcus aureus*, *Clostridium perfringens*, etc.

VI. EFFECTS OF PHYSICAL AND CHEMICAL AGENTS ON MICROBES

From the dawn of history, human beings have used various physical and chemical methods in preservation of food without knowing the actual mechanism of inactivation of microbes in various foods. After the discovery of microbes and the studies on various types of microbes in foods, scientists began to understand the rationales behind the success of some physical and chemical treatment of foods. The following are brief descriptions on the effects of important physical and chemical agents on microbes from a practical standpoint on food microbiology. More detailed discussions on specific physical and chemical effects on microbes in poultry will be presented elsewhere in this book.

A. Drying

This is probably the most widely used method of food preservation in the world. Many cereals, meat, fish, and fruit are dried and preserved for long periods of time. Drying can be accomplished by controlled dehydration or by sun drying. Controlled dehydration provides good sanitation and economy of space but is more expensive than sun drying. Drying removes water from food so that microorganisms cannot grow.

This process, however, does not sterilize the food. When moisture leaves the food, enzymatic activity ceases. However, when the food is rehydrated, microbial growth resumes and may result in spoilage of the food in question. Spores of bacteria and molds can survive long periods of time in dried form.

There are several ways of dehydrating foods. Hot air is commonly used as a medium to remove moisture from foods. Methods include cabinet, tunnel, kiln, and spray drying. Moisture can also be removed by direct contact on hot surfaces such as drum, vacuum, and heat drying.

B. Low-Temperature Effects on Microbes

Chemical reactions in general follow the $Q_{10} = 2$ principle, i.e., for every 10°C increase of temperature there is a doubling of chemical reaction rate. Conversely, for every 10°C decrease of temperature, the chemical reaction is reduced by half. Since microbial activities are closely related to chemical and enzymatic reactions, low-temperature effects on microbes are essentially a retardation process. Microbial activities will be greatly reduced at refrigeration temperature. The four common low-temperature preservations are chilling temperature (10–15°C), refrigeration temperature (0–10°C), slow freezing (reduced to −20°C within 3–72 hr), and quick freezing (reduce to −20°C within 30 min). Most microbes will not multiply in refrigerated temperatures but will survive for a long period of time and will start to grow again when placed in a warmer temperature. Freezing temperatures will kill some microbes but will preserve a variety of commonly occurring bacteria. Some microbes can grow slowly in chilling temperatures and refrigeration temperatures. These groups of microbes, the psychrotrophs, can grow and metabolize under these temperature ranges and can spoil foods. Microbes in general do not grow in freezing temperature. Slow freezing is more detrimental to microbes than quick freezing, because of formation of large, extracellular ice crystals which disrupt cell membranes as well as draw solute out of the cells. Quick freezing results in formation of small, intracellular crystals exerting minimum damage to the cells. Low-temperature preservation can be achieved by direct contact, immersion, nitrogen freezing, or air-blast freezing.

In low-temperature treatment, there is usually a sudden death of part of the population (10%), but thereafter, the microbes die off slowly. Characteristically, gram-positive cocci are more resistant to low temperatures than are gram-negative rods, and spores are more resistant than vegetative cells. Viruses (such as foot-and-mouth disease virus), nematodes (such as *Trichinella*), and protozoa cannot survive freezing.

C. Freeze-Drying

Freeze-drying (or lyophilization) is a relatively new technology. This process depends on the unique property of water—the triple point of water. Food is first frozen, then a vacuum is applied to the frozen liquid until sublimation occurs, and the water is evaporated. The resultant food retains its shape and porosity. The advantage of freeze-drying is that the food need not be refrigerated in storage, since microbes cannot grow in the absence of water. Upon hydration, however, the food will be subjected to the same type of spoilage as before. Although it kills some organisms, freeze-drying is the most efficient way of preserving bacterial cultures. The cost of freeze-drying is still high compared with other forms of food preservation and only a few products (such as freeze-dried coffee) are commercial successes.

D. High Temperatures

High temperature kills microorganisms by denaturation of protein and renders enzymes inactive. The purpose of high-temperature preservation of food is to kill pathogenic and spoilage organisms without destroying nutritive and other desirable properties of foods.

Pasteurization of food usually means 63°C (145°F) treatment of food for 30 min, or 72°C (161°F) for 15 sec. This time and temperature will destroy most vegetative cells, especially *Mycobacterium tuberculosis* and *Coxiella burnetti* in milk. A group of organisms called thermoduric organisms will survive pasteurization and will later be able to spoil the pasteurized food. Therefore, pasteurized food should be kept refrigerated and used as soon as possible.

Temperatures higher than pasteurization temperatures are called very high temperature (VHT) or ultrahigh temperature (UHT) treatments. These usually mean treatments of food at 130°C or higher for 1 sec or more. The problem of these high-temperature treatments is that flavor defects in products such as "sterile" milk. The advantage is that the products can be kept at room temperature for an extended period of time (42–90 days).

2. Types of Microorganisms

In order to achieve sterilization temperature, food is cooked under pressure in sealed containers. This is the practice of most commercial canning. The aim of commercial canning is to have time and temperature combinations such that *Clostridium botulinum* spores are destroyed. To date, commercial canning has a good record in the lack of outbreaks of botulism. Most of the problems are home canning of food using less than ideal temperature and time combinations. Many more people are killed by botulism due to improper home canning compared with commercial canning.

E. Radiation Effects on Microorganisms

There are two types of radiation to be considered—ionizing radiation and nonionization radiation. Ionizing radiation has the ability to ionize compounds along its pathway, thereby creating highly reactive free radicals. Nonionizing radiation does not have enough energy to achieve ionizing of molecules in its pathway, but it can cause vibration of molecules and generate heat in the process.

Alpha, beta, gamma rays and X-rays are able to destroy microorganisms. Alpha and beta particles can be stopped by paper and aluminum, respectively. Due to their poor penetration power, they are not used for food preservation. Gamma rays and X-rays can penetrate aluminum and can only be stopped by lead. These have good penetration power and have been tested for food preservations.

The killing effects of ionization particles can be attributed to breaking of chemical bonds of essential macromolecules such as DNA (the target theory) or the ionization of water which results in the forming of highly reactive radicals such as HO^-, H_2O^-, etc. (the free radical theory). These free radicals split carbon–carbon bonds of macromolecules in living organisms, thereby killing the organisms. Since no heat is generated in this form of destruction of microorganisms, radiation sterilization of food is commonly known as "cold sterilization." Radiation can also cause mutation of microorganisms in sublethal doses. The problem of radiation preservation of food includes formation of undesirable odors, high costs, and the potential of radiation hazards for personnel. A detailed treatment of radiation is given in Chapter 8.

Ultraviolet radiation (UV)—a nonionizing radiation—is useful only for surface sterilization of foods, since it has poor penetration power. The most effective wavelength is 2600 Å, which affects protein and nucleic acids. UV can cause mutation of organisms as well as destruction of organisms.

Microwaves are a form of nonionization radiation. When applied to foods, the waves, usually at 2540 MH_z, bounce back and forth through the food and create a tremendous vibration of asymmetric, dielectric molecules such as water. This vibration causes friction, and friction generates heat. It is the heat that is most responsible for killing microorganisms as well as for cooking food. There remain some questions as to the possibility of microwave killing effects in addition to mere heating (Fung and Cunningham, 1980). Since microwaves generate heat from the inside of the food, it has a peculiar killing pattern of microorganisms, i.e., more microorganisms are killed first in the center of the food compared with the outer surface of foods (Fung and Kastner, 1983). This is the reverse in conventional heating, in which organisms are killed first from the outside then the inside of the food. Microwave cooking is now a common method in homes and institutions for food preparations.

F. Chemical Effects on Microorganisms

Chemicals can kill microorganisms (bactericidal) or prevent the organisms from growing (bacteriostatic). Some are used to treat equipment and the environment (sanitizers, disinfectants, strong acids and bases, halogens, etc.), while others are for human and animal uses, such as antimetabolites, antibiotics, alcohols, and food additives (see Chapter 9).

Important points to consider in using chemical agents to destroy microorganisms include concentration (usually the higher the better), contact time (usually the longer the better), temperature (usually the higher the better), active ingredients, mode of action, costs, mode of application, and availability. From a practical standpoint, a technician must always follow instructions carefully and must have a continuous program to suppress harmful microorganisms in the particular environment.

The specific sites of destruction of microorganisms are quite varied, depending on the chemicals in question. Antimetabolites are used to block the metabolic pathways of microorganisms so that essential macromolecules cannot be formed or so nonfunctional molecules are formed instead of the normal molecules. For example, sulfa drugs will compete with a compound called para-aminobenzoic acid in the microorganisms and will create a nonfunctional folic acid. Without a functional folic acid many bacteria will die. Since human beings do not synthesize folic acid, sulfa drugs will not affect humans but will kill bacteria. Antibiotics act very specifically in different sites of microorganisms. For example, penicillin acts on cell walls; polymyxins act on cell membranes of bacteria; polyenes act on cell membranes of yeasts and molds; mitomycin prevents replications of new DNA; and chloramphenicol, tetracycline, and streptomycin disrupt translation of amino acids into functional polypeptides. For a detailed description of antibiotics on microorganisms, readers are recommended to read the book by Gale et al. (1972).

Heavy metals such as silver, mercury, copper, etc., have been used to combat microorganisms. These heavy metals bind sulfhydryl (—SH) groups in proteins and render enzymes inactive. Basically, this is a bacteriostatic action and is reversible in the presence of high H_2S levels. Carbolic acid has been used to destroy microorganisms. This compound coagulates protein and inhibits synthesis of DNA and RNA. This is one of the first compounds to be used as antisepsis in surgery rooms. Alcohols denature microorganism protein by dehydration. Although vegetative cells are effectively destroyed, spores are not affected by alcohol treatments. One advantage of alcohol treatment is that after treatment the compound evaporates from the surface such as human skin.

Halogens, such as chlorine, iodine, bromine, etc., are strong oxidizing agents. These agents are able to oxidize proteins and irreversibly damage enzymes in living systems.

Gases, such as ethylene and propylene oxides, bind reactive side groups, such as NH_2, SN, OH, and COOH, of macromolecules by a process called alkylation and kill the microorganisms. These gases are good in treatment of disposable items such as plastic petri dishes, pipettes, bags, etc., but are not used in foods because of residuals of these compounds.

Food additives are compounds added to foods for a variety of purposes such as flavor, texture, rheology, color, pH changes, water-holding capacity, emulsification, etc. Some of these compounds may directly affect microbiology of foods. Compounds allowed to be used in foods and controlled by governmental agencies are listed under the generally recognized as safe (GRAS) list. The newest list was recently published by Oser et al. (1984). The positive and negative aspects of these compounds are listed in other chapters in this book.

VII. CONCLUSIONS

The intent of this chapter is to present some general characteristics of important microorganisms relative to poultry and poultry products. Related issues concerning growth and death of microorganisms and some important physical and chemical effects on microorganisms are also presented. This information is presented so that a reader can appreciate more fully the impact of microorganisms on fresh and processed poultry products from the standpoint of food spoilage and foodborne disease.

For more detailed discussions on food microbiology readers are advised to consult the following texts: Banwart (1979), Frazier and Westhoff (1978), Jay (1978), Ayres et al. (1980), and Roberts and Skinner (1983). The books "Microbial Ecology of Foods, Vol. 1: Factors Affecting Life and Death of Microorganisms" and "Vol. II: Food Commodities," issued by the International Commission on Microbiological Specification for Foods (1980a,b) are particularly important references.

ACKNOWLEDGMENTS

Contribution No. 85-353-B, Department of Animal Sciences and Industry, Kansas Agricultural Experiment Station, Kansas 66506.

REFERENCES

Ayres, J. C., Mundt, J. O., and Sandine, W. E. 1980. "Microbiology of Foods." W. H. Freeman and Comp., San Francisco.
Bailey, W. R. and Scott, E. G. 1974. "Diagnostic Microbiology." C. V. Mosby Comp., St. Louis.
Banwart, G. J. 1979. "Basic Food Microbiology." AVI Publishing Co., Westport, CT.
Frazier, W. C. and Westhoff, D. C. 1978. "Food Microbiology," 3rd ed. McGraw-Hill, New York.
Fung, D. Y. C. 1983. Microbiology of batter and breading. In Suderman, D. R. and Cunningham, F. E. eds. "Batter and Breading Technology." AVI Publishing Co., Westport, CT. pp. 106–119.
Fung, D. Y. C. 1984. Rapid methods for determining the bacterial quality of red meats. J. Environ. Health 46(5): 226.
Fung, D. Y. C. and Cunningham, F. E. 1980. Effects of microwave cooking on microorganisms in foods. J. Food Protect. 43: 641.
Fung, D. Y. C. and Kastner, C. L. 1983. Microwave cooking and meat microbiology. Am. Meat Sci. Asso. Reciprocal Meat Conference Proceed. 35: 81.
Fung, D. Y. C., Kastner, C. L., Hunt, M. C., Dikeman, M. E. and Kropf, D. H. 1980. Mesophilic and psychrotrophic bacterial populations on hot-boned and conventionally processed beef. J. Food Protect. 43: 547.
Gale, E. F., Cundliffe, E., Reynolds, P. E., Waring, M. J., and Richmond, M. N. 1972. "The Molecular Basis of Antibiotic Action." John Wiley and Sons, N.Y.

Gerhardt, P., Murray, R. G. E., Costilow, R. N., and Wester, E. W., Wood, W. A., Krieg, N. R., and Phillips, G. B. 1981. "Manual of Methods for General Bacteriology." Am. Soc. Microbiol. Washington, D.C.

Holt, J. G. 1984. "Bergey's Manual Systematic Bacteriology," Vol I, II, III, IV. Williams and Wilkins. Baltimore, MD.

International Commissions on Microbiological Specifications for Foods. 1980a. "Microbial Ecology of Foods Vol I: Factors Affecting Life and Death of Microorganisms." Academic Press, N.Y.

International Commissions on Microbiological Specifications for Foods. 1980b. "Microbial Ecology of Foods. Vol. II: Food Commoditie." Academic Press. N.Y.

Jay, J. M. 1978. "Modern Food Microbiology," 2nd ed. Van Nostrand Reinhold. N.Y.

Lennette, E. H., Balows, A., Hausler, W. J., Jr., and Truant, J. P. 1980. "Manual of Clinical Microbiology," 3rd ed. Am. Soc. Microbiol. Washington, D.C.

Oser, B. L., Ford, R. A., and Bernard, B. K. 1984. 13. GRAS Substances. Food Technol. 38(10): 66.

Roberts, T. A. and Skinner, F. A. 1983. "Food Microbiology: Advances and Prospects." Academic Press, N.Y.

CHAPTER 3

Types of Microorganisms Associated with Poultry Carcasses

F. E. Cunningham

Department of Animal Sciences
Kansas State University
Manhattan, Kansas 66506

The quality of poultry meat is considered optimum immediately after processing, and maintenance of acceptable quality depends on initial microbial levels and measures taken to minimize growth of organisms. The two major concerns are control of spoilage organisms which cause consumers to reject the product due to odor or flavor and minimization of pathogenic organisms which may, under faulty handling, lead to a health hazard.

Poultry products have caused few foodborne illness outbreaks in recent years. In 1976, of 438 outbreaks reported, 18 were attributed to poultry products, 11 of which were of unknown etiology. The others were evenly divided among *Clostridium perfringens, Salmonella,* and *Staphylococcus.* In 1974 and 1975, 15 of 956, and 27 of 497 outbreaks, respectively, were attributed to chicken and turkey.

In most instances, careless handling of foods before consumption has been primarily responsible for the foodborne illnesses. Nevertheless, the poultry industry has a great responsibility in assuring that poultry and poultry products remain safe, wholesome, and highly acceptable.

INCIDENCE OF MICROORGANISMS

Many reports have been presented on numbers and kinds of different microorganisms on poultry meat; several studies were reported in an early bulletin by Dawson and Stadelman (1960). Several reviews (Ayres,

1966; Barnes, 1972; Brune and Cunningham, 1971) have presented information on the incidence and distribution of different organisms on processed poultry. As is true with several foods, microorganisms of importance on poultry may be nonpathogenic spoilage types or pathogens.

A. *Salmonella*

Because salmonellae are a public health hazard, it is essential that efforts be directed toward their elimination from eviscerated poultry. However, the entry for slaughter of salmonellae-contaminated birds (carriers) makes the elimination of salmonellae from these carcasses a difficult task. Galton et al. (1955) and Magwood et al. (1967) isolated salmonellae from flocks submitted for slaughter and reported that the organisms became widely disseminated in a plant during processing. Morris and Ayres (1960), however, found no buildup of salmonellae in inspected eviscerating plants.

Poultry is still considered to be the single most important source of salmonellae, ever since Edwards' report in 1939. In 1975, chickens and turkeys accounted for about 28% of the incidences of *Salmonella* in domestic animals and their environment, but only about 19% of the incidence of human foodborne isolations (Center for Disease Control, 1976). In the survey of turkey processing operations and turkey products by Bryan (1968a,b) the predominant serotypes changed from time to time in the plants. Of 23 serotypes recovered, *Salmonella anatum* and *S. san diego* were the most frequent isolates. In general, the most common isolates of *Salmonella* have been *S. typhimurium, S. heidelberg, S. enteriditis, S. infantis,* and *S. newport.* Sadler and Corstvet (1965) also found these serotypes as well as several others in a market survey of chickens and turkeys. *Salmonella typhimurium* and *S. san diego* were the most common isolates from a turkey flock. A total of more than 2000 turkeys yielded 108 positive *Salmonella* isolations (5.25%); for about 2800 chicken fryers, the incidence was only about 1%, and *Salmonella* was recovered from 7 (0.54%) of about 1280 chicken hens.

Bryan (1968a) also pointed out that the incidence of pullorum disease and fowl typhoid has decreased considerably, because of testing programs for these flock diseases. In 1956, about 70% of isolations from chickens and turkeys were *S. pullorum* and *S. gallinarum,* whereas in 1965, these serotypes decreased to about 6% of isolations from poultry.

Incidence of contamination of poultry by salmonellae in processing plants and retail markets was reported by Bryan (1968a) to vary from about 1 to 50% in various surveys made by different investigators. Processing operations, particularly defeathering by machine, are important sources for spread of salmonellae in the plant.

The persistence of salmonellae in flock contamination was shown by Dougherty (1976) to drop significantly during growth of broilers until processing at 8 or 9 weeks. Initial contamination of chicks was 37.5%, but at the time of processing, infection with salmonellae had decreased to only about 2.5%.

Campbell *et al.* (1983) reported that a 1979 survey of chicken eviscerating plants showed that chickens did not have a statistically significant reduced incidence of salmonella compared to chickens analyzed in 1969. In 1979, in nine plants surveyed, the incidence of salmonella on chickens at the exit of the chillers was 11.6%. In 1979, there was a 45% reduction of the incidence of salmonella in eviscerated chickens in the nine plants studied compared to the incidence of salmonella in eviscerated chickens in 1969.

From the reports reviewed, it appears that contamination of incoming poultry to processing plants is relatively low, but that spread of salmonellae occurs during processing. Market poultry becomes contaminated to varying degrees, undoubtedly due to a marked extent by differences in sanitation practices in processing plants as well as later handling and retailing conditions.

B. *Staphylococcus*

Although the main reservoir of *Staphylococcus aureus* is humans, poultry products have been incriminated in outbreaks of staphylococcus food poisoning (Bryan, 1968b). Poultry meat made into salads may be implicated in staphylococcus food poisoning. Since processed poultry is handled considerably, it may be expected that the bacteria may be contributed by processing-line workers as well as occurring as a "natural" contaminant of the skin and feathers of birds. Barnes (1972) stated that food poisoning from multiplication of *S. aureus* in poultry meat was relatively rare and generally came from the human food-handler. This implies that the poultry itself may not be as important a cause of staphylococcus food poisoning as is the worker, whether on the processing line or in a food service establishment. A number of staphylococcus strains isolated from poultry were of similar phage types as those from humans.

In a survey of eight processing plants in which about 2500 poultry livers were sampled, Genigeorgis and Sadler (1966) found that 22 of 35 coagulase-positive and 2 of 122 coagulase-negative strains of staphylococci were typable with human phage types. In addition, 7 of the 14 coagulase-positive strains tested were enterotoxigenic. Only about 1.6% of the livers demonstrated presence of coagulase-positive staphylococci.

In 1967, daSilva obtained swab and rinse samples of meat, equipment, and personnel in three federally inspected turkey processing plants producing rolls and roasts. Coagulase-positive staphylococci were isolated from several locations in the plants and from uncooked final products, but cooked rolls did not yield *S. aureus*. Hands of workers, the defeathering operations, and chill tank water contributed to increases in incidence of *S. aureus* on the raw poultry meat.

In the study made by Salzer et al. (1964), coagulase-positive staphylococci were recovered from 62 of 360 turkey livers before washing, for an incidence of 17%, but on only 3 after washing (less than 1%). The importance of water use in plant management procedures will be described in a later section, but is worthy of notice here with regard to adequate sanitation measures.

Minor and Marth (1972) reviewed several reports of incidence of staphylococci in barbecued chicken, "commercial" chicken, precooked frozen chicken, turkey pies, chicken salad, and other products. *Staphylococcus aureus* was isolated in low numbers from chicken during processing and from chill tank waters by Surkeiwicz et al. (1969). Bryan and McKinley (1974) did not recover *S. aureus* from cooked turkeys, but found that about 20% of 24 raw turkey carcasses harbored the organism; one isolate produced enterotoxin D.

Bryan (1976) also mentioned that healthy poultry tissues do not support prolonged growth of staphylococci, but bruised tissue will allow persistence of these organisms.

Bryan (1968a) reported that the natural flora of chicken, turkey, and beef pies may inhibit growth of *S. aureus*, although this organism can grow in chicken gravy at 10°C (50°F) but not at 5°C (41°F). Reports of isolations of staphylococci from poultry products are numerous; however, the significance of these findings in relation to foodborne disease potential of the products is not well defined.

An interesting observation of Scheusner and Harmon (1971) was that enterotoxin was not detected in pot pies in which good growth of inoculated *S. aureus* occurred. Mere presence of enterotoxigenic staphylococci does not prove that the toxin was produced in a food.

C. *Clostridium perfringens*

Clostridium perfringens is found in the intestinal tract of humans and animals, including chickens and turkeys (Banwart, 1969). Poultry processing operations can spread the organism (Barnes, 1972), which has been isolated from carcasses at various stages of processing. In England, in a survey reported in 1970 of clostridia isolated from poultry during processing, 91% were *C. perfringens* from chickens, and only 56% from turkeys.

In Walker's (1975) review, different surveys of poultry products are described in which *C. perfringens* was isolated from cooked poultry products as well as from birds during processing operations (Lillard, 1971, 1973). Boiled chicken contaminated with *C. perfringens* was reported by McClung (1945) to have caused an outbreak of food poisoning.

Bryan and McKinley (1974) recovered *C. perfringens* from surfaces of equipment contacting raw turkey and from 3 of 13 cooked turkeys. They stated that a greater number of raw turkeys (10 of 24) were contaminated with *C. perfringens* than with *S. aureus* (5 of 24). These investigators also gave recommendations for insuring safety of turkey products in school food service, e.g., cooking to a minimum internal temperature of 74°C and refrigerating rapidly after cooking as well as before.

D. *Clostridium botulinum*

Raw agricultural commodities, including poultry, have not usually been considered as important causes of botulism, but their heat-processed, low-acid food counterparts may be hazardous in certain instances. Tompkin and Christiansen (1976) reviewed outbreaks and means of prevention of botulism. Perishable foods such as poultry meat which are dependent on refrigerator freezing for protection are rarely involved in botulism outbreaks.

E. Coliforms

Because coliforms are facultative anaerobes, they may outgrow psychrotropic, gram-negative rods on refrigerated, vacuum-packaged poultry (Arafa and Chen, 1977). *Enterobacter* species may predominate over *Pseudomonas* and other typical spoilage types under conditions imposed by vacuum packaging.

F. *Bacillus* Species

Walker (1976) reviewed the role of *Bacillus* species in foods and mentioned spoilage of duckling by *B. subtilis*. The duckling carcasses were moist during refrigeration at 10°C, whereas 4°C storage controlled odor and surface slime.

An extensive review of *B. cereus* by Goepfert et al. (1972) indicated that the organism has usually been considered to be a harmless saprophyte but also may be implicated in foodborne diseases of humans. Whether a fatal infection of a zoo tiger by *B. cereus* can be extrapolated to poultry meat is debatable, but the organism has been isolated in food poisoning outbreaks involving meats. Chicken soup contaminated with high numbers of bacilli caused food poisoning in Italy. The high incidence of spore-formers in spices may be related to *B. cereus* food poisoning in highly seasoned meat dishes.

G. Coryneforms

Saprophytic coryneform bacteria have been found in poultry and poultry deep litter. Such organisms have been termed "diphtheroids" when found on the surface of freshly killed chickens. Genera include *Corynebacterium* and *Microbacterium*. In England, Barnes (1960) recovered psychrophilic *Corynebacterium* spp. from birds entering a processing plant. A large percentage of the organisms isolated by Salzer et al. (1964) from turkey giblets was classified as *Corynebacterium* and *Brevibacterium*. Many of the organisms could grow at 5°C, but their role in spoilage of poultry meat was not apparent.

H. *Campylobacter*

Campylobacter fetus ssp. *jejuni* (Smibert, 1974) has recently been recognized as an important microorganism causing diarrheal disease in humans. These bacteria, also known as "related vibrio's" (King, 1957) are highly motile, small, spiral-shaped or curved gram-negative rods belonging to the genus *Campylobacter* of the family Spirillaceae. *Campylobacter fetus* ssp. *jejuni* can be distinguished from other *C. fetus* subspecies by the ability to grow in culture media under microaerophilic conditions at 42–43°C, whereas multiplication at 25°C does not occur.

Campylobacter fetus ssp. *jejuni* is known as the causative agent of vibrionic hepatitis in poultry (Voute and Grimbergen, 1959) and "winter dysentery" in sheep (Firehammer, 1965). Since a selective culture technique for detection of these bacteria in human feces is available (Skirrow, 1977), the enteropathogenic character of this campylobacter subspecies became evident; their presence mostly as the only enteric pathogen was demonstrated in 3–14% of the fecal samples from diarrheic patients. In some comparative studies the campylobacter isolation rate was even higher than salmonellae isolation rates (Blaser et al., 1979a; Bruce et al., 1977; Butzler, 1978; Pearson et al., 1977; Severin, 1978; Skirrow, 1977; Telfer Brunton and Heggie, 1977). The organism is seldom found in healthy persons (Severin, 1978; Skirrow, 1977).

In many instances the course of campylobacter enteritis is more severe and of longer duration than that of samonellosis. The incubation period varies from 2 to 7 days, averaging 3 days. The prodromal symptoms of fever, malaise, headache, arthralgia, and abdominal pains are followed by profuse, watery diarrhea with blood, pus, or mucus, accompanied by abdominal pain. Sometimes symptoms are so severe that admission to a hospital is required. The diarrhea lasts from 1 to 3 days, patients are away from work for 10–14 days. Most cases recover spontaneously; severe cases are treated with antibiotics, preferably erythromycin (Severin, 1978; Skirrow, 1977). Only during a short period after their recovery are patients still excreting campylobacters.

Campylobacter fetus ssp. *jejuni* is found widely spread in nature and is isolated from wild and domestic animals as well as from the environment (Knill et al., 1978). Oosterom (1980) found 61% of 182 intestinal content samples of normal slaughtered pigs contaminated, whereas the organism was isolated from 14–91% of cecal and rectal samples and carcasses of healthy chickens (Bruce et al., 1977; Grant et al., 1980; Riberiro, 1978; Simmons and Gibbs, 1977, 1979).

Contact with infected animals may play an important role in transmission of *C. fetus* ssp. *jejuni* to humans (Blaser et al., 1979a, 1980). Also, person-to-person infection, especially among young children, is reported (Severin, 1978; Skirrow, 1977).

Ingestion of untreated water, probably contaminated from animal sources, was incriminated in some campylobacter enteritis outbreaks (Skirrow, 1981; Tiehan and Vogt, 1978). *Campylobacter* may survive in fresh water for 11 days to 4 weeks at 4°C and for 2 days at room temperature (Blaser et al., 1980; Robinson et al., 1979).

Though multiplication of *C. fetus* ssp. *jejuni* in foodstuffs stored at normal temperatures is unlikely, transmission by way of contaminated foods happened in several outbreaks. Campylobacter enteritis caused by ingestion of unpasteurized milk was noticed by Blaser et al. (1979b), Robinson et al. (1979), Taylor et al. (1979), and McNaughton et al. (1981). Milk may be contaminated by the fecal route or by infected udders; severe campylobacter mastitis could be initiated by intramammary inoculation of only few bacteria (Lander and Gill, 1979). *Campylobacter* may survive in milk for 21–164 days at 4°C and for 61 days at room temperature (Blaser et al., 1980; Robinson et al., 1979), but its survival after proper pasteurization processes is unlikely.

Campylobacter fetus ssp. jejuni was also isolated from commercially processed poultry (Bruce et al., 1977; Simmons and Gibbs, 1977; Smith and Muldoon, 1974). The organisms may survive commercial processing procedures, including chlorination (Simmons and Gibbs, 1979), as well as storage for 2–5 days at 3°C and 1–2 months at about −20°C (Robinson et al., 1980; Smith and Muldoon, 1974). Contact with contaminated raw chicken in the preparation stage and ingestion of undercooked, contaminated poultry products are suggested as causes of campylobacter enteritis (Brouwer et al., 1979; Severin, 1978; Skirrow, 1977). Normal heating procedures in preparation of chicken will kill campylobacters.

A study of the occurrence of C. fetus ssp. jejuni in chicken parts from retail outlets and poultry slaughterhouses (Hartog et al., 1981) found that chilled chicken livers at retail outlets were heavily contaminated with C. fetus ssp. jejuni: 19 out of 52 samples (36%) contained 10^3 colony-forming units per gram (CFU/g); 7.4×10^4 CFU/g was the highest campylobacter colony count. Other chilled chicken parts and further-processed poultry products were contaminated at a lower degree: 3 of 143 samples (2%) contained 10^3 CFU/g. The number of viable campylobacters in chicken livers is reduced considerably by freezing and frozen storage. The organisms could be isolated from only 2 out of 44 frozen chicken liver samples (4%).

Campylobacter fetus ssp. jejuni was found to be widespread in four poultry slaughterhouses; 4 out of 60 chicken carcasses (73%) harbored the organism. Cooling-water seems to be an important reservoir of these bacteria: 100–3000 CFU/ml were demonstrated, and survival over long periods at low temperature is possible. The organisms were isolated from "air samples" in two slaughterhouses. Equipment was found contaminated as well (Hartog et al., 1981).

Campylobacter fetus ssp. jejuni strains isolated from human patients, poultry products, and poultry slaughterhouses were indistinguishable. Reliable serotyping systems are not available at present, but it seems reasonable to assume that C. fetus ssp. jejuni strains isolated from poultry are human enteropathogens. Shanker et al. (1982) reported a high incidence of Campylobacter jejuni on broiler carcasses in Australia. Campylobacter jejuni was isolated from 18 of 40 processed broiler carcasses and 134 of 327 cloacal swabs obtained at four processing plants in Sydney, Australia. Three of four flocks examined carried C. jejuni; 82% of the chicken and 98% of the human isolates from the area were of identical biotypes. Therefore, more data on the behavior of these bacteria in foods are needed. Measures to lower the contamination level of chicken livers need to be taken.

I. *Yersinia enterocolitica*

Yersinia enterocolitica is a gram-negative, non-spore-forming, facultatively anaerobic bacterium that can multiply at refrigeration temperatures (0–4°C). The organism is associated with a variety of human illnesses (Bottone, 1977), of which enterocolitis is the most frequent manifestation of a *Y. entercolitica* infection. There is a steady increase in the number of isolations of *Y. enterocolitica* in recent years, not only from clinical materials but also from water and foods. *Yersinia enterocolitica* is a common organism in many types of food such as milk (Hughes, 1979; Severin, 1978), meat (Hanna *et al.*, 1976, 1980; Inou and Kurose, 1975; Leistner *et al.*, 1975; Schiemann, 1979), seafoods (Kapperud and Jonsson, 1976; Peixotto *et al.*, 1979), and also in water (Keet, 1974; Lassen, 1972).

Though yersiniosis is frequently suspected to be a foodborne disease, a possible link between *Y. enterocolitica* in foods and human infections is not yet clear. There has been only one well-documented outbreak of foodborne infection of *Y. enterocolitica* caused by consumption of contaminated chocolate milk (Black *et al.*, 1978). More epidermiological studies are needed to obtain a better knowledge of possible modes of transmission and ecology of *Y. enterocolitica*.

Some authors distinguish between "clinical" and "environmental" *Y. enterocolitica* strains. The clinical strains belong to some serotypes, of which serotypes 0:3 and 0:9 are isolated frequently in Europe, Japan and Canada, while 0:8 is the most prevalent clinical serotype in the United States (Stern and Pierson, 1979). These strains display a proven pathogenicity for humans, while the environmental strains probably do not have any pathogenic significance in human beings. It is suggested that swine in particular are one of the main reservoirs of the pathogenic *Y. enterocolitica* serotype 0:3 (Hanna *et al.*, 1980; Wauters *et al.* 1976).

Yersinia enterocolitica strains can be arranged according to their biochemical reactions into groups, or "biotypes." The classification scheme according to Nilehn and Wauters (1979) mentions five different biotypes. Most of the food isolates are of biotypes different from the typical clinical strains; the former strains usually belong to biotype 1 (Wauters, 1973).

It is difficult to detect small numbers of *Y. enterocolitica* among large numbers of other microorganisms. Several enrichment systems are being tested.

Lee et al. (1980), using selenite media for the recovery of Y. enterocolitica from meats, concluded that restriction of the sample size of a blended meat suspension to 0.2 g of meat per 100 ml of selenite broth was essential.

From the fact that a relatively large number of media is in use, one may conclude that an optimal enrichment and isolation medium has not yet been found. Some media do not recover all clinically important serotypes. Several authors recommend the simultaneous use of two or more enrichment and isolation media (Hanna et al., 1980; Stern and Oblinger, 1980).

Phosphate-buffered saline solution with and without addition of 1% sorbitol and 0.15% bile salts, modified Rapaport medium, and selenite broth were used as enrichment media for isolation of Y. enterocolitica from poultry products (deBoer and Hartog, 1981). Seventy-three of 108 (68%) samples of poultry products contained Y. enterocolitica. It was isolated most frequently using phosphate-buffered saline solution with sorbitol and bile salts. Nearly all Y. enterocolitica strains were isolated on MacConkey agar after KOH treatment of the enrichment media just before plating. Serotyping showed that the isolated strains belonged to the category of the nonpathogenic, so-called environmental strains.

REFERENCES

Arafa, A. S., and Chen, T. C. 1977. Ascorbic acid dipping as a means of extending shelf life and improving microbial quality of cut-up broiler parts. Poultry Sci. 56: 1345. S.A.A.S. Abstr.

Ayres, J. C. 1966. Microbial implications in the processing of poultry. Proceedings: 2nd Intern. Congr. Food Sci. Technol., Aug. 22–27, Warsaw, Poland.

Banwart, G. J. 1969. Rapid detection of Salmonella in turkey rolls and on fresh chicken parts. Poultry Sci. 48: 1529.

Barnes, E. M. 1960. Bacteriological problems in broiler preparation and storage. Royal Soc. Health J. 80: 145.

Barnes, E. M. 1972. Food poisoning and spoilage bacteria in poultry processing. Vet. Rec. June 24, 720.

Black, R. E., Jackson, R. J., Tsai, T., Medvesky, M., Shayegani, M., Feeley, J. C., Macleod, K. I. E., and Wakelee, A. M. 1978. Epidemic Yersinia enterocolitica infection due to contaminated chocolate milk. New Eng. J. Med. 297:76.

Blaser, M. J., Berkovitz, J. D., LaForce, F. M., Cravens, J., Reller, L. B., and Wang, W. L. L. 1979a. Campylobacter enteritis: Clinical and epidemiologic features. Ann. Intern. Med. 91: 179.

Blaser, M. J., Cravens, J., Powers, B. W., LaForce, F. M., and Wang, W. L. L. 1979b. Campylobacter enteritis associated with unpasteurized milk. Am. J. Med. 67: 715.
Blaser, M. J., Hardesty, H. L., Powers, B. W., and Wang, W. L. L. 1980. Survival of *Campylobacter fetus* ssp. *jejuni* in biological milieus. J. Clin. Microbiol. 11:309.
Bottone, E. J. 1977. *Yersinia enterocolitica*: A panoramic view of a charismatic microorganism. Crit. Rev. Microbiol. 5: 211.
Brouwer, R., Mertens, J. J. A., Siem, T. H., and Katchaki, J. 1979. An explosive outbreak of *Campylobacter enteritis* in soldiers. Antonie van Leeuwenhoek 45: 517.
Bruce, D., Zochowcky, W., and Ferguson, J. R. 1977. *Campylobacter enteritis*. Brit. Med. J. pp. 1219.
Brune, H. E., and Cunningham, F. E. 1971. A review of microbiological aspects of poultry processing. World's Poultry Sci. J. 27:223.
Bryan, F. L. 1968a. What the sanitarian should know about staphylococci and salmonellae in non-dairy products. II. Salmonella. J. Milk Food Technol. 31: 131.
Bryan, F. L. 1968b. What the sanitarian should know about staphylococci and salmonellae in non-dairy products. I. Staphylococci. J. Milk Food Technol. 31: 110.
Bryan, F. L. 1976. *Staphylococcus aureus*. pp. 12–128. *In:* DeFigueiredo, M. P., and Splittstoesser, D. F. (eds.), "Food Microbiology: Public Health and Spoilage Aspects." Avi Publishing Co., Westport, Connecticut.
Bryan, F. L., and McKinley, T. W. 1974. Prevention of foodborne illness by time-temperature control of thawing, cooking, chilling and reheating turkeys in school lunch kitchens. J. Milk Food Technol. 37: 420.
Butzler, J. P. 1978. Infection with *Campylobacter*. pp. 214–239. *In* J. D. Williams (ed.) "Modern Topics in Infection." W. Heinemann, London.
Campbell, D. F., Johnston, R. W., Campbell, G. S., McClain, D., and Macaluso, J. F. 1983. The microbiology of raw, eviscerated chickens: A ten year comparison. Poultry Sci. 62: 437.
Center for Disease Control. 1976. "Salmonella Surveillance. Annual Summary, 1975." U.S. Department of Health, Education and Welfare, Center for Disease Control, Atlanta, Georgia.
daSilva, G. A. N. 1967. Incidence and characteristics of *Staphylococcus aureus* in turkey products and processing plants. PhD Thesis, Iowa State University, Ames, Iowa.
Dawson, L. E., and Stadelman, W. J. 1960. Microorganisms and their control on fresh poultry meat. Tech, Bull. 278, North Central Regional pub. 112. Michigan State University, East Lansing, Michigan.
deBoer, E., and Hartog, B. J. 1981. Occurrence of *Yersinia enterocolitica* in poultry products. *In* Proceedings of the Fifth European Symposium "Quality of Poultry Meat," Spelderholt Institute for Poultry Research, Beekbergen, the Netherlands.
Dougherty, T. J. 1976. A study of salmonella contamination in broiler flocks. Poultry Sci. 55: 1811.
Edwards, P. R. 1939. Incidence of salmonella types in fowl in the U.S. Proc. Worlds Poultry Congr. pp. 271.
Firehammer, B. D. 1965. The isolation of vibrios from ovine feces. Corness Vet. 55: 482.
Galton, M. M., Mackel, D. C., Lewis, A. C., Haire, W. C., and Hardy, A. V. 1955. Salmonellosis in poultry and poultry processing plants in Florida. Amer. J. Vet. Res. 16: 132.
Genigeorgis, C., and Sadler, W. W. 1966. Characterization of strains of *Staphylococcus aureus* isolated from livers of commercially slaughtered poultry. Poultry Sci. 45: 973.
Goepfert, J. M., Spira, W. M., and Kim, H. U. 1972. *Bacillus cereus* food poisoning organism. A Review. J. Milk and Food Technol. 35: 213.

Grant, I. H., Richardson, J. J., and Bokkenheuser, V. D. 1980. Boiler chickens as potential source of *Campylobacter* infections in humans. J. Clin. Microbiol. 11: 508.

Hanna, M. O., Zink, D. L., Carpenter, Z. L., and Vanderzant, C. 1976. *Yersinia enterocolitica*-like organisms from vacuum packaged beef and lamb. J. Food Sci. 41: 1254.

Hanna, M. O., Smith, G. C., Hall, L. C., Vanderzant, C., and and Childers, A. B. 1980. Isolation of *Yersinia enterocolitica* from pig tonsils. J. Food Prot. 43: 23.

Hartog, B. J., Severin, W. P. J., and deBoer, E. 1981. Epidermiology of campylobacteriosis and occurrence of *Campylobacter fetus* ssp. *jejuni* in poultry products. *In* Proceedings of the Fifth Symposium "Quality of Poultry Meat," Spelderholt Institute for Poultry Research, Beekbergen, the Netherlands.

Hughes, D. 1979. Isolation of *Yersinia enterocolitica* from milk and a dairy farm in Australia. J. Appl. Bacteriol. 46: 125.

Inou, M., and Kurose, M. 1975. Isolation of *Yersinia enterocolitica* from cow's intestinal contents and beef meat. Jpn. J. Vet. Sci. 37: 91.

Kapperud, G., and Jonsson, B. 1976. *Yersinia enterocolitica* in brown trout (*Salmo trutta* L.) from Norway. Acta Pathol. Microbial. Scand. 84B: 66.

Keet, E. E. 1974. *Yersinia enterocolitica* septicema. Source of infection and incubation period identified. N.Y. State J. Med. 1974: 2226.

King, E. O. 1957. Human infection with *Vibrio fetus* and a closely related vibrio. J. Infect. Dis. 101: 119.

Knill, J., Suckling, W. G., and Pearson, A. D. 1978. Environmental isolation of heat-tolerant *Campylobacter* in the Southampton area. Lancet ii: 1002.

Lander, M. K., and Gill, K. P. W. 1979. *Campylobacter* mastitis. Vet. Rec. 105: 333.

Lassen, J. 1972. *Yersinia enterocolitica* in drinking. Scand. J. Infect. Dis. 4: 125.

Lee, W. H., Harris, M. E., McClain, D., Smith, R. E., and Johnson, R. W. 1980. Two modified selenite media for the recovery of *Yersinia enterocolitica* from meats. Appl. Environ. Microbiol. 39: 205.

Leistner, L., Hechelmann, H., Kashiwazaki, M., and Albertz, R. 1975. Nachweis von *Yersinia enterocolitica* in Faeces und Fleisch von Schweinen, Rindern und Geflugel. Fleischwirtschaft 11: 1599.

Lillard, H. S. 1971. Occurrence of *Clostridium perfringens* in broiler processing and further processing operations. J. Food Sci. 34: 1008.

Lillard, H. S. 1973. Contamination of blood system and edible parts of poultry with *Clostridium perfringens* during water scalding. J. Food Sci. 38: 151.

McClung, L. S. 1945. Human food poisoning due to growth of *Clostridium perfringens* (*C. welchii*) in freshly cooked chicken: preliminary note. J. Bacteriol. 50: 229.

McNaughton, R. D., Leyland, R., Dicon, J. M. S., and Mueller, L. 1981. *Campylobacter* enteritis associated with the consumption of raw milk. Alberta. C.D.W.R. 7: 20.

Magwood, W. E., Rigby, C., and Fung, P. H. 1967. *Salmonella* contamination of the product and environment of selected Canadian chicken processing plants. Can. J. Comp. Med. Vet. Sci. 31: 88.

Minor, T. E., and Marth, E. H. 1972. *Staphylococcus aureus* and staphylococcal food intoxications. A review. IV. Staphylococci in meat, bakery products and other foods. J. Milk Food Technol. 35: 228.

Morris, T. G., and Ayres, J. C. 1960. Incidence of salmonellae on commercially processed poultry. Poultry Sci. 39: 1130.

Oosterom, J. 1980. The presence of *Campylobacter fetus* subspecies *jejuni* in normal slaughtered pigs. Tijdschr. Diergeneesk. 105: 49.

Pearson, A. D., Suckling, W. G., Ricciardi, I. L. D., Knill, M., and Ware, E. 1977. *Campylobacter* associated diarrhea in Southampton. Brit. Med. J. 2: 955.

Peixotto, S. S., Finne, G., Hanna, M. O., and Vanderzant, C. 1979. Presence, growth and survival of *Yersinia enterocolitica* in oysters, shrimp and crab. J. Food Prot. 42: 974.
Ribeiro, C. D. 1978. *Campylobacter enteritis*. Lancet ii: 270.
Robinson, D. A., Edgar, W. J., Gibson, G. L., Matchett, A. A., and Robertson, L. 1979. *Campylobacter enteritis* associated with consumption of unpasteurized milk. Brit. Med. J. 1: 1171.
Robinson, D. A., Gilbert, and Skirrow, M. B. 1980. *Campylobacter enteritis*. Environ. Health 88: 140.
Sadler, W. W., and Corstvet, R. E. 1965. Second survey of market poultry for salmonella infection. Appl. Microbiol. 13: 348.
Salzer, R. H., Kraft, A. A., and Ayres, J. C. 1964. Bacteria associated with giblets of commercially processed turkeys. Poultry Sci. 43: 934.
Scheusner, D. L., and Harmon, L. G. 1971. Temperatures range for production for different enterotoxins by *Staphylococcus aureus* in BHI broth. Bacteriol. Proc. p. 18.
Schiemann, D. A. 1979. Synthesis of a selective agar medium for *Yersinia enterocolitica*. Can. J. Microbiol. 25: 1298.
Severin, W. P. J. 1978. *Campylobacter* enteritis. Ned. T. Geneesk. 122/(15): 499.
Shanker, S., Rosenfield, J. A., Davey, G. R., and Sorrel, T. C. 1982. *Campylobacter jejuni*: Incidence in processed broilers and biotype distribution in human and broiler isolates. Appl. & Environ. Microb. 43: 1219.
Simmons, N. A., and Gibbs, F. J. 1977. *Campylobacter* enteritis. Brit. Med. J. ii: 264.
Simmons, N. A., and Gibbs, F. J. 1979. *Campylobacter* spp. in oven-ready poultry. J. Infect. 1:159.
Skirrow, M. B. 1977. *Campylobacter* enteritis: A "new" disease. Brit. Med. J. 2: 9.
Skirrow, M. B. 1981. *Campylobacter* enteritis in a golfer. Communicable Dis. Rep. 80(53): 3.
Smibert, R. M. 1974. Genus II. *Campylobacter* Sebald and Veron 1963. pp. 207–212. In Buchanan, R. E. and Gibbons, N. E. (eds.) "Bergey's Manual of Determinative Bacteriology," 8th ed. Williams and Wildins, Baltimore.
Smith, M. V., II, and Muldoon, P. J. 1974. *Campylobacter fetus* subspecies *jejuni* (*Vibrio fetus*) from commercially processed poultry. Appl. Microbiol. 27: 995.
Stern, N. J., and Pierson, M. D. 1979. *Yersinia enterocolitica*: a review of the psychrotrophic water and foodborne pathogen. J. Food Sci. 44: 1736.
Stern, N. J., and Oblinger, J. L. 1980. Recovery of *Yersinia enterocolitica* from surfaces of inoculated hearts and livers. J. Food. Prot. 43: 706.
Surkiewicz, B. F., Johnson, R. W., Moran, A. B., and Krumm, G. W. 1969. A bacteriological survey of chicken eviscerating plants. Food Technol. 23: 1066.
Taylor, P. R., Weinstein, W. M., and Bryner, J. H. 1979. *Campylobacter fetus* infection in human subjects. Am. J. Med. 66: 779.
Telfer Brunton, W. A., and Heggie, D. 1977. Campylobacter-associated diarrhea in Edinburg. Brit. Med. J. 2: 956.
Tiehan, W., and Vogt, R. L. 1978. Waterborne campylobacter gastroenteritis-Vermont. Morbid. Mortal. Weekly Rep. 27(25): 207.
Tompkin, R. B., and Christiansen, L. N. 1976. *Clostridium botulinum*. pp. 156–169. In DeFigueiredo, M. P., and Splittstoesser, D. R. (eds.) "Food Microbiology: Public Health and Spoilage Aspects." AVI Publishing Co., Westport, CT.
Voute, E. J., and Grimbergen, A. H. M. 1959. *Vibrio* hepatitis bij kippen. Tijdschr. Diergeneesk. 84: 1380.
Walker, H. W. 1975. Foodborne illness from *Clostridium perfringens*. Crit. Rev. Food Sci. Nutr. 7: 71.

Walker, H. W. 1976. Aerobic and anaerobic spore forming bacteria and food spoilage. pp. 356–386. *In* DeFigueiredo, M. P., and Splittstoesser, D. F. (eds.) Food microbiology: Public health and spoilage aspects. AVI Publishing Co., Westport, CT.

Wauters, G. 1973. Improved methods for the isolation and the recognition of *Yersinia enterocolitica*. Contrib. Microbiol. Immunol. 2: 68.

Wauters, G., Pohl, P., and Stevens, J. 1976. Portage de *Yersinia enterocolitica* par le porc de boucherie. Med Mal. Infect. 6: 484.

CHAPTER **4**

Standard and Experimental Methods of Identification and Enumeration

John R. Chipley

U.S. Tobacco Company
Research and Development Division
Nashville, Tennessee 37202

I. INTRODUCTION

Perhaps the most basic concept of food microbiology lies in the detection, isolation, enumeration, and characterization of microorganisms associated with foods. Because of the widespread recognition of the role of foods in spreading disease, numerous microbiological methods have been developed for their analysis.

This chapter attempts to outline several of the established and experimental methods for the microbiological examination of foods. Where appropriate, advantages and disadvantages associated with each method will be mentioned, and examples of foods or food products analyzed by each method will be cited. Many of these methods have not yet been thoroughly evaluated for use with poultry products. Therefore, foods other than those of poultry and studies involving analyses of some biological or clinical specimens have also been included.

The choice of a particular method or methods must be thoroughly evaluated on an individual basis by the food microbiologist. It is hoped that the reader will determine from this chapter that there is no one "perfect" method of analysis. Many factors such as the acceptance of the method by regulatory agencies, feasibility of use for food, costs, time, personnel, equipment, space, accuracy, and reproducibility must all enter into the final decision. If this chapter can inform the food microbiologist of the preceding, then it will have served its purpose.

II. DETECTION AND ENUMERATION

A. Direct Microscopic Count

The direct microscopic count is a method for estimating the number of microorganisms per gram or milliliter of food by direct microscopic examination of a stained slide prepared from an aqueous suspension of a food or from a liquid food product. This method has been used to enumerate bacteria, yeasts, and fungi in milk, fermented foods, egg products, and processed fruits and vegetables. A known quantity of aqueous food suspension or liquid food product is spread evenly over a prescribed area of a microscope slide. The film is dried, fixed, and stained, and microorganisms in a given number of fields are counted. The number of fields counted can be translated into the amount of product examined, and the number of microorganisms per gram or milliliter can be determined.

Pusch et al. (1976) listed several advantages and disadvantages of the direct microscopic count. Advantages included (1) speed, (2) convenience, (3) a minimum requirement for equipment, (4) some information towards subsequent microbial identifications, and (5) storage of slides for future reference. The disadvantages are that (1) it is suited only for foods which contain large numbers of microorganisms (at least 10^5/ml or g), (2) precision is limited since only small specimens (0.01 or 0.001 ml) are examined, (3) presence of debris may make microbial detection difficult, (4) analyst fatigue occurs and reduces precision, and (5) it is generally not possible to distinguish accurately between viable and nonviable cells.

A thorough outline for analysis of several food products by direct microscopic count has been provided by Pusch et al. (1976) and by Koch (1981).

B. Plate Count Techniques

Agar pour plate procedures were among some of the earliest methods developed by microbiologists for the routine detection and enumeration of microorganisms. Koch introduced this procedure in 1880 when he developed agar media, and by 1895 it was a recognized procedure. Analysis of this procedure in 1916 resulted in the development of procedures currently in use. Several variations of this technique have been developed and are described by Koch (1981).

The basic assumption of plate count techniques is twofold: (1) one viable cell will form one visible colony, and (2) all types of microorganisms from a sample being analyzed for total count will grow on a single agar medium incubated under one set of conditions. It should be obvious to any reader having experience in microbiology that these assumptions contain serious deficiencies! Plate counts do not necessarily detect the actual total numbers of viable microorganisms per aliquot of analyzed sample, since microbial cells can occur singly and in pairs, chains, clusters, or clumps. Likewise, special requirements involving nutrients, atmospheric gases, incubation temperatures, injured cells, etc., make it imperative that samples be plated on more than one nonselective medium and be incubated under more than one condition. However, once optimum procedures and conditions for a particular food are determined, this technique can be quite useful for routine microbiological analysis of the food.

Much effort in research has been aimed at finding a better method than the pour plate, primarily because it is both time-consuming and expensive. As alternatives, the drop plate and the spread plate are currently used for detection of mesophilic aerobes. Both methods offer cost savings compared to the pour plate method. Van Soestbergen and Lee (1969) reported that the pour plate method was more precise than the spread plate method but generally gave lower counts. Clark (1967) found 70–80% higher colony counts from chicken meat when surface plating was compared to pour plating. This difference was related to both genus and species of bacteria. Klein and Wu (1974) reported that heterotrophic organisms in water samples were susceptible to the transient stress of agar at 45°C used in the standard methods pour plate procedure. Significant increases were noted in the numbers of cells recovered from water samples when surface plating (spread plating) techniques were used.

A study of plate preparation, inoculum volume, and technique of spreading in the surface plate method showed that predrying plates to a weight loss of 3–4 gm per 15-ml plate allowed the use of inoculum volumes of up to 1 ml without affecting recovery of bacteria (Clark, 1971). Several factors affect moisture loss from agar plates during incubation (Alexande and Marshall, 1982).

The components of the plate medium are of utmost importance. Under some circumstances, normally essential components may actually inhibit the detection of a desired microorganism. For example, blood supplementation is recommended for isolation of *Campylobacter jejuni* in enrichment broths. However, when this organism was isolated from poultry and red meats (Stern et al., 1985), it was found that supplementation was inhibitory to isolates from ground beef but was required for isolates from chicken.

A comparative study was made of five methods (pour plate, surface spread plate, surface drop, agar droplet, and microdilution) for the enumeration of microorganisms in 100 samples of food products. A variation in count between the methods of <0.5 $\log_{10}$ cycles was given by 98% of the samples. The authors (Kramer and Gilbert, 1978) reported that substantial savings were achieved with the microdilution and agar droplet techniques, compared with conventional pour plate and spread plate methods.

In a more recent survey, the pour plate, surface spread, agar droplet, and spiral plate methods were used in parallel with the surface drop method for enumeration of microrganisms in foods (Greenwood et al., 1984). Good agreement was obtained between all surface methods of enumeration, but there was poor agreement between molten-agar methods and the surface drop method.

A thorough discussion of the many advantages and disadvantages associated with plate count techniques as well as detailed procedures for analysis of several food products using this method has been provided by Thatcher and Clark (1973), Gilliland et al. (1976), and Koch (1981). A random-number sampling method for estimation of bacteria of particular interest when they occur in complex mixtures on plates has been described (Ordonez, 1979).

A modification of the plate count technique, the plate loop method, has been developed. This method was evaluated for determining total counts of orange juice (Murdock and Hatcher, 1976). The authors reported savings in time and equipment but cautioned that analysts using this technique must be specially trained to maintain their degree of accuracy.

C. Spiral Plate Count

The Bureau of Foods of the U.S. Food and Drug Administration has developed the Spiral Plate Maker (marketed by Spiral Systems Marketing, Bethesda, Maryland). This instrument distributes a continuously decreasing volume of liquid on the surface of a rotating petri plate while moving from the center toward the edge in the form of an Archimedes' spiral (Gilchrist et al., 1973). This method was compared to the pour plate procedure using several bacterial cultures in water and milk. No statistically significant differences were observed when concentration of 600 to 12×10^5 bacteria per milliliter were used, but the spiral plate method gave counts that were higher than counts obtained by the pour plate method.

The spiral plate method has been used to enumerate microorganisms in foods (Jarvis et al., 1977; Kramer et al., 1979; Zipkes et al., 1981) and in raw and pasteurized milk (C. B. Donnelly et al., 1976). A comparison of the media and equipment required for the examination of 100 food samples using pour plate and spiral plate techniques has recently been reported (Konuma et al., 1982). Use of the spiral plate system, along with a laser bacterial colony counter, has been reported to cut the cost of microbial analysis by 50% (S. B. Newton, 1978). Several advantages regarding the use of the Spiral Plate Maker have been reported (Hedges et al., 1978). It can be used in conjuction with the Ames test (Couse and King, 1982).

D. Most Probable Number

The most probable number (MPN) technique has been used for detecting microbial growth for approximately 50 years. It is based on the assumption that microorganisms are normally distributed in products and will grow when conditions are favorable. With this technique, if the number of organisms in a sample is large, the differences between samples will be small; all the individual results will be closer to the average. If the number is small, the differences, relatively speaking, will be larger.

Food samples for MPN analysis are blended and serially distributed in an appropriate diluent, and three serial aliquots or dilutions are inoculated into 9 or 15 tubes of appropriate medium for the 3- or 5-tube method, respectively. The tubes are incubated, and the numbers of organisms in the original sample are extrapolated from the number of tubes containing microbial growth by the use of standard MPN tables. This method provides a statistical determination of viable cells, and MPN results are generally higher than standard plate counts.

The MPN technique has been used for several years for the detection of coliforms and other organisms in water, dairy products, and several foods, including broiler carcasses (Rigby, 1982) and giblets (Christopher et al., 1982). Small numbers (less than 1 cell per gram) of *Campylobacter jejuni* were recovered from inoculated poultry products using a 3-tube MPN method (Wesley et al., 1983). Most probable number analysis was also employed for detecting small numbers (10 cells per 100 ml) of *Campylobacter* from water (Bolton et al., 1982). By using a fluorogenic MPN procedure, *Escherichia coli* was detected in oysters approximately 4 days earlier than with standard techniques (Koburger and Miller, 1985). *Clostridium perfringens* was easily detected in food products by using an iron milk MPN procedure (Abeyta et al., 1985).

Statistical analysis of this technique can be automated (Hurley and Roscoe, 1983; Clarke and Owens, 1983; Russek and Colwell, 1983; MacDonell, 1983). A number of references contain detailed reviews of MPN procedures and discussions of the advantages and disadvantages associated with this technique. They include Thatcher and Clark (1973), deFigueredo and Jay (1976), Mayou (1976), and Koch (1981).

E. Spectrophotometric and Colorimetric Methods

Spectrophotometric and colorimetric methods have been employed for several years for the detection of microbial growth. Basically, these procedures involve inoculation of a suitable solution of growth medium, incubation, and periodic determination of the increase in absorbance of resulting microbial suspensions compared to that of an uninoculated control solution. At the intervals in which the absorbance is determined, aliquots are often withdrawn for serial dilution and standard plate count determination. In this manner, standard plate counts can be plotted against changes in absorbance to yield a linear relationship during the phase of logarithmic growth. Absorbance of microbial suspensions is usually determined at 400–465 nm.

Spectrophotometers and colorimeters used in detecting microbial growth operate in the following manner. Cuvettes containing microbial suspensions are interposed between a unit source of light and a photoelectric unit attached to a galvanometer. The reading on the galvanometer depends on the passage of light from the unit source. Of the total light from the unit source, the percentage transmitted through the tube will be diminished in proportion to the turbidity (number of cells).

The Klett-Summerson colorimeter (Klett Manufacturing Co., New York) is perhaps the best known of these instruments.

These methods are subject to a variety of errors due to variations in cell morphology, presence of clumps or clusters, particulates in the media, and nonviable cells or fragments. However, they are one of the quickest and simplest detection methods and are reasonably accurate. Optical density has been used to estimate concentrations of *Penicillium* and *Geotrichum* spores in aqueous suspensions (Morris and Nicholls, 1978) and to measure yeast growth in grape juice (Monk and Costello, 1983).

Differential light-scattering techniques have been used for studying bacterial morphology. These methods consist of recording the angular distribution of light scattered from a suspension of bacteria illuminated by a plane parallel beam of monochromatic light. Differential light scattering was used to determine the effects of heat on suspensions of *Staphylococcus epidermidis* (Berkman and Wyatt, 1970) and in morphological studies of *Staphylococcus aureus* (P. J. Wyatt, 1970). A *Differential I* light-scattering photometer (Science Spectrum, Inc., Santa Barbara, California) was used in these studies. This technique has been proposed for use in the rapid identification of various microorganisms (Jenkins et al., 1980; Salzman et al., 1982). A thorough review of these techniques has been published by Koch (1981) and by Wood (1981).

F. Direct Count

1. Particle Counting

In particle-counting instruments, operation is based upon differences in electrical conductivity between particles and diluents. Particles act as insulators, whereas diluents act as conductors. Particles, suspended in an electrolyte, are forced through a small aperture through which an electrical current path has been established. As each particle displaces electrolyte in the aperture, a pulse proportional to the particle volume is produced. The assumption is made that changes in electrical resistance are due to microbial cells. The disadvantages with this system lie in the difficulty of recognizing relatively few organisms in a "sea" of particles (dust, fibers, food particles, etc.). Other disadvantages are discussed in detail by Koch (1981).

Particle-counting instruments are used most extensively in clinical analyses (hematology and urology) (Dow et al., 1979); however, application to microbiological analyses of foods may be possible, provided extracts can be prepared which contain a minimum of particulate material. A commercial source for this type of instrument is the Coulter counter (Coulter Electronics, Inc., Hialeah, Florida). A rapid method to differentiate between total and viable bacteria using the Coulter counter has been reported (Seydel and Wempe, 1980). Several types of microorganisms can be sized with this instrument (Poole, 1982).

Modifications in particle counting, through the addition of a pulse-height analyzer, can be used to rapidly detect microorganisms in body fluids. Upon addition of specific antisera, microorganisms can be identified by changes in pulse heights and areas when agglutination reactions occur (Curbey and Gall, 1977; Gall et al., 1978; DeCastro and Widomski, 1978).

2. Immunological Methods

Several automated and/or rapid methods have been reported in which immunological techniques can be employed to detect and to identify microorganisms.

Thomason et al. (1975) evaluated a semiautomatic system for the direct fluorescent antibody detection of salmonellae in various samples, including eggs, poultry, and sausage. They concluded that the system worked reasonably well, but that research was needed to improve enrichment procedures for samples to be processed by the system. In a later evaluation, Munson et al. (1976) described a method in which this automated fluorescent antibody (FA) procedure was successfully used to detect Salmonella in foods. The automated system added a sample of enrichment broth to membrane filters supported on plastic slides, removed spent broth, washed cells, and stained them with polyvalent FA conjugate at the rate of 120 slides/hr. The degree of fluorescence was automatically determined in a slide reader at the rate of 360 slides/hr. After some modification, the authors were able to obtain excellent correlation with conventional methods, finding only 4% false-positive FA results with 116 frog leg samples. A semiautomatic method for Salmonella detection in foods and feeds has also been reported (Barrell and Paton, 1979).

A membrane filter-disc immunoimmobilization technique has been developed for rapid detection of salmonellae in food (Swaminathan et al., 1978a,b). This method, when used to detect salmonellae in raw meats and poultry, was found to have good correlation with conventional cultural methods. Fluorescent antibody and conventional methods have been compared for the detection of salmonellae in several agricultural and food products (Gibbs et al., 1979).

A thorough outline for detection of salmonellae in several food products by FA techniques has been provided by Thomason (1976). Procedures for the identification of species of *Clostridium* (Batty and Walker, 1966; Koch et al., 1978) and enteric microorganisms (Ayres, 1967) have also been published. Control of nonspecific staining will increase the acceptance of this technique (Swaminathan et al., 1978a; Tharrington et al., 1978).

A serological method for rapid identification of *Vibrio parahaemolyticus* from marine samples has been proposed (Shinoda et al., 1982). The authors found the method useful, especially for samples in which there were many related organisms, because many biochemical tests could be omitted. The production of a common antigen by members of the *Enterobacteriaceae* may provide a rapid method of classification (Ramia et al., 1982).

An indirect enzyme-labeled antibody technique was developed by Krysinski and Heimsch (1977) to detect *Salmonellae* in foods. This method involved the use of a 47-mm membrane filter, on which 14 or more samples could be tested simultaneously. A double-antibody enzyme immunoassay that used a microtechnique was developed for detecting staphylococcal enterotoxin A in hot dogs and mayonnaise (Saunders and Bartlett, 1977). Assay sensitivity ranged from 0.4 ng (20-hr test time) to 3.2 ng (1- to 3-hr test time) of toxin per milliliter of prepared sample. Separation and detection of enterotoxin from spiked food products ranged between 72 and 98% of the amount added. An immunosorbent assay for *Yersinia enterocolitica* has been developed in which lipopolysaccharide could be detected in minute quantities (0.5 ng/ml or more) (Gripenberg et al., 1979). Other assays have been developed for *C. botulinum* type A, B, and E toxins (Notermans et al., 1979; Dezfulian and Bartlett, 1985), for aflatoxins (Pestka et al., 1980), and for staphylococcal and *E. coli* enterotoxins (Freed et al., 1982; C. Edwin et al., 1984; M. R. Thompson et al., 1984).

A commercially available enzyme immunosorbent assay (ELISA) was recently compared with conventional methods for detection of salmonellae in meat and poultry products (Emswiler-Rose et al., 1984). The authors reported 100% agreement between the methods. Identification of contaminated products was accomplished in 2–3 days, compared to the 4–6 days required for conventional methods. An enzyme immunoassay (EIA) which could be completed within 27 hr has also been developed for the detection of salmonellae in foods (Aleixo et al., 1984; Mattingly and Gehle, 1984). A rapid ELISA technique to detect *E. coli* enterotoxins by using bacterial cell sonicates has been proposed (Olsvik and Berdal, 1982). These immunoassays have been outlined and compared in a recent publication (Salzman and Gregg, 1984).

Conditions for detection and quantitation of *Clostridium perfringens* enterotoxin by counter-immunoelectrophoresis have been described by Naik and Duncan (1978). As little as 0.2 µg of enterotoxin per milliliter could be detected. The test was reported to be rapid (approximately 1 hr), sensitive, and specific. Detection of this enterotoxin was increased (down to 1.0 ng/ml for meat products) by using an ELISA technique (Olsvik et al., 1982).

Using a radiolabeled antibody staining technique, Benborough and Martin (1976) detected *Bacillus subtilis* spores, *Serratia marcescens*, and *Francisella tularensis* within 1 hr. Microwells and microporous filtration were used to concentrate samples. The technique was most successful when tritium was used as the label. As few as 7000 *B. subtilis* spores, 9000 cells of *S. marcescens*, and 50,000 cells of *F. tularensis* could be detected. A similar technique was used to quantitate the growth of streptococci (Ling et al., 1982). A radioimmunoassay for detection of lipopolysaccharide has been developed (Munford and Hall, 1979). Procedures for the preparation of labeled staphylococcal enterotoxin A containing a high specific activity have also been reported (Niskanen and Lindroth, 1976). Recently, the production of monoclonal antibodies for mycotoxin and *Salmonella* detection has been outlined (Hunter et al., 1985; Robison et al., 1983).

A comparison of several serological techniques for sensitivity and analysis times has been reported (Jay, 1978). Techniques for several immunoassays have been published (Lagone and Van Vunakis, 1982).

3. Membrane Filtration

In order to improve the accuracy of quantitative microbial analyses of certain foods or food ingredients, it may be necessary to use relatively large samples. Large volumes of liquid foods or solutions of dry foods which can be dissolved and can be passed through a bacteriological membrane filter (pore size 0.45 µm) may be analyzed for microbial content by the membrane filtration method. This method may be quite useful for samples containing low numbers of bacteria.

Membrane filtration (sometimes called "direct count") procedures have been reported for beverages, soft drinks, water testing, fecal coliforms, osmophiles, spores, and stressed cultures and are outlined in the "Compendium of Methods for the Microbiological Examination of Foods" (Speck, 1976). A detailed procedure has been provided by Gilliland et al. (1976). Detection of enterococci by a membrane-filter-FA test has also been reported (Pugsley and Evison, 1975).

Two membrane filter methods were compared with an MPN procedure for enumerating *E. coli* biotype I in foods (Sharpe et al., 1983a). The results were available in 24 hr by both membrane filter methods, compared with 10–14 days by the MPN procedure. Both of these methods gave significantly higher recoveries than the MPN procedure.

Advantages of this method include the capability of analyzing samples containing low numbers of microorganisms and the ability to both concentrate and characterize these organisms with a single procedure. The disadvantages are that clogging of the membrane with food particles often occurs. Food samples must be completely dissolved for effective filtration. It has also been observed by the author that certain types of compounds, especially polyphenolics, may chemically react with the membranes and result in clogging.

Several of these disadvantages appear to have been eliminated with the development of the Iso-Grid hydrophobic grid–membrane filter system (Tilton, 1982). The incorporation of a wire-cloth prefilter in this system prevents deposit of particulate matter on the membrane filter surface, and the application of an appropriate enzyme or surfactant treatment allows several foods to be filtered. In some cases, food particles can be rinsed completely from unincubated filter surfaces without affecting subsequent counts (Sharpe et al., 1978). The limited counting range of conventional membrane filters was also eliminated with the introduction of the hydrophobic grid–membrane filter (HGMF). These modifications may make membrane filtration a viable alternative approach to classical methods for the microbiological analysis of foods (Sharpe et al., 1979a,b; Cousins et al., 1979; Pettipher et al., 1980; Peterkin and Sharpe, 1980, 1981).

An automated HGMF technique was compared to MPN and spread plate methods for isolation and enumeration of several indicator organisms in foods (Brodsky et al., 1982). Recovery efficiencies for this technique compared to conventional methods ranged from 80 to 88%. The authors suggested that the automated system was a viable alternative to conventional methods. However, modifications of the system were proposed for optimum recovery of injured cells.

By coupling membrane filtration with ELISA, a procedure was developed for rapid enumeration of *Staphylococcus aureus* in foods (Peterkin and Sharpe, 1984). The test required 3 hr to complete and allowed direct enumeration of confirmed *S. aureus* in foods within 27 hr.

A ready-to-use, disposable system for simultaneous membrane filtration and culturing of microorganisms has been developed (Nalge Company, Rochester, New York, and Millipore Corp., Bedford, Massachusetts).

The type of filter membrane selected for use in membrane filtration studies is critical and may affect subsequent recovery of bacteria (Green *et al.*, 1975; Lorenz and Tuovinen, 1980; Farber and Sharpe, 1984). Drying of microbial cells on these membranes may induce injury (Asada *et al.*, 1979). However, injured cells can be repaired and detected if proper procedures are followed (Brodsky, 1982).

Membrane filtration and epifluorescent microscopy have been used for rapid enumeration of bacteria from cream (Griffiths *et al.*, 1984), milk (Pettipher *et al.*, 1983; Pettipher, 1982), and food (Pettipher and Rodrigues, 1982). A technique for the isolation of *Y. enterocolitica* from water allowed presumptive identification within 50 hr. (Morrison *et al.*, 1982). Membrane filtration has also been used to monitor cannery cooling water (Rey *et al.*, 1982) and to detect *C. botulinum* spores in honey (Hauschild and Hilsheimer, 1983). The accuracy of this method can be substantially improved (by 20–50%) without a large increase in cost when the proper degree of replication at each level is performed (Kirchman *et al.*, 1982).

Verification of membrane filter total coliform counts from drinking water was increased 87% by testing for the presence of β-galactosidase and cytochrome oxidase, compared with verification by determining gas production in lauryl tryptose broth (LeChevallier *et al.*, 1983b). Several factors, such as media preparation and storage and sample holding times, were found to affect the detection sensitivity of this technique (Hsu and Williams, 1982).

G. Measurement of Microbial Growth

1. Changes in Electrical Impedance

Whenever microorganisms are metabolically active and change the chemical composition of their supporting medium, there are small changes in the electrical impedance of the medium. Instruments are available to measure these impedance changes and thus record the growth rates of microorganisms. Samples are inoculated into media-filled containers (modules or bottles), each equipped with impedance-sensing electrodes, and are incubated. Impedance ratio measurements are then made on a pair of wells (reference and sample) at specified time intervals, and the entire testing cycle can be repeated. Data can be either displayed on the face of the instrument or recorded. One commercial source for this type of instrument is Bactomatic, Inc., Palo Alto, California.

The major disadvantage with this technique would be the initial cost of instrumentation. Background knowledge of the microflora associated with the samples to be analyzed is also required in order to select media for optimum growth. However, for most high-count samples, results can be obtained in a fraction of the time required for conventional techniques (Firstenberg-Eden and Eden, 1984).

Changes in electrical impedance have been used for determining microorganisms in milk (Cady et al., 1978; Firstenberg-Eden and Tricarico, 1983), the shelf life of milk (Martins et al., 1982), and the bacterial content of frozen vegetables (Hardy et al., 1977) and raw meat (Firstenberg-Eden, 1983). In these studies, accuracies of from 88 to 93% (compared with standard plate counts) were achieved when detecting samples with specified numbers of microorganisms per gram or milliliter. A rapid (24-hr) procedure has been described for the quantitation of coliforms in ground meat using electrical impedance (Martins and Shelby, 1980). This technique can also be used to detect the growth of yeasts (Fleischer et al., 1984).

An automated system has been recently evaluated for estimating total and selective counts of fish (D. M. Gibson et al., 1984). Total counts were completed in one-fifth of the time taken for conventional methods, and selective counts were completed in one-third of the time. Correlations between the methods were good. This system has also been evaluated for detection of bacteria and yeasts in frozen concentrated orange juice (Weihe et al., 1984). Detection times were rapid (10.2 hr for bacteria and 15.8 hr for yeasts), and more than 96% of the samples were correctly classified as acceptable or unacceptable compared to plate counts. Procedures by which samples are prepared and diluted, however, may affect the subsequent estimation of the number of microorganisms when using this system (Firstenberg-Eden, 1983).

Another detection system, based upon electrical conductance of bacterial suspensions, has been proposed (Baynes et al., 1983). However, detection time and suspending medium affect the precision and accuracy of this system. Its usefulness for food analysis remains to be seen.

2. Radiometry

The radiometry technique involves the detection of microbial growth by measuring metabolites of radiolabeled substrates. Usually, $^{14}CO_2$ production from ^{14}C-labeled carbohydrates and amino acids is monitored. The time required for detection of this gas is related to inoculum density, growth rate, and metabolic activity, but is generally more rapid (6–18 hr) than conventional culture techniques (Johnston Laboratories, Towson, Maryland). Disadvantages would include the cost of initial instrumentation and the use of radiolabeled compounds. This method has been utilized to detect microorganisms in frozen concentrated orange juice (Hatcher et al., 1977), coliforms (Bachrach and Bachrach, 1974; Previte et al., 1977), Clostridium (Evancho et al., 1974), Salmonella, Staphylococcus, and heat-shocked Clostridium spores (Previte, 1972), and several other bacteri of significance in foods (Waters, 1972).

Radiometric methods have been used as a rapid screening method for assessing the quality of cooked, frozen foods (Rowley et al., 1978a), for detecting injured fecal coliforms (Rowley et al., 1979), for detecting microbial contamination in liquid foods (Mafart et al., 1978), and for estimating the replication time of bacteria (Buddemeyer et al., 1978). Radiometric methods have also been used to determine the effect of osmotic stabilizers, such as sucrose, glycerol, and ethylene glycol, upon cell wall–damaged bacteria (Martinez and Malinin, 1979). A rapid method for detection of Salmonella in foods has been reported (Stewart et al., 1980).

3. Microcalorimetry

Small changes in heat production resulting from bacterial metabolism and growth in liquids and food systems can be measured in calorimeters (Forrest, 1972; Sachs and Menefee, 1972). Methods have been reported whereby microorganisms can be identified on the basis of characteristic thermograms (Boling et al., 1973; Mor, 1976). Lampi et al. (1974) used a combination of radiometry and microcalorimetry for rapid detection of foodborne microorganisms. These authors reported a high degree of correlation between viable cell count and radiometric measurements as meat loaf underwent spoilage. In ground meat, significant correlations were observed between microcalorimetric measurements and plate counts with all experimental conditions (Gram and Sogaard, 1985). Estimation of the bacterial levels (10^5-10^8 CFU/g) could be made in less than 24 hr.

In addition to the preceding applications, microcalorimetry has also been used for the identification of some *Enterobacteriaceae* (Herman et al., 1980) and lactic acid bacteria (Fujita et al., 1978), estimation of bacteria in milk (Cliffe et al., 1973; Monk, 1979), bacterial classification (Monk and Wadso, 1975), and characterization of yeasts (Beezer et al., 1979; H. Wang et al., 1978; Brettel et al., 1980; Perry et al., 1983).

Microcalorimetry can be used to study the effects of antimicrobial agents on metabolism (Perry et al., 1980), aerobic and anaerobic metabolism, membrane biochemistry and transport, etc. The advantage of this process is that it lends itself well to automation and computer interpretation, but the disadvantage may be the cost of initial instrumentation. One commercial source is the LKB BioActivity Monitor (LKB Instruments, Inc., Rockville, Maryland).

4. Measurement of Bioluminescence and Chemiluminescence

Measurements of bioluminescence [bacterial adenosine 5'-triphosphate (ATP) interaction with firefly luciferase] and chemiluminescence (bacterial porphyrin reaction with luminol) (Ewetz and Thore, 1978) have been used for the detection of microorganisms in clinical and biological specimens, water, and food samples (Sanville, 1985; Van Dyke, 1985). Rapid detection and enumeration of brewery microorganisms, bacteria in foods, yeasts, and phospholipase and lipase using bioluminescence have been reported (Hysert et al., 1976; Sharpe et al., 1970; L. F. Miller et al., 1977; Ulitzur and Heller, 1978; Ulitzur, 1979). This method has also proven useful for quantitation of total microbial growth and for determination of lipopolysaccharide (Ng et al., 1985b; Dhople and Hanks, 1973; Chappelle and Levin, 1968; Ulitzur et al., 1979; Vanstaen, 1980). It has been proposed for use as an early indication of spoilage in fish (Barak and Ulitzur, 1980) and as a bioassay for mycotoxins (Yates and Porter, 1982). In ground beef, the bioluminescent assay time for a given sample was less than 1 hr, and correlation coefficients between microbial ATP and plate counts ranged from .86 to .99 (Kennedy and Oblinger, 1985). In raw meat, this method was used to give a rapid (20-25 min) estimate of microbial contamination (Stannard and Wood, 1983). However, in the analysis of foods, one must be aware of the limitations of bioluminescence (Williams, 1971; Jakubczak and LeClerc, 1980). For example, ATP content in bacterial cells was found to vary with the conditions of sample storage (Kaneko et al., 1984).

A chemiluminescent immunoreaction for rapidly identifying and quantitating small numbers of microorganisms has been developed (Halmann et al., 1977). With this method, as few as 30–300 bacterial cells could be detected. A chemiluminescent immunoassay has been developed for the determination of staphylococcal enterotoxin B (Velan and Halmann, 1978) and for rapid detection of viruses (Pronovost et al., 1982).

Commercial sources for bioluminescence instrumentation include the Lumac System (3M, St. Paul, Minnesota), the Packard Picolite system (Packard Instrument Co., Downers Grove, Illinois), and the Dupont Luminescent Biometer (Dupont Instruments Division, Wilmington, Delaware). All of these systems are rapid and relatively sensitive. For example, with the Lumac System, results can be obtained within 1 min for samples containing at least 10^3 microorganisms per milliliter. Disadvantages include somewhat complex and expensive detection equipment, the requirement that any nonmicrobial ATP or porphyrin compounds be removed prior to microbial analysis, and variations in the ATP content of microbial cells.

5. Measurement of Deoxyribonucleic Acid

The determination of deoxyribonucleic acid (DNA) base compositions can be used as an aid in bacterial taxonomy. However, this method is time-consuming, relatively complex, and outside of the capabilities of a routine diagnostic laboratory. Significant improvements in this procedure were made by the use of reversed-phase liquid chromatography (Wickstrom and Mischke, 1980). Following isolation and digestion of DNA, 1–2 nmol of each of the deoxynucleosides of DNA could be detected in 25 min. Another rapid method, whereby five bacterial samples could be analyzed simultaneously, has been reported (Flossdorf, 1983). Several strains of *Streptococcus mutans* and other bacteria have been identified by DNA base analysis (Coykendall and Lizotte, 1978; MacDonell et al., 1983).

DNA hybridization can be used for detecting *Shigella* and *E. coli* (Boileau et al., 1984) and *Y. enterocolitica* and *Salmonella* in food (Hill et al., 1983a; Fitts, 1985; Fitts et al., 1983). This procedure can detect approximately 100 cells of *E. coli* per gram of food (W. E. Hill, 1981; Hill et al., 1983b).

Donkersloot et al. (1972) described a rapid method for the determination of intact DNA in both gram-positive and gram-negative bacteria. The method is based upon fluorometric determination of DNA with ethidium bromide after alkaline digestion of bacteria. Assays required less than 2 hr to complete and could detect at least 0.2 μg of DNA per 4 ml of solution. This method was used to determine the specific growth rate of *E. coli* and could be used for the estimation of total bacterial growth. Analysis of adenosine phosphates and guanine phosphates can also be used to determine microbial growth (Bostick and Ausmus, 1978; Karl, 1978).

Flow cytometry, coupled with appropriate staining techniques, permits the measurements of cellular DNA and protein in bacteria, algae, fungi, and yeasts (Hutter and Eipel, 1978, 1979; Boye et al., 1983; Van Dilla et al., 1983). Living and dead cells can be differentiated with this technique. Selective adsorption of fluorescent dyes can be used to characterize a wide variety of bacteria (Shelly et al., 1983). Advantages, disadvantages, and procedures associated with microbial detection by flow cytometry have been recently published (Van Dilla et al., 1985).

6. The *Limulus* Lysate Assay

The *Limulus* amebocyte lysate assay is used to detect endotoxins of gram-negative bacteria. The lysate is prepared from the circulating amebocytes of the horseshoe crab (*Limulus polyphemus*). This technique evolved from the observation by Bang (1956) that gram-negative infections of the horseshoe crab resulted in fatal intravascular coagulation. Levin and Bang (1964) later demonstrated that this clotting was a result of the action between endotoxin and a clottable protein in the circulating amebocytes of the crab. Following development of a suitable anticoagulant for *Limulus* blood, lysates from washed amebocytes have served as extremely sensitive indicators for the presence of endotoxins (Sullivan et al., 1976).

The sensitivity and specificity of this assay have been reported (Jorgensen and Smith, 1973), as well as have other factors, such as magnesium, which affect the sensitivity of the assay (Sullivan and Watson, 1974; Tsuji and Steindler, 1983; Ho, 1983). Although this assay has been applied primarily to clinical analyses, it has also been used for rapid detection of endotoxin in drinking water (Jorgensen et al., 1976), milk and dairy products (Hansen et al., 1982), and ground beef (Jay, 1977). Jay (1977) showed a good correlation of increasing *Limulus* lysate test titers to the actual content of gram-negative bacteria in ground beef.

This technique has been used to assess the quality of fish (Sullivan et al., 1983). Endotoxin levels agreed with aerobic plate counts and chemical indices of spoilage. Correlation between levels of endotoxin and levels of total volatile bases was highly significant. In vacuum-packed, cooked turkey meat, there was a linear correlation between titers and numbers of Enterobacteriaceae (Dodds et al., 1983). Ten bacterial cells per gram of meat could be detected with this assay.

A rapid, automated method for the performance of the *Limulus* assay has been developed by using the Abbott MS-2 Microbiology System (Abbott Laboratories) (Jorgensen and Alexander, 1981). Up to 176 samples could be assayed during a 1-hr period. Coupling this assay to a robotic system allowed three samples and a standard to be analyzed in duplicate in 48 min (Tsuji et al., 1984).

A draft guideline listing acceptable conditions for using the lysate assay has recently been published (Anonymous, 1983). This procedure may be used in lieu of the currently official rabbit pyrogen test.

7. Reduction of Dyes

Most actively growing anaerobic and facultatively anaerobic microorganisms bring about a lowered oxidation–reduction potential in their medium (Costilow, 1981). Several oxidation–reduction dyes have been used to estimate the numbers of bacteria present in media. They include litmus, methylene blue, tetrazolium salts, neutral red, and resazurin. Methylene blue and resazurin have been used in determining the bacteriologial quality of dairy products and meats (Espejo, 1977; Sudarsanam et al., 1978; Barton, 1980; Labadie and Doumbia, 1984).

A method for rapid identification of enterococci based upon reduction of litmus in litmus milk has been reported (Blazevic and Schierl, 1981). Nitroblue tetrazolium reduction by strains of *Staphylococcus epidermitis*, *E. coli*, and *Pseudomonas aeruginosa* was found to be proportional to the numbers of bacteria present (Urban and Jarstrand, 1979). This technique may be useful for measuring the influence of bactericidal, bacteriostatic, or growth-stimulating factors on bacteria. Reduction of benzyl viologen has been proposed as an aid for identifying strains of *Campylobacter* species (Goodman and Hoffman, 1983).

A method has been devised which permits quantitative determination of injured cells in populations of *Saccharomyces cerevisiae* (Bonora and Mares, 1982). Injury could be determined by measuring the absorbance (at 664 nm) of suspensions of cells treated with methylene blue without affecting cell integrity. This method was based on the observation that methylene blue could be taken up by dead or severely damaged cells but not by living cells. A similar method for determining viability of yeast cells based upon reduction of tetrazolium salts and other dyes has been reported (Trevors, 1983; Trevors et al., 1983). Cells of *Staphylococcus aureus* maintain their ability to replicate when stained with the fluorescent dye rhodamine 123 (Matsuyama, 1984).

The application of dye reduction tests to foods other than milk has not been very successful, because many food components (such as oxidoreductase enzymes, etc.) may interfere with microbial reduction of the dyes. These tests provide only estimates of microbial activity, since reduction times are influenced not only by microbial numbers but also by the growth phase and types of microorganisms present at the time of testing (National Academy of Sciences, 1985).

H. Measurement of Microbial Products

1. Enzymatic Activities

Certain enzymes may be highly characteristic of certain genera and species of microorganisms. The ability of a microorganism to ferment or hydrolyze simple carbohydrates such as glucose, sucrose, and lactose; to utilize as hydrogen acceptors such substances as nitrate; to hydrolyze gelatin, casein, complex polysaccharides, and lipids; and to utilize amino acids such as arginine, ornithine, and lysine are but a few of the many enzymatic activities that can be assayed. The secretion into media of several of these enzymes by microorganisms has been reviewed (Lampen, 1966).

Clarke and Steel (1966) were among some of the earliest researchers to describe some rapid and simple biochemical tests involving enzyme assays for bacterial identification. Results could be obtained within approximately 2 hr for most assays. Bascomb (1976) described an automated continuous-flow analytical system through which several different classes of microbial enzymes could be detected. This system could assay 18 different enzymes in approximately 1 hr and was used to detect several members of the Enterobacteriaceae.

Other automated and rapid methods for measuring enzyme activities of bacterial suspensions have been described (Rollof et al., 1984; Bascomb and Spencer, 1980; Waitkins et al., 1980; Dobrogosz, 1981). Farber and Idziak (1982) reported that a fourfold increase occurred in the activity of glucose dehydrogenase during spoilage of chilled beef. They attributed this increase to the increase in numbers of Pseudomonas during spoilage. A rapid enzymatic procedure has been proposed for measuring accumulation of ethanol in tissues as a chemical indicator of fish spoilage (Kelleher and Zall, 1983). A similar procedure has been proposed for acetate (Clarke and Payton, 1983) and ethanol detection when microorganisms are growing on agar plates (Jacobs et al., 1983).

Detection of tetrathionate reductase is of value in the differentiation of Enterobacteriaceae (Richard, 1977), as is the presence of β-galactosidase (LeMinor, 1979) and aminopeptidase (Peterson and Hsu, 1978; Cerny, 1978). Detection of staphylococcal thermonuclease provides a convenient method of screening foods for possible enterotoxin production (Park et al., 1978). Detection of phosphatase, catalase, thiaminase, and proteolytic activities can be of value in taxonomic studies (Pacova and Kocur, 1978; Satta et al., 1979; Giammanco et al., 1982; Chester, 1979; Edwin et al., 1978). Detection of phospholipase C activity has been proposed as an applicable method for estimating viable Clostridium perfringens in long-time, low-temperature processed food products (Foegeding and Busta, 1980).

An agar medium was developed which permitted the detection of bacteria with β-glucuronidase activity (Kilian and Bullow, 1979). It was used for the rapid identification of Enterobacteriaceae and was found to have several advantages compared to conventional techniques. A fluorogenic assay procedure, based upon β-glucuronidase activity, has been proposed for rapid detection and confirmation of E. coli in foods (Robison, 1984). Fluorescence of inoculated plates was indicative of the presence of E. coli; extensive biochemical confirmation was not necessary. When compared with classical methods for detecting E. coli in foods, total agreement between the two methods was 94.8%. There was a false-positive rate of 4.8% and no false-negatives. Another assay for this enzyme, utilizing headspace gas chromatography, has been proposed (Koeppen and Dalgaard, 1984). An extensive review of appropriate enzymes used in the identification of gram-negative bacteria has been published (Richard, 1978).

It should be noted that detection of a particular enzyme from an unknown organism should only serve as an aid, and not the basis, for subsequent identification of that organism. For example, thermonuclease production, extensively used for the identification of S. aureus, has been found in Bacillus and enterococci (Bissonnette et al., 1980; Batish et al., 1982). An attempt was made to correlate coagulase, DNase, and enterotoxin production from 200 strains of staphylococci isolated from various foods (Sankaran and Leela, 1982). However, no correlation could be seen between coagulase and enterotoxin or DNase and enterotoxin production. In fact, a number of coagulase-negative strains were found to produce enterotoxins.

Enzymatic assays have been used to determine whether or not some food products have been properly pasteurized, i.e., phosphatase and amylase assays in milk and egg products, or sterilized (J. E. Anderson et al., 1983). However, as in the case of egg products, both the heat-processing conditions and the food components may affect the stability of the enzyme and may thus limit the usefulness of these assays.

In addition to microbial detection and identification, enzymatic assays can be used to monitor metabolic activities in relation to changes in media or environment. For example, Tomlins et al. (1971) assayed several enzymes, including those of the tricarboxylic acid (TCA) cycle, to determine the effects of thermal injury upon *Staphylococcus aureus* and *Salmonella typhimurium*. Decreases in the specific activity of several key enzymes were observed. Enzymatic assays have also been conducted to determine the effects of microwave irradiation and conventional heating upon *Staphylococcus aureus* (Dreyfuss and Chipley, 1980) (Table I) and to

TABLE I

Effects of Microwave Irradiation and Conventional Heating upon Enzymatic Activity of *Staphylococcus aureus*[a]

Enzyme	Specific activity[b]			Ratios	
	Control[c]	Irradiated[d]	Heat-treated[e]	Irradiated/control	Heat-treated/control
Glucose-6-phosphate dehydrogenase	6.16	4.57	8.51	0.74	1.38
Malate dehydrogenase	3.60	9.82	6.64	2.72	1.84
α-Ketoglutarate dehydrogenase	29.34	51.82	7.10	1.76	0.24
Lactate dehydrogenase	2.89	2.81	1.50	0.97	0.52
Alkaline phosphatase	16.27	16.31	14.61	1.00	0.90
Cytochrome oxidase	0.45	0.55	—[f]	1.23	—
ATPase[g]					
cell lysate	0.81	1.26	1.35	1.55	1.66
cell walls	0.28	0.29	—	1.06	—

[a] From Dreyfuss and Chipley (1980).
[b] Expressed as milliunits per milligram of total protein.
[c] Cells incubated at 34°C.
[d] Cells irradiated for 20 sec; internal temperature of media, 46°C.
[e] Cells conventionally heat-treated for 39 sec; internal temperature of media, 46°C.
[f] Not assayed.
[g] Expressed as moles of inorganic phosphate liberated per gram of total protein.

TABLE II

Enzymatic Activity of *Aspergillus flavus* following 65 krad X-ray Treatment[a]

Enzyme	Culture age (days)	Level of activity[b]		Difference (%)	Order[c]
		Test	Control		
Lactate dehydrogenase	6	0.0046	0.0045	2	T > C
	7	0.0835	0.0068	1128	T > C
	8	0.0024	0.0010	140	T > C
Glycerol dehydrogenase	6	0.0103	0.0065	59	T > C
	7	0.0048	0.0042	14	T > C
Glucose-6-phosphate dehydrogenase	6	0.0035	0.0056	38	T < C
Malate dehydrogenase	6	0.4000	0.5000	20	T < C
	7	0.0160	0.0145	10	T > C
	8	0.0024	0.0028	14	T < C
Peroxidase	6	0.0047	0.0062	24	T < C
	7	0.0206	0.0222	7	T < C
	8	0.0055	0.0072	24	T < C

[a] From Chipley (1981).
[b] Expressd as change in absorbance/min/mg of protein.
[c] T, test treatment, C, control treatment.

determine the effects of high-energy irradiation (X-ray) upon *Aspergillus flavus* (Chipley, 1981) (Table II). Addition of salt to laboratory media significantly stimulated the release of penicillinase from *S. aureus* (Kim et al., 1979).

It is not possible to provide a detailed review of the numerous procedures developed for enzymatic analyses. For details regarding specific enzymes, selected references (Table A-II) such as Speck (1976) should be consulted.

2. Microbial Toxins

Bacterial contamination of food is the most frequent cause of foodborne disease. Many bacterial pathogens (salmonellae, some shigellae, and some enteropathogenic strains of *Escherichia coli*) conveyed by foods invade the intestinal mucosa, causing true infections. Illness may be due to the presence of endotoxins. Others (*Vibrio cholerae*, some enteropathogenic *E. coli*) release enterotoxins during growth or lysis of microbial cells or during sporulation (*Clostridium perfringens*) in the gut. Other bacteria (*Clostridium botulinum* and *Staphylococcus aureus*) produce exotoxins as they grow within a food, and, when the food is eaten, cause an intoxication (Sanders et al., 1976).

Staphylococcal intoxication, salmonellosis, and *Clostridium perfringens* gastroenteritis have been and are currently the foodborne diseases reported most frequently in the United States (Bryan, 1974). However, other agents of increasing significance during the past 15 years have also been reported (Roberts, 1982; Bryant, 1983). These include *Bacillus cereus*, enteropathogenic *E. coli*, *Vibrio parahaemolyticus*, *V. vulnificus*, *Yersinia enterocolitica*, *Listeria monocytogenes*, and *Campylobacter jejuni*. *Campylobacter jejuni*, for example, is the subject of an entire reference (Butzler, 1984). Hence, the testing of foods incriminated in outbreaks should not be limited solely to those methods that are selective for the commonly recognized etiological agents (Sanders et al., 1976).

The following sections on toxins are not intended to be comprehensive reviews. Rather, the more recent methods of detection will be briefly discussed. For more detailed information, the reader should consult the selected references cited in each section.

a. Toxins from Gram-Negative Bacteria. Gram-negative bacteria have been characterized by their production of endotoxins. Endotoxins consist of a lipopolysaccharide layer of the cell envelope complex. This layer is pyrogenic and responsible for many of the symptoms that accompany infections caused by gram-negative bacteria (Jay, 1978).

Endotoxins have been extensively studied over the past 20 years. Excellent reviews of earlier research involving endotoxins are available (Braude, 1964; Asselineau, 1966; Nowotny, 1966; Kass and Wolff, 1973; Bernheimer, 1976; DeFigueiredo and Splittstoesser, 1976).

Methodologies for the detection and quantitation of endotoxins have been rather slow in developing. This may be related to the complexity of structure and to the methods employed for isolation of endotoxins (Chipley, 1972, 1974; Darveau and Hancock, 1983).

In food analyses, an indirect approach towards solving this problem has been through quantitation of gram-negative bacteria by plating procedures employing classical bile salts–containing media. However, this method is time-consuming and depends upon the capacity of the cells to produce colony-forming units under the cultural and incubation conditions employed.

The advantages of a direct assay method for endotoxins are obvious from the standpoints of savings in labor, material, and time, and the ability to detect both viable and nonviable cells. An early study by Davis et al. (1969) involved the use of gas–liquid chromatography to analyze sugars in endotoxins. Several strains of *Salmonella*, *Escherichia*, and *Proteus* were rapidly analyzed by this technique. Other methods for endotoxin detection have been outlined by Speck (1976) and/or discussed in other sections of this chapter.

Currently, one of the most rapid and sensitive assays for detection of endotoxins is the *Limulus* lysate test, which was discussed in Section II,G,6.

Although the source of endotoxin has been identified as the surface of the bacterial cell, circulating endotoxin, free of cells, can also be detected during infections. Endotoxins can be found in cell-free filtrates of cultures of gram-negative bacteria. In fact, the release of such "free endotoxin" does not require cell lysis but is a consequence of normal growth (Russell, 1976).

Recent research has indicated that a variety of extracellular products may be produced by gram-negative bacteria during their normal cycle of growth. In addition to endotoxins, these products may also be toxigenic in nature. Enterotoxins from *E. coli* have been isolated, and modified bioassays have been developed (Stavric and Jeffrey, 1977; Waldman *et al.*, 1984). Purification and characterization of an exotoxin from *Pseudomonas aeruginosa* has been reported (Callahan, 1974), and its mechanism of action elucidated (Pavlovskis *et al.*, 1978). An enterotoxin from *Vibrio cholerae* has also been extensively studied (Bennett and Cuatrecasas, 1975). Preformed enterotoxins have been detected in cultures of *Yersinia enterocolitica* (Velin and Emody, 1982) and *V. vulnificus* (Gray and Kreger, 1985). Enterotoxins from several gram-negative bacteria have similar properties and modes of action (Gemmell, 1984).

The role of the Enterobacteriaccae in producing other toxins, such as histamine, has also been recognized (Cattaneo and Cantoni, 1978). In ground beef, significant correlations were noted between the concentrations of amines and total microbial counts (Sayem El Daher *et al.*, 1984; Sayem El Daher and Simard, 1985). It was suggested that amines might serve as indicators of beef quality. However, in fish, no direct correlation was found between the levels of histidine-decarboxylating bacteria and levels of histamine (Karolus *et al.*, 1985). A simple medium has been developed for detection of amines produced by the Enterobacteriaceae (Taylor and Woychik, 1982).

b. Toxins from Gram-Positive Bacteria. Several gram-positive bacteria possess the ability to produce exotoxins. Exotoxins are protein in nature and are generally released by cells during their normal cycle of growth. A major exception occurs in *Clostridium perfringens* enterotoxin, which is released during sporulation.

Because exotoxins can be obtained and purified in reasonable quantities from cell-free extracts, methodologies for detection and quantitation have proceeded at a somewhat quicker pace compared to those for gram-negative endotoxins.

In food microbiology, several serological techniques offer potential for the rapid detection and quantitation of minute amounts of exotoxins. Enterotoxins from *Staphylococcus aureus* and *Clostridium perfringens*, as well as exotoxins from *Clostridium botulinum*, have been analyzed by these techniques. A comparison of several serological techniques for sensitivity and analysis times has been reported (Jay, 1978).

A rather unique direct skin test for rapid detection of staphylococcal enterotoxin B in foods has been proposed (Scheuber *et al.*, 1983). This test involved the use of highly sensitized guinea pigs. The entire assay could be finished within 20 min, with a sensitivity of 10–100 pg of enterotoxin per milliliter of food sample.

Excellent reviews regarding isolation, detection, and characterization of exotoxins are available (Bernheimer, 1976; deFigueiredo and Splittstoesser, 1976; Speck, 1976; Ajl *et al.*, 1970, and subsequent volumes; Stephen and Pietrowski, 1981; Alouf *et al.*, 1984). Detailed, excellent reviews of many techniques used for the identification of staphylococcal toxins in clinical and food microbiology laboratories are also available (Sperber, 1977; Freer and Arbuthnott, 1982).

A guide to aid in the detection of exotoxins, as well as other toxic agents, according to symptoms and incubation periods has been proposed by Sanders *et al.* (1976). Detection methods for specific exotoxins are discussed in other sections of this chapter.

c. Mycotoxins. A mycotoxin may be defined as a toxic, secondary metabolite produced during the growth of a fungus. Toxicity syndromes resulting from ingestion of foodstuffs contaminated with mycotoxins are referred to as "mycotoxicoses." These toxins have been the object of very intensive research, since it was discovered that some of them are the most potent biologically produced carcinogens known. Since their discovery approximately 30 years ago, hundreds of reports have been published annually involving isolation, detection, purification, and characterization techniques. Several pertinent references have been published and are cited in a timely review of the occurrence and control of mycotoxins and mycotoxicoses (Rodricks, 1976).

Mycotoxins are extremely diverse in structure, and procedures for their detection are equally diverse. Thin-layer chromatography was one of the earliest detection methods to be employed, and a relatively simple screening procedure for the detection of 18 mycotoxins was described by P. M. Scott *et al.* (1970). Stoloff *et al.* (1971) developed a detection method for 5 major mycotoxins based upon selective extraction with acetonitrile–water, and subsequent transfer to chloroform. In 1976, a significant improvement in detection procedures became available with the application of high-performance liquid chromatography (HPLC) to this field. Pons (1976) developed an HPLC procedure to resolve and detect 4 major aflatoxins (B_1, B_2, G_1, and G_2) in 15 min. Peanut butter samples, as well as other agricultural products, could be rapidly analyzed with significant savings of time and solvent consumption. A modification of this procedure was developed for HPLC detection of 7 mycotoxins (Schweighardt and Leibetseder, 1981). Aflatoxin M_1 can be rapidly detected in milk and dairy products by HPLC (Chambon et al., 1983), and T-2 toxin can be detected in cereals with this technique (Schmidt and Dose, 1984). Other detection procedures have been reported. They include spectrofluorometric and gas chromatographic assays for sterigmatocystin and *Fusarium* toxins (Maness et al., 1976; Vanyi et al., 1982), Sephadex LH-20 column chromatographic assays for *Penicillium* mycotoxins (Bouhet et al., 1976), and fluorometric assays for aflatoxins (Chakrabarti, 1984). Enzyme-linked immunosorbent assays have been developed for mycotoxins, such as T-2 toxin (Gendloff et al., 1984). Several immunoassays are available for mycotoxin detection (Chu, 1984).

Many of the mycotoxins exhibit carcinogenic, mutagenic, and teratogenic properties. Therefore, bioassays have been recommended as integral portions of detection programs (Vesely et al., 1983). Hundreds of reports have been published in which ducklings, rainbow trout, rats, mice, rabbits, chicken embryos, dogs, monkeys, guinea pigs, hamsters, sheep, and cattle served as bioassay systems. These systems have been used to detect both native mycotoxins and their metabolites, for example, the tissue distribution and metabolism of labeled aflatoxins in broiler and layer chickens (Chipley et al., 1974). However, some authors feel that bioassays have given little valuable information in the surveillance of food and foodstuffs (Watson and Lindsay, 1982).

With the introduction of the Ames procedure (Ames et al., 1973), a new system was developed for detecting carcinogenic compounds. This system was based on the observation that in bacteria, carcinogens acted as mutagens. Another detection system was developed using recombination-deficient mutant cells of *Bacillus subtilis* (Ueno and Kubota, 1977; Mazza, 1983). Several mycotoxins were tested, and a good correlation between mutagenic effects in this organism and carcinogenic effects in animals was reported. Detection systems based on induction of bacteriophages from lysogenic cells of *Bacillus megaterium* with aflatoxin B_1 (Whittaker and Chipley, 1979) and inhibition of protein synthesis by T-2 toxin (Thompson and Wannemacher, 1984) have also been reported. Toxicity of mycotoxins, as well as other microbial products, can be determined within 30 min with a commercially available bacterial bioassay system (Microtox System, Beckman Instruments, Carlsbad, California).

Detailed reviews of toxigenic fungi, mycotoxins, assay procedures, and control mechanisms have been published by Rodricks and Lovett (1976), Egan et al. (1982), and Naguib et al. (1983).

3. Gas–Liquid Chromatography

Although originally employed to identify anaerobic bacteria of medical and veterinary importance, gas–liquid chromatography has been gaining the approval of food microbiologists over the last 5–10 years. The detection of both volatile and nonvolatile metabolites in culture media and foods has been reported.

Clostridium perfringens can be identified within 35 hr after inoculation of specimens by the use of gas–liquid chromatography (GLC) (Anderson and Fugate, 1977). *Clostridium botulinum* has also been detected in contaminated food by the use of GLC (Mayhew and Gorbach, 1975). In this procedure, the authors used butyric and other short-chain fatty acids as presumptive indicators of *C. botulinum*. Analysis of fatty acids has also been used to identify other anaerobes (Sullivan et al., 1978) and yeasts (Moss et al., 1982). Complete fatty acid profiles of bacteria, including *Streptococcus faecalis*, can be easily obtained (Moss et al., 1980; Teixeira et al., 1983).

Richter et al. (1982) were able to differentiate *Salmonella gallinarum* from *S. pullorum* based upon GLC measurement of succinic acid production from dulcitol media. GLC has also been used for the identification of *Achromobacter* (Dees and Moss, 1978), *Bacillus cereus* (Niskanen et al., 1978; Larsson et al., 1984), *C. perfringens* (Larsson et al., 1978a,b), anaerobes (Guillou and Chevrier, 1979; Rizzo, 1980), and *Streptococcus* (Drucker et al., 1982). A GLC technique has been developed for detection of β-nitropropionic acid produced by certain fungi (Gilbert et al., 1977) and for detection of aflatoxins (Rosen et al., 1984). Microbial hydrolysis of hippurate and conversion of fumarate to succinate have been detected with GLC (Kodaka et al., 1982). Headspace gas chromatography has been employed in microbial identification procedures for the analysis of amines and other volatile bacterial products (Larsson et al., 1978a,b; Van Vuuren et al., 1978; Lee et al., 1979; Snygg et al., 1979; Britz and Steyn, 1979; Schafer et al., 1982; Salveson and Bergan, 1981). This technique can be automated for greater sensitivity (Larsson and Holst, 1982).

Pyrolysis gas–liquid chromatography (PGLC) was first used by Reiner (1967) to anlyze more than 1500 microbial species and strains. With this technique, complex components of a bacterial cell are thermally degraded into volatile constituents that are monitored by recorders and integrator/analyzers. The term "fingerprinting" was first used by Reiner (1967) to describe the resulting pyrograms. PGLC has been subsequently used to differentiate gram-positive and gram-negative bacteria (Wasserfallen and Rinderknecht, 1978; Gutteridge and Norris, 1979; Dahlen and Ericsson, 1983), *Salmonella* serotypes (Emswiler and Kotula, 1978), *Staphylococcus* (Magee et al., 1983), *Bacillus* species (O'Donnell et al., 1980), proteolytic and nonproteolytic strains of *C. botulinum* (Reiner and Bayer, 1978), and fungi (Blomquist et al., 1979). Differences in growth conditions can dramatically alter the resulting pyrograms (Gutteridge and Norris, 1980; Drucker and Gibson, 1982). Stepwise discriminant analysis of PGLC data has been used to differentiate between invasive and non-invasive strains of *Yersinia enterocolitica* (Stern et al., 1980) and to correctly identify 12 strains of *B. cereus* (Stern, 1981b). However, PGLC does not appear to be capable of objectively identifying all bacteria with the methods presently available (Stern, 1982c).

4. Other Chemical Methods

There are many reports of diverse chemical methods and procedures whereby microorganisms or their metabolites can be identified. For example, HPLC has been used in the presumptive identification of *E. coli* (Shihabi and Wasilauskas, 1979) and *Salmonella* (Hirsh and Martin, 1983a,b). Other techniques include density centrifugation (Basel et al., 1983), continuous-flow analysis of enzymatic actvities, gel electrophoresis, affinity and circular thin-layer chromatography, chemical profile analysis of antibiotic sensitivities, atomic absorption spectrophotometry, pyrolysis/mass spectrometry, direct-probe mass spectrometry (Puckey et al., 1980), and acidometric and dip-slide methods. References pertaining to most of these methods may be found in an excellent review by Southern (1979). Completely automated systems exist for some of these analyses. They include the AutoMicrobic (Vitek Systems, Inc.), the MS-2 (Abbott Laboratories, Dallas, Texas), and the Sceptor (Johnston Laboratories, Towson, Maryland) systems.

I. Rapid Methods of Identification

Several corporate, federal, and educational institutions have accelerated their research on development of mechanized and automated microbiological procedures. The need for this type of research was clearly voiced by Dr. Robert Angelotti of the U.S. Food and Drug Administrations Bureau of Foods. Speaking at a seminar on the need for automating microbiological procedures, he stated that "Robert Koch could enter a modern microbiology laboratory and find very little that would be new or strange. Microorganisms today are cultured, purified, plated, and counted much as they were 100 years ago."

This situation, however, is changing rapidly with new instruments and systems being introduced on a yearly basis for rapid and automated detection, isolation, quantitation, and characterization of all types of microorganisms (Stanndard et al., 1982; Tilton, 1982; Nelson, 1985). This portion of the current chapter on methods of identification will attempt to highlight some of these developments. A format for the teaching of automation and rapid methods in food microbiology has been proposed (Fung, 1980).

1. Biochemical Systems or Kits

Traditionally, batteries of biochemical tests have offered the best means for specific identification of isolated microorganisms. For example, flow diagrams based upon typical biochemical reactions have served as the basis for identifying important members of the Enterobacteriaceae (Ewing, 1970; Ewing and Davis, 1970; Ewing et al., 1970; Burdash et al., 1978).

During the past 15–20 years, these tests have been miniaturized and placed in small, disposable growth chamber devices or kits. Although each brand of kit is somewhat unique, they all contain prepackaged combinations of substrates and tests based upon relatively conventional systems of biochemical differentiation. Large numbers of bacterial isolates could thus be identified with significant savings of time, money, space, labor, and material. A reference entitled "Miniaturized Microbiological Methods" (Hartman, 1968) cited several of the earliest published works in this field.

Currently, there are a number of these kits being marketed in the United States (Table III). They provide about 90–95% accuracy compared with conventional techniques, and some utilize special codes and

TABLE III

Miniaturized Microbiological Identification Systems

System	Number of tests performed	Supplier
API-20 E	22	Analytab Products, Plainview, New York
API-50 E	50	
API-Staph-Ident	19	
Enteric 20	20	Inolex Corporation, Glenwood, Illinois
Spectrum-10	20	Austin Biological Laboratories, Austin, Texas
Enterotube II	15	Roche Diagnostics Division, Hoffman-LaRoche, Inc., Nutley, New Jersey
Micro-ID	15	General Diagnostics Division, Warner-Lambert Co., Morris Plains, New Jersey
Minitek	20	BioQuest Division, Becton, Dickinson and Co., Cockeysville, Maryland
Pathotec	13	General Diagnostics Division, Warner-Lambert Co., Morris Plains, New Jersey
R/B	14	Corning Medical Microbiology, Corning Glass Works, Rosyln, New York
Enteric-Tek	14	
Oxi/Ferm	10	Roche Diagnostics Division, Hoffman-LaRoche, Inc., Nutley, New Jersey
Biostix/Mycostix	1/1[a]	Ames Co., Division, Miles Laboratories, Inc., Elkhart, Indiana
Nalgene Nutrient Pad Kits	13[a]	Nalgene Labware Department, Nalge Co., Rochester, New York
RSBA	3[b]	Akro-Medic Engineering, Denville, New Jersey
Salmonella Fluro-kit	1[c]	Clinical Sciences, Inc., Whippany, New Jersey
Salmonella Bio-Enzabead Screen Kit	1[d]	Litton Bionetics, Charleston, South Carolina
SeroSTAT Staph	1[e]	Scott Laboratories, Inc., Fiskeville, Rhode Island
Lysostaphin Test Kit	1[f]	Remel, Lenexa, Kansas

[a] Detection and quantitation of bacteria and fungi.
[b] Rapid sequential bacteriological analyzer for total bacterial counts, total coliform counts, and *Salmonella*.
[c] Fluorescent antibody detection method for *Salmonella*.
[d] Enzyme immunoassay for *Salmonella*.
[e] Agglutination test for *Staphylococcus aureus*.
[f] Enzymatic differentiation of staphylococci from micrococci.

a computer to analyze the results (Guthertz and Okoluk, 1978). When these kits were made commercially available, they were first evaluated in studies involving bacteria of clinical significance, especially members of the family Enterobacteriaceae (Rutherford et al., 1977; McCarthy et al., 1978; Barry et al., 1979a,b). However, several publications have appeared involving their use in food microbiology (Fung, 1982; Jarvis, 1982; Tilton, 1982). Some of these kits have been extensively evaluated for a number of years (Oberhofer, 1983). Others have been evaluated either at instructional workshops or by experienced personnel (Fung et al., 1984; Cox et al., 1984).

Holmes et al. (1977) and Leighton and Little (1983) compared the API 20E, Enterotube, and Pathotec systems (see Table III) for identification of Enterobacteriaceae. There was little or no difference in kit reproducibility and a probability of error of approximately 1.7%. The API and Enterotube systems also gave overall agreements of 88.5 and 75.8% for the confirmation of *E. coli* from dairy products (Cooke et al., 1977). Guthertz and Okoluk (1978) compared four systems with classical methods to identify 134 strains of Enterobacteriaceae. Little difference was observed when identifying genera (92.7–96.9% agreement with conventional methods), but the API system was able to identify 95.2% of the organisms to the species level, while Minitek, Pathotec, and Inolex were able to speciate 88.0, 43.2, and 36.0% of the organisms, respectively. Results comparable to conventional systems have also been reported using a Minitek system for identifying Enterobacteriaceae (Hanson et al., 1978) and *Lactobacillus* (Gilliland and Speck, 1977), and API and Oxi/Ferm systems to identify nonhemolytic streptococci (Waitkins et al., 1980; Ruoff and Kunz, 1983) and nonfermentative bacteria (Isenberg and Sampson-Scherer, 1977; Oberhofer et al., 1977; Oberhofer, 1979; Hofherr et al., 1978; Rosenthal et al., 1978; Holmes et al., 1979; Appelbaum and Leathers, 1984).

A miniaturized technique involving the use of microtiter plates for the microbiological examination of poultry skins for total counts and *Pseudomonas* correlated well with results obtained by standard methods (LaHellec and Colin, 1982). The authors also claimed that the cost for one biochemical identification of a *Salmonella* strain was five to six times less with a miniaturized method than with a classical one. Micromethods have been reported for other gram-negative, nonfermentative bacteria (Gibson et al., 1978), *Salmonella* (Guinee et al., 1983), coagulase-negative staphylococci (Brun et al., 1978), and micrococci from meat and dairy products (Delarras et al., 1979). Rapid slide agglutination tests for *Staphylococcus aureus* are available (Pennell et al., 1984). However, they may be less dependable for detecting antibiotic-resistant strains of this organism than for detecting sensitive strains (Woolfrey et al., 1984).

The Spectrum-10, a newly developed minaturized system, was evaluated for accurate and rapid identification of Enterobacteriaceae by use both of fresh isolates from raw and frozen foods and of stock cultures (Cox et al., 1985). In comparison with the Micro-ID and the API-20E systems, the Spectrum-10 identified 95–96% of the stock cultures to genus and species, whereas 93% of the fresh isolates were identified to genus, and 82% to species.

The advantages of miniaturized systems are low cost, flexibility of tests (except for commercially prepared kits), mass production of data for many isolates, speed of reactions, and savings of space and time. In some cases, identification can be made in 4 hr (Izard et al., 1984). Their disadvantages include the need for manual dexterity, experience, and training (Fung, 1982). Before these systems can be used, preliminary taxonomic grouping of isolates is necessary (Mossel et al., 1983). Pure cultures must be used for inoculation of these systems as well as for inoculation of standard systems. Commercial systems can be adapted for rapid identification of bacteria in foods (Cox et al., 1982).

An excellent review of rapid biochemical testing procedures for Enterobacteriaceae in foods, including poultry products, has been published (Cox and Mercuri, 1979a). Performance data for several systems were given, and pertinent references were cited in this article. Several advantages, disadvantages, and precautions regarding the use of these rapid procedures were also emphasized. Many systems can be programed for computer-assisted identification (Kelley and Kellogg, 1978; Kellogg, 1979; Phillips and Amsterdam, 1977; Lee et al., 1982). Multitest systems for rapid identification of yeasts are also available (Qadri and Nichols, 1978; Cooper et al., 1978; Gille and Guinet, 1978; Land et al., 1979).

Three of the kits described above have been adopted as official first action as alternatives to the Association of Official Analytical Chemists (AOAC) biochemical tube system for presumptive generic identification of foodborne *Salmonella* and for screening and eliminating non-*Salmonella* isolates (Poelma et al., 1978). Use of the API system has also been proposed for differentiation between emetic and other strains of *B. cereus* (Logan et al., 1979), for identification of *Bacillus* strains (Logan and Berkeley, 1984), for characterization of staphylococci (Hofherr and Lund, 1979; Kloos and Wolfshohl, 1982), and for identification of anaerobic bacteria (Karachewski et al., 1985).

The practical value of these kits or systems for rapid identification of microorganisms has been appraised (Duke, 1980), and a bibliography of research involving their use in microbiology and immunology during 1976–1980 has been published (Palmer, 1981). A cost-comparison study revealed that for the total direct costs of identifying an organism, kits were cheaper than conventional techniques ($4.30–4.96 versus $5.66) (Matsen and Bale, 1981).

2. Methods for Replication and Counting

Techniques for microdilution in both broth and agar are currently available for identifying microorganisms. These systems utilize replicators for simultaneous inoculation of several media by many bacterial strains (Burman and Ostensson, 1978; Buckwold *et al.*, 1979), and some systems have also been linked to computer-based, automatic identification modules (Ellner and Myers, 1981; Staneck *et al.*, 1983). Although originally designed for identifying members of the family Enterobacteriaceae, they have been modified for nonfermentative bacilli, anaerobes, and others (Southern, 1979; Lennox and Ackerman, 1984). A rapid microdilution method for total bacterial counts in foods correlated well with the standard plate count method (Kramer, 1977; Zavanella *et al.*, 1980).

A considerable savings in time and analyst fatigue has been realized by the development of automated colony counters (for example, the BioTran, New Brunswick Scientific Co., New Brunswick, New Jersey). Several studies have been conducted to evaluate the precision and accuracy of these instruments. In one such study (DeScrilli *et al.*, 1976), it was possible to count 240–720 plates per hour compared to approximately 40 plates per hour using a manual colony counter. However, in a more recent evaluation (Peeler *et al.*, 1982), the authors stated that the use of these counters might be unacceptable to regulatory agencies, when compared to manual counting.

3. Instruments for Automated Microbiological Analyses

Currently, the food microbiologist has available new instruments, such as the AutoMicrobic system, specifically designed for the automatic detection, enumeration, identification, and characterization of microorganisms. Decisions regarding the purchase of these instruments must take into account such factors as speed, precision, laboratory space, costs, personnel requirements, acceptance of the method by regulatory agencies, feasibility of use for foods, ease and availability of repair, etc. Two of these instruments have been evaluated for use in clinical situations (P. B. Smith *et al.*, 1978; Brown and Washington, 1978; Isenberg *et al.*, 1979, 1980; Tilton, 1982; Dudley *et al.*, 1983). The Abbott MS-2 system

(Abbott Laboratories, Dallas, Texas) provided rapid (5–6 hr) and accurate identification of Enterobacteriaceae (Izard et al., 1982). However, in one study, problems were encountered with identification of *Salmonella* and *Shigella* strains (Chomarat et al., 1982). The AutoMicrobic system has been used to identify several genera of gram-positive bacteria (Ruoff et al., 1982). It also correctly identified 98% of the Enterobacteriaceae and all of the *Salmonella* isolated from food and feed samples (Bailey et al., 1985).

In addition to the instruments described in this chapter, discussions of several others may be found in excellent review articles by Goldschmidt and Fung (1978, 1979) and Fung (1984). An outline of pertinent instrumentation literature is included in these articles.

J. Detection and Enumeration of Injured Microorganisms

Perhaps no other concept has had more profound impact on the field of food microbiology than has the recognition of sublethally injured microorganisms. Within the last 25 years, literally hundreds of published reports have all emphasized the significance of this phenomenon, although it had been recognized decades earlier and reviewed by Nelson (1943, 1944).

Sublethal injury may be defined as a loss of the reproductive and/or metabolic capability of microbial cells following exposure to some nonsterilizing treatment. This loss is evidenced by the failure of injured cells to grow or reproduce under cultural conditions that are satisfactory for unexposed cells. Injury may be classified into two categories (Ray and Speck, 1973): (1) cells may be metabolically injured, i.e., able to multiply and form colonies on a nonselective complete agar medium but not on a minimal salts–glucose–agar medium, and (2) cells may be structurally injured, i.e., able to multiply and form colonies on a nonselective complete agar medium, but not on a selective complete agar medium. These classifications are depicted in Fig. 1.

Injured cells, when placed in a suitable environment, may repair their damaged components and may subsequently grow and reproduce in a normal manner. This should always be considered when monitoring food processing, preservation, and storage in order to ensure the safety of foods and protection of consumers. Other essential considerations could include use of injury to enhance the lethal action of a processing treatment (such as heating with the addition of salt or preservatives), minimizing injury to cultures used in food fermentations, predicting

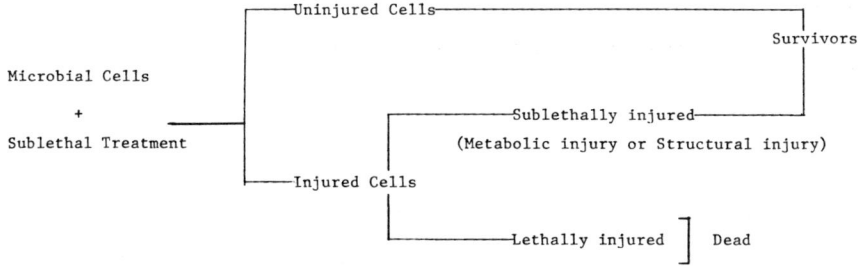

Fig. 1. Effects of sublethal injury upon microorganisms. Adapted from Ray and Speck (1973).

effects of product formula modification on subsequent microbial injury and survival, and eliminating inadequacies within existing and proposed methodologies to detect and enumerate specific microorganisms (Busta, 1976).

Addition of steps which will ensure the recovery of injured microorganisms to methodologies used to detect and enumerate specific microorganisms cannot be overemphasized. Most current methodologies for specific microorganisms utilize growth media that contain inhibitory or selective agents. Because injured cells may not grow in the presence of these agents, a "false security" about the safety of a food product may result. For example, a review, by the author, of data from several published reports in which injured microorganisms were allowed to recover indicated that the numbers of viable cells detected prior to recovery was lower than the numbers detected after recovery by factors generally ranging from 0.5 to 1.5 log cycles. Incorporation of a recovery step was found to be essential when microorganisms were monitored in several food service systems (Chipley and Cremer, 1980).

A survey of the literature pertaining to sublethal injury indicates that the methods by which microbial cells sustain injury are diverse. Likewise, the ways in which these injuries can be manifested are diverse, as can be seen from Table IV. In addition to general changes which are relatively easy to detect, specific changes may occur within the cellular components of injured microorganisms. Specific changes are somewhat more difficult to detect and may be manifested in rather unique ways. For example, mycotoxin production was significantly stimulated when spores of *Aspergillus* were first exposed to X-ray or gamma irradiation and were then allowed to germinate in enriched and minimal media (Chipley, 1981). Progeny of repaired cells of *Staphylococcus aureus* retained their ability to produce enterotoxin, although there were large variations in the amount produced by individual colony-forming units (Smith *et al.*, 1984).

TABLE IV

Summary of Methods Known to Cause Microbial Injury and Some Characteristics of Injured Cells

Method of injury	Characteristics of injured cells
freezing	general changes
Refrigeration	loss of ability to multiply
freeze-drying	extended lag phase
low water activity	increased need for nutrients
frying	increased sensitivity to other treatments
aerosolization	increased sensitivity to surfactants, inhibitory compounds, and salts
temperature alternation (freeze-thawing)	leakage of cellular components
sublethal heating	increased changes in morphology
irradiation	specific changes (may or may not be observed)
high-energy irradiation	spores
dyes and inhibitory compounds	cell walls and membranes
salts	nucleic acids
osmotic pressure (high ionic strength)	proteins, peptides, and amino acids
	enzyme activity
certain food additives	respiration and electron transport
pH extremes (low or high)	motility
sanitizing compounds	agglutination
	genetic stability
	decrease in growth temperature rangs
	ability to recover in simple rather than complex media
	stimulation in production of some metabolites

A number of recovery methods have been proposed for various organisms (Jay, 1978). They range in complexity from short periods of incubation in commercial broths to plating on complex, nutritionally complete agars with or without added nutrients and enzymes. However, in some cases, these recommendations have been questioned. For example, in the recovery of various *Salmonella* serotypes from chilled and frozen broiler carcasses and meat products, it was found that direct enrichment procedures were adequate for detecting low levels of salmonellae. Preenrichment procedures could be omitted (Cox and Mercuri, 1979b; Cox et al., 1980).

Regardless of the types of injury incurred, in most cases, repair is rapid under conditions appropriate for recovery and precedes cellular multiplication, and repaired cells respond normally to the selective agents present in the medium used for their detection. However, notable exceptions do occur as in the case of the "Phoenix phenomenon," which is associated with repair of metabolic injury (Koburger and

Dargan, 1985). Procedures that promote maximum repair must be evaluated to minimize or eliminate multiplication of competing or noninjured organisms (Ordal et al., 1976).

Excellent reviews involving sublethal injury, recovery mechanisms, and procedures may be found in articles by deFigueiredo and Jay (1976), Busta (1976, 1978), Ordal et al. (1976), Jay (1978), Perlman (1978), Andrews (1986), and Ray (1986). In addition, two more recently published references provide extensive information regarding these topics (Andrew and Russell, 1984; Hurst and Nasim, 1984).

III. SAMPLING TECHNIQUES

Changes in the composition of food products may depend to some extent on biochemical activities of microorganisms associated with the products. Sampling for microbial numbers and types is a prerequisite for assessing contamination and its possible effects on food products. Various sampling and screening systems have been developed, and methodologies for sampling have been reviewed in several publications (Dawson and Stadelman, 1960; Sharf, 1966; Dam et al., 1970; Brune and Cunningham, 1971). Since microbial contamination and spoilage occur primarily on the surfaces of many food products, several techniques have been designed to remove microorganisms from exposed surfaces for subsequent enumeration or identification. Several of these procedures are nondestructive in nature so that products may be analyzed during routine quality control operations (Patterson, 1972; Dawson et al., 1979). Basically, these methods fall into one of two categories; (1) elution of organisms by rinsing and assaying the microbial suspension according to accepted microbiological procedures or (2) removal of organisms by direct contact with solid materials that either allow microbial growth directly or are used to transfer organisms to growth-supporting media (Fincher, 1965).

A. Swab or Swab-Rinse Method

The use of cotton swabs for sampling is one of the oldest and most widely used techniques for detection of microorganisms on the surfaces of foods, utensils, equipment, walls, and floors in food plants and institutions.

Quantitative samples can be taken by using a premoistened swab and swabbing a defined area within a sterile template of paper or metal laid over the area to be sampled. The swab is transferred to a buffer solution and shaken, and the resulting microbial suspension is serially plated in a suitable growth medium (American Public Health Association, 1972). This method has been extensively used for microbial analysis of poultry carcasses (Thomson et al., 1976).

When cotton swabs are used, not all of the recovered organisms are released from the cotton into the diluent. In addition, most researchers have also found that approximately 50% of the surface organisms are recovered when good swabbing techniques are used. With the introduction of calcium alginate swabs, all recovered organisms are released into the diluent upon dissolution of the alginate by sodium hexametaphosphate. In recovery studies, most investigators have found alginate superior to cotton swabs. However, when cotton and alginate swabs were compared in a study involving poultry carcasses, no differences were observed in total, Enterobacteriaceae, or *Salmonella* counts (Notermans et al., 1976). When swabbing was compared to tissue removal combined with blending, blending provided higher and more representative values of the microbial flora on beef carcasses (Lazarus et al., 1977; deZutter et al., 1982) and yielded higher total and Enterobacteriaceae counts from broiler skins (Cox et al., 1976).

Angelotti et al. (1964) found the swab-rinse method was better for porous, irregular, and greasy film surfaces, while the RODAC plate method was better for smooth, nonporous, flat surfaces free of greasy films. An alternative to the swab-rinse method is the cellulose sponge swab technique. Cellulose sponges are cut into $2 \times 2 \times 2$ inch pieces individually bagged and steam-sterilized. A sterile sponge is removed, moistened with rinse solution, and used for swabbing. After sampling, the sponge is placed in a sterile plastic bag where it may be subjected to a variety of analytical tests. Procedures for use of swabs and cellulose sponges have been outlined by Gabis et al. (1976).

A rapid technique has been described for sampling surfaces of meats (Olgaard, 1977). Using cotton wool swabs, bacteria are transferred directly to the surface of a segment of an agar plate and incubated. The number of colonies is estimated using a comparator disc and then converted into an average number of colonies per segment, using a table. Results are reported as relative bacterial levels.

In an evaluation of several different types of swabs, it was found that the survival of bacteria on dry swabs varied from 0 to 20% after 4 hr (Mandler and Sfondrini, 1977). Cotton wool swabs markedly reduced survival of bacteria; alginate or albumin swabs were less toxic for gram-positive bacteria, and charcoal swabs less toxic for gram-negative bacteria. Transport systems allowed organisms to remain viable for as long as 24–48 hr. The authors made the following recommendations. Cotton wool swabs should be used only when cultures are made immediately after collection. Dry swabs—charcoal, albumin-coated, or calcium alginate—should be used only when cultures are to be made within 4 hr. For longer periods, only transport systems should be employed.

Storage of swab samples can also affect the subsequent numbers of bacteria recovered. Results of a study involving swab sampling of beef carcasses revealed that both phosphate-buffered solutions and peptone broths were equally effective as diluents for swab samples (Marriott *et al.*, 1978). However, refrigerated storage of swab samples resulted in recovery of fewer microorganisms.

B. Rinse Method

This technique is useful for sampling entire surfaces of relatively small, irregularly shaped objects by immersing in diluent within a closed container and shaking thoroughly. The diluent can then be serially plated in a suitable growth medium.

Although higher average percentage of recoveries may be obtained with surface rinsing compared to swabbing, in some cases the swab method may be more convenient or more logically used. For example, thorough rinsing of large carcasses or meat portions would be difficult, because of the amount of sterile diluent needed and because of problems encountered in thoroughly shaking large containers. Because of these problems, the alginate swab was the method of choice when several techniques for quantitative determination of bacterial contamination on poultry carcasses were compared (Fromm, 1959). However, Mallmann *et al.* (1958) stated that the rinse method was more accurate than the swab method for predicting the shelf life of dressed poultry. A low-volume water rinse technique was successfully used to detect *Salmonella* on broiler carcasses (Cox *et al.*, 1981). Several advantages of using smaller volumes of rinsing fluid were listed. When air-chilled broiler carcasses were surveyed for *Salmonella*, it was found that counts obtained by the rinse method were significantly higher than those obtained by the neck skin maceration method (Dezeure-Wallays and Van Hoof, 1980). More bacteria were removed by the rinse method than by either the swab-rinse or imprint methods when meat surfaces were evaluated (Lame, 1978). The effects of different procedures upon *Salmonella* recovery from catfish were studied by Wyatt *et al.* (1980). Both swabbing and rinsing were the least effective of the methods examined. Immersion of whole fish yielded the highest number of *Salmonella*—positive samples. A modification of the rinse method, the wipe–rinse technique, has been evaluated for the bioassay of large inert surfaces (Kirschner and Puleo, 1979). Results indicated a significantly higher recovery efficiency for polyester-bonded cloth than for cotton swabs when used on these surfaces.

Blending, rinsing, and a technique involving pressing (dispersion) in plastic bags followed by diluent addition were compared for recoveries of bacteria from frozen vegetables (Rey *et al.*, 1985). The pressing technique was found comparable to blending and superior to rinsing techniques for these products.

Detection of *Salmonella* in thaw water, pericloacal skin, and whole-carcass rinse samples from naturally contaminated broiler chickens was compared (D'Aoust *et al.*, 1982b). Recovery with the whole-carcass rinse technique (93%) was significantly greater than that obtained with thaw water (63%) and skin (74%) methods. The authors stated that thaw water does not provide a reliable index of *Salmonella* contamination and therefore is not a suitable method for the nondestructive analysis of poultry carcasses.

Two relatively unique methods were compared for sampling beef and pork slices (Goulet *et al.*, 1983). A procedure involving the use of abrasive discs and a spray gun technique were tested on inoculated meat slices incubated at 4°C for varying time periods. The disc method was slightly less efficient than the spray gun in recovering bacteria during the early stages of incubation but was much more efficient during the later stages.

Techniques for surface rinsing of containers and equipment have been described by deFigueiredo and Jay (1976) and Gabis *et al.* (1976a).

C. RODAC Plate Counting

The RODAC (replicate organism direct agar contact) plating system has been extensively used in determining the sanitary quality of utensil, equipment, and floor surfaces in dairy and food processing plants and in hospitals. Each prepared RODAC plate contains a convex or raised nutrient agar bed for application directly to the surface being tested. When sampling surfaces which have been chemically sanitized, appropriate neutralizing compounds should be incorporated into the medium. The sampling procedure (Gabis et al., 1976b) consists of removing the plastic lid and carefully pressing the agar surface to the surface being sampled, making certain that the entire agar meniscus contacts the surface. The lid is replaced, plates are incubated in an inverted position, and colonies are counted. Ideally, the RODAC plate method should be used on previously cleaned and sanitized flat, impervious surfaces. It should not be used for pervious, creviced, or irregular surfaces (Gabis et al., 1976b). Cotton swab and RODAC methods were compared for the recovery of *Bacillus subtilis* spore contamination from stainless steel surfaces (Angelotti et al., 1964). The swab method recovered an average of 47% of the spores, while the RODAC method recovered 41%. However, results from the RODAC method were more reproducible than those from the swab technique.

RODAC plate and calcium alginate swab techniques were compared for detecting bacterial contamination of surfaces (Scott et al., 1984). High levels of contamination were demonstrated more frequently with swabs, but for some sites where low numbers of organisms were present, higher recovery rates were obtained with RODAC plates. Although results from the two methods were correlated, the authors recommended the plate method for sampling large numbers of sites.

Two other direct agar contact methods, the agar syringe and the agar sausage, have been described by deFigueiredo and Jay (1976). Both the swab and the agar syringe techniques were found to be less reliable for sampling carcass surfaces compared to an excision method (Nortje et al., 1982).

An adhesive tape method has been developed for estimating the microbial load on meat surfaces (Fung et al., 1980). With red meat, this method was found to be significantly correlated with conventional sampling methods. Swab, RODAC, and Mylar adhesive tape methods were evaluated for estimating surface bacteria on pork carcasses (Cordray and Huffman, 1985). The authors suggested that the tape method was a good alternative for estimating surface bacterial loads.

A membrane filter contact technique for sampling of moist surfaces has been described (Craythorn et al., 1980). Using this method, the recovery of *S. aureus* from beef muscle surfaces was significantly higher than the recovery with RODAC plates.

Decreasing the sampling variance in bacteriological analyses of meat surfaces may be simply a matter of rubbing the meat surfaces together prior to sampling (Anderson et al., 1980).

An excellent comparison of several of the surface sampling methods just described has been published by Baldock (1974). Advantages, disadvantages, and limitations of each method are discussed.

D. Stomacher

The stomaching technique was developed by Sharpe and Jackson (1972) for preparation of food samples for microbiological analysis. With this technique, a sterile plastic bag containing sample and diluent is tightly clamped into an instrument called the "Stomacher." The action of two reciprocating paddles produces sponging and shearing forces on the sample. These forces effectively release even deep-seated bacteria. After samples are taken for analysis, the bag and its remaining contents may be discarded. The effectiveness of this technique for analysis of chicken skin has been evaluated by scanning electron microscopy (Thomas and McMeekin, 1980).

Emswiler et al. (1977b) compared two techniques (Stomaching versus blending) for the recovery, isolation, and identification of several selected types of bacteria from meat products. Their data (Table V) indicated that differences between the two techniques were small and nonsignificant. The stomaching method was favored because of the reduction of labor involved. Advantages and disadvantages of this method were discussed.

Modification of the Stomacher bag has been reported by Konuma et al. (1982). When tested in combination with the spiral plating technique, use of the modified bag prevented large particles of food samples from blocking the hole of the stylus tube of the spiral plating machine. Both the Stomacher technique and the spiral plate method are recommended for the microbiological analysis of meat (Gerats and Snijders, 1978). In general, food samples prepared by blending or stomaching yielded coliform and *Salmonella* counts which were not statistically different (Andrews et al., 1978a). Addition of Tween 80 did not result in a consistent increase in coliform counts. However, stomaching yielded aerobic plate counts which were lower compared to blending (Andrews et al., 1978b). A scrape-sampling device for the microbiological examination of poultry carcasses yielded only slightly lower counts than did samples which were stomached (Adams et al., 1980). Skin tissue was removed from five different sites on broiler carcasses and was stomached (Klinger et al., 1981). No significant differences among plate counts were found among the different sites.

Five techniques were evaluated for recovering *B. subtilis* spores from inoculated beef (Kreiger et al., 1984). The combined use of a tissue homogenizer with ionic dispersants was more effective than washing, blending, sonicating, or stomacher procedures for quantitative recovery.

TABLE V

Bacterial Counts from Meat Products Homogenized by Stomacher and Blender[a,b]

Meat type	Aerobic plate count (20°C)				S. aureus			
	Stomacher		Blender		Stomacher		Blender	
	With Tween	Without Tween	With Tween	Without Tween	With Tween	Without Tween	With Tween	Without Tween
Ground beef	1.3×10^7	1.0×10^7	1.6×10^7	1.5×10^7	13	42	32	24
Frankfurters	2.0×10^6	1.3×10^6	2.3×10^6	1.6×10^6	0	0	0	0
Round steak	1.2×10^5	7.0×10^4	1.3×10^5	1.4×10^5	1	1	2	2
Chicken skin	1.3×10^5	1.7×10^5	2.3×10^5	1.8×10^5	133	24	75	18
Chicken meat	3.8×10^4	3.3×10^4	4.8×10^4	4.5×10^4	6	6	3	7

Meat type	Coliforms				Fecal coliforms				E. coli			
	Stomacher		Blender		Stomacher		Blender		Stomacher		Blender	
	With Tween	Without Tween	With Tween	Without Tween	With Tween	Without Tween	With Tween	Without Tween	With Tween	Without Tween	With Tween	Without Tween
Ground beef	3311	4572	5126	3251	98	60	38	138	72	40	25	32
Frankfurters	0	1	0	2	0	0	0	0	0	0	0	0
Round steak	12	8	10	12	2	2	3	3	1	1	2	3
Chicken skin	559	584	753	1076	317	298	353	508	77	67	223	183
Chicken meat	93	50	80	58	18	17	17	14	9	10	14	8

[a] From Emswiler et al. (1977b).
[b] Each value is the average of 8 samples, in bacterial counts per gram.

E. Atmospheric Sampling

In addition to the surface sampling techniques just described, one must also be aware of the consequences that airborne contamination may have upon products or processes. An excellent discussion of methods and instruments for air sampling has been published by Fincher and Mallison (1967). Centrifugal air samplers are available with a variety of media for detection of several types of airborne microorganisms (Hycon, Division of Seralc, Inc., Miami, Florida). Studies comparing some of these methods have been published (Johnston et al., 1978; Anonymous, 1978; Delmore et al., 1981; Lundholm, 1982; Henningson et al., 1982).

F. Sampling—General Considerations

Sampling might be defined as the number of individual portions, units, or containers randomly taken from an overall population or quantity to which analytical tests will be applied in order to satisfy the requirements of a predetermined plan. Any sampling plan must be based on random samples taken throughout the population or quantity. It is also based upon the hypothesis that any defect that is to be analyzed is distributed randomly throughout.

Sampling is necessary, because it is usually not practical or possible to subject entire quantities of food ingredients or products to microbiological examination. Results from microbial analyses of samples are therefore used to draw conclusions about the entire population or quantity.

There are several areas of concern inherently associated with any sampling program for microbiological analysis. They include the following:

1. Hazard. Without question, this should be of paramount concern. What are the types and numbers of microorganisms present? How hazardous are they? Are laboratory analyses to be based upon qualitative or quantitative techniques?

2. Characteristics of the product. What ingredients are used for formulation? What is the effect of each upon the microorganisms present? Is the product perishable? Is the final product uniform in nature, e.g., a liquid with all ingredients thoroughly dissolved and evenly distributed throughout?

3. If ingredients from outside suppliers are used, what is their case history?

4. What are the critical control points for the formulation of the product? Are there procedures within the formulation, processing, storage, and distribution which might be favorable for microbial contamination or growth?

5. Is the sampling plan feasible? How many samples should be taken? What quantity should they be? How often should they be taken and by whom? What facilities are available for storage and/or transportation to the laboratory for analysis?

From these questions, it should be obvious that a correct sampling plan requires careful attention to details. At stake are the health and safety of the consumer as well as the reputation and economic health of the manufacturer. For details, the reader is referred to sampling procedures outlined by Speck (1976) and to benchmark references by Thatcher and Clark (1974) and the National Academy of Sciences (1985) involving principles and specific proposals for sampling and sampling plans. Statistical aspects of sampling have been discussed by Cowell and Morisetti (1969) and Malcolm (1984). A procedure for random sampling of frozen chicken to determine the occurrence of salmonellae has been proposed (Hildebrandt et al., 1977). Several sampling schemes have been reviewed for use within the meat industry (Kilsby, 1982).

IV. EVALUATION OF INDICATOR MICROORGANISMS AS A MEASURE OF FOOD QUALITY

Methods used to detect pathogenic microorganisms are not as effective whenever the pathogens occur in low numbers in a food. This is particularly true when the food also contains large numbers of other organisms. In order to overcome these problems, many food microbiologists have suggested the use of groups or types of organisms which are more easily detected. Their presence in foods may indicate that the foods were exposed to conditions favorable for the introduction and growth of pathogenic organisms. These groups or types of organisms are called "indicator" organisms. They have been extensively used to assess both the microbial quality and the safety of many foods over the past three decades.

When choosing which organisms to use as indicators, Buttiaux and Mossel (1961) established an early set of guidelines which are still currently appropriate. These include (1) specificity of occurrence—ideally, the bacteria of choice should be found only in intestinal environments; (2) large population—they should occur in very high numbers in feces, so that they can still be detected in high dilutions; (3) resistance to environment—they should be highly resistant to the environment in order to be effective indicators of fecal pollution, and (4) easy and reliable detection—they should be easily detectable even when present in very low numbers. An assessment of the microbiological safety and keeping quality of foods by enumeration of indicator organisms has been published (Doyle, 1979). However, during the processing of broiler chickens at sealed temperatures of less than 60°C, none of several bacterial groups tested was found to be suitable as an index of fecal contamination during subsequent processing (Notermans et al., 1977).

The following section is not intended to be a comprehensive review of indicator microorganisms. Rather, the more significant characteristics of those indicator systems currently in use or proposed for use will be briefly discussed. For more detailed information regarding indicators, the reader should consult Thatcher and Clark (1973), deFigueiredo and Jay (1976), Speck (1976), Jay (1978), and the National Academy of Sciences (1985). An excellent early review of this subject is also available (Slanetz et al., 1963). A European symposium involving indicator organisms in food and water (*Antonie van Leeuwenhoek* 48, 609-644) and a symposium entitled "Indicator Organisms: A Current Look at Their Usefulness" have recently been published (*Food Technol.* 37, 105–117).

A. Total Count

Total plate counts are a measure of the number of aerobic, mesophilic microorganisms in a food product. There is no attempt to further characterize the microbial population as to specific types, classes, or genera. Total plate counts (also called "aerobic plate counts") do not necessarily detect the total microbiological population in a food sample. Instead, they measure only that fraction of the population that is able to produce visible colonies in the medium used under the conditions in which the plates are incubated (National Academy of Sciences, 1985).

Total counts have been used for many years to evaluate the sanitary quality of dried or frozen foods and milk (Silliker, 1963). However, perishable items such as fresh meats cannot be accurately evaluated by this technique alone. The growth of different types of microorganisms in and on meats will cause different types of spoilage, depending upon the conditions of treatment and storage. Several workers have demonstrated that, during the spoilage of meats, the growth patterns of specific organisms, such as *Pseudomonas*, are more important than knowing the total numbers present.

There are instances in which foods (with the exception of fermented products) might be regarded as unwholesome when they contain large numbers of organisms, even though the organisms may not be known pathogens and they have not noticeably altered the character of the food (Thatcher and Clark, 1973). For example, high total counts often indicate contaminated raw materials, unsatisfactory sanitation, or unsuitable time or temperature conditions during formulation, processing, or storage. Several common bacteria not normally regarded as food pathogens (such as fecal streptococci, *Proteus*, *Pseudomonas*, and others) have been suspected of causing food poisoning outbreaks when present in large numbers in food. High total counts may also indicate the likelihood of spoilage, since many foods contain 10^6 to 10^8 microorganisms per gram at the time when spoilage becomes evident.

Thatcher and Clark (1973) also pointed out several important limitations to the value of total counts. For instance, in foods naturally fermented by large numbers of bacteria, high total counts have practically no significance. In dried, frozen, or heat-processed foods, viable counts may not accurately reflect the true sanitary history, and other detection methods, such as direct microscopic examination, may be required. If incubation temperatures of inoculated plates do not match storage temperatures of the food under examination, then total counts may be of little value in predicting shelf life.

It should be noted that foods cannot always be assumed to be microbiologically safe just because they contain low total counts. Monford and Thatcher (1961), in an early study of commercial frozen-egg preparations, isolated *Salmonella* and coliforms from several samples containing total counts of less than 5,000/g. The possibility also exists that foods may have low total counts but contain toxins (either endotoxins or exotoxins) that remain stable to conditions which may not favor continued survival of viable cells.

Levine (1961) reviewed the relative merits of using coliforms, enterococci, and total plate counts as indices of sanitary quality for water and foods. He suggested that total plate counts would probably give as much information as any other suggested microbial index for assessing processing and storage of food products. In 1976, two reports by Solberg et al. and Miskimin et al. compared the total, coliform, and E. coli counts for evaluating the safety of a large number of raw and ready-to-eat foods. The authors stated that it is not the number of pathogens with respect to the number of indicator organisms that is important but rather the level of indicator organisms which may represent the presence of any pathogen. None of the indicator tests provided any useful knowledge relating to the safety of raw foods. Although more meaningful with respect to ready-to-eat foods, the indicators were still unable to provide a measure of safety, regardless of the level established as the acceptance standard. The authors concluded that the total count was the most suitable method for evaluating the microbial quality of ready-to-eat foods. They also concluded that indicator tests for foods had nothing to do with safety, but only indicated improper handling. Likewise, in a study of vacuum-packaged sliced bologna, Paradis and Stiles (1978) found no relationship between total microbial load and indicator organisms or pathogenic bacteria. When samples of ice cream were analyzed, appreciable numbers of indicator organisms were detected, but no salmonellae were found (Tamminga et al., 1980). A similar lack of correlation has been observed in frozen chicken (Norberg, 1981).

An evaluation of the sanitary quality of swine and beef carcasses by the use of indicator microorganisms has been reported (Wojton and Kossakowska, 1977). In this study, a double swab method was used for aerobic and facultatively anaerobic organisms. A correlation between the level of contamination by coliforms, enterococci, and the total number of bacteria was found, except for *Clostridium*. However, in a later study involving red meat carcasses, no correlation between indicators and total counts could be found (Roberts et al., 1980).

Gerba et al. (1980) evaluated the use of bacterial indicators as related to enterovirus contamination of oysters. They reported a moderate correlation between viruses in water and total coliforms in oysters. However, the presence of viruses in water was not correlated with the presence of viruses in oysters. They concluded that current bacteriological standards do not reflect the occurrence of enteroviruses in marine waters. In a similar study, results indicated no positive correlation between indicator bacteria in the water and the number of viruses in oysters (Ellender et al., 1980).

B. Coliforms, Fecal Coliforms, and *E. coli*

The family Enterobacteriaceae contains the coliform bacteria as well as such known pathogens as *Salmonella*, *Shigella*, and *Yersinia*. The coliforms differ from most of the other members of this family because of their ability to ferment lactose, with the production of acid and gas occuring within 48 hr. Coliforms are actually comprised of four genera, *Citrobacter, Klebsiella, Escherichia,* and *Enterobacter*. However, *Escherichia coli* and *Enterobacter aerogenes* are the most widely recognized members of these genera.

The natural habitat of *E. coli* is the gastrointestinal tract of warm-blooded animals, including humans. *Enterobacter*, on the other hand, occurs primarily as a saprophytic agent of vegetation but may be found occasionally in the intestine. Over 90 years ago, researchers noted that since *E. coli* was so uniformly present in the intestinal tract, its presence outside the intestines might be regarded as evidence of contamination with fecal discharges of humans or animals. This observation marked the beginning of the use of the coliforms as indicators of pathogens in water, followed by their use in the food industries (Jay, 1978). However, the value of their use as indicators by the food industry has been questioned (Peabody, 1963).

The detection of large numbers of these organisms in foods and water may be indicative of fecal pollution or contamination. The existence of pollution indicates the possibility that pathogenic organisms may also be present. However, many researchers feel that the presence of intestinal organisms in foods indicates a lack of cleanliness, not safety. Safety can be determined only by examining for the presence of pathogens (McCoy, 1961).

In processed foods, the detection of coliforms in large numbers does not indicate contamination in the sense of demonstrating recent contact with fecal matter. Rather, it indicates poor practice in the formulation, processing, and storage of foods and raises the possibility that pathogens might have entered the food product through the same route. A much more important issue than the presence or absence of coliforms is their relative numbers compared to the rest of the microbial population and whether they are of fecal or nonfecal origin (Thatcher and Clark, 1973).

Fecal coliforms are a group of organisms selected by incubating inocula derived from a coliform enrichment broth at elevated temperatures (usually 44–45.5°C) (Thatcher and Clark, 1973). Fecal coliforms produce gas in enrichment broths at these temperatures, whereas nonfecal coliforms generally do not. Thus, detection of *E. coli* and fecal coliforms in foods may have special sanitary significance. In a study involving refrigerated market milk, Asperger and Brandl (1979) reported that the coliform count could be used as a criterion for good manufacturing practice in the dairy or in samples stored for a short period at low temperature. It was not possible, however, to draw this conclusion for market milk, which is stored under market conditions. For assessing meat quality, Newton (1979) stated that coliforms have no value as indicators of potentially harmful contamination, and he suggested that coliform tests should not be applied to fresh meat and meat products. In sausage processing, Kovacs and Takacs (1979) found that salmonellae were considerably more resistant than coliforms to changes taking place during ripening. These authors also stated that a negative coliform test was not conclusive of the absence of *Salmonella* contamination. Hood *et al.* (1983a) reported that fecal coliform and total counts did not correlate with those for vibrios in fresh oysters. However, strong correlations were observed in oysters stored for 7 days. The authors suggested that these indicators might be useful in monitoring oyster quality when meats are stored for a limited time as shell stock. Hood *et al.* (1983b) found that low fecal coliform levels in both fresh and stored oysters were good indicators of the absence of *Salmonella,* but that high levels were somewhat limited in predicting the presence of *Salmonella. Escherichia coli* levels correlated very strongly with fecal coliform levels.

From reviewing numerous studies, Buttiaux and Mossel (1961) concluded that various pathogens might persist after *E. coli* is destroyed in processed foods and treated waters. Only in acidic foods did *E. coli* have particular value as an indicator organism, due to its relative resistance to low pH values. Nevertheless, *E. coli* is a good indicator of fecal contamination in most foods; coliforms other than *E. coli* are good indicators of unsatisfactory processing or sanitation, and foods containing fecal coliforms have a higher probability of containing organisms of fecal origin (Thatcher and Clark, 1973).

In spite of some of the aforementioned disadvantages, coliform detection has been widely used in assessing sanitary quality, with standards being either established or recommended for a wide variety of foods.

C. Enterococci

The enterococci are members of the genus *Streptococcus,* which contains gram-positive, catalase-negative cocci in short or long chains. *Streptococcus faecium* (subspecies *durans*) and *S. faecalis* (subspecies *faecalis, liquefaciens,* and *zymogenes*) constitute the enterococcus group. They are all members of Lancefield's serologic group D streptococci. Enterococci may be differentiated from other group D streptococci by their ability to grow in the presence of 6.5% NaCl, at a pH of 9.6, at 10°C and 45°C, and to withstand a temperature of 60°C for 30 min. The taxonomy of these organisms has been extensively reviewed (Hartman *et al.*, 1966).

The enterococci are found primarily in the intestinal tract of humans and other animals and tend to exhibit some degree of host specificity. *Streptococcus faecalis* and its subspecies are more commonly associated with the intestinal tract of humans, while *S. faecium* occurs predominantly in swine (Jay, 1978). In addition to their fecal origins, enterococci are widely distributed in nature, existing on plants and insects and in soils, dusts, slaughterhouses, and curing rooms.

A large number of investigators have detected enterococci in foods such as frozen seafood, dried whole egg powder, raw and pasteurized milk, commercially frozen fruits, fruit juices, vegetables, and processed meats (deFigueiredo and Jay, 1976). Several of these workers feel that enterococci are a better index of food sanitary quality than are coliforms, especially for frozen foods.

In an early study of more than 400 commercially prepared frozen chicken pies, Hartman (1960) reported that enterococcus counts were more closely related to total counts than to coliform counts, while coliform counts were more closely related to enterococcus counts than to total counts. He also found that enterococci occurred in greater numbers than coliforms. This latter finding has also been reported by several other investigators who analyzed frozen foods.

The greater resistance of enterococci to adverse environmental conditions may tend to make these organisms better indicators of food sanitary quality than are the coliforms. However, this increased resistance may also be the reason for the unreliability of enterococci as general indicators of fecal contamination (Thatcher and Clark, 1973). Because of increased relative survival, their numbers may have little relationship to the hazard posed by less durable, but recognized pathogens such as *Salmonella*, which even if deposited simultaneously with the enterococci, would probably not have survived.

Stiles *et al.* (1978) compared counts of group D streptococci with coliform, *E. coli*, and *Enterobacteriaceae* counts in raw and processed meats but found no meaningful relationships. They stated that their results did not support the use of group D streptococci as alternative indicator organisms for meats. However, the association of these organisms with packing plant contamination may prove to be of value. Kielwein (1978) stated that the demonstration of *Streptococcus durans* in milk and dairy products is an indicator of inadequate sanitation of dairy equipment. The function of enterococci as indicator organisms in cheese is strongly limited according to the proliferation of the organisms in these products.

Jay (1978) listed several characteristics which could be used to compare the relative merits of enterococci and coliforms as indicators of food sanitary quality. These are presented in Table VI.

TABLE VI

Comparison of Coliforms and Enterococci as Indicators of Food Sanitary Quality[a]

Characteristic	Coliforms	Enterococci
Morphology	Rods	Cocci
Gram reaction	Negative	Positive
Incidence in intestinal tract	10^7–10^9/g feces	10^5–10^8/g feces
Incidence in fecal matter of various animal species	Absent from some	Present in most
Specificity to intestinal tract	Generally specific	Generally less specific
Occurrence outside of intestinal tract	Common in low numbers	Common in higher numbers
Ease of isolation and identification	Relatively easy	More difficult
Response to adverse environmental conditions	Less resistant	More resistant
Response to freezing	Less resistant	More resistant
Relative survival in frozen foods	Generally low	High
Relative survival in dried foods	Low	High
Incidence in fresh vegetables	Low	Generally high
Incidence in fresh meats	Generally low	Generally low
Incidence in cured meats	Low or absent	Generally high
Relationship to foodborne intestinal pathogens	Generally high	Lower
Relationship to nonintestinal, foodborne pathogens	Low	Low

[a] From Jay (1978).

D. Enterobacteriaceae

While total counts, coliforms, and enterococci have been extensively utilized for several decades as indicator systems, other organisms have also been proposed. In 1963, Mossel et al. suggested the collective use of all members of the family Enterobacteriaceae as a means of detecting fecal contamination.

The advantages of using this system include (1) avoiding problems dues to the ill-defined taxonomy of coliforms, (2) eliminating the possibility of obtaining atypical strains as the predominant enteric contaminants, (3) increasing the reliability of monitoring by adding direct detection of pathogens, and (4) detecting lactose-negative members, such as *Salmonella* and *Shigella,* which would have more direct public health significance than the coliforms (Thatcher and Clark, 1973; Mossel et al., 1978b).

Use of this detection system may also result in significant savings of time, supplies, and numbers of samples to be monitored. For example, Drion and Mossel (1972) reported that when one or two 1-g aliquots of dried foods are tested and no Enterobacteriaceae can be detected, the same degree of consumer protection is afforded as when 60 25-g amounts of dried foods are tested for salmonellae and all these samples are found to be salmonellae-negative. The authors based their conclusions upon a minimum ratio of Enterobacteriaceae to salmonellae of 1000:1.

Many more studies are needed to fully evaluate the potential of this system as a food sanitary indicator. As cited previously, there appear to be distinct advantages associated with its use. However, some of the genera within this family (*Serratia, Proteus,* and *Erwinia*) have natural habitats other than the intestinal tract of humans and animals. This would make them unsatisfactory for use as fecal indicators.

E. Fungi

Two fungi have received somewhat limited attention as indicator systems for food quality. *Geotrichum candidum,* frequently called "machinery mold," may grow on moist surfaces of food-contact equipment in fruit and vegetable processing plants, especially tomato canning plants, and bottling plants. Mycelial fragments then may enter bottled drinks and fruit and vegetable products from the processing equipment. The amount of *G. candidum* in a product is determined by direct microscopic count of fragments of mycelium with the characteristic feathery branching habit of this organism (Pusch et al., 1976). *Geotrichum* is also sometimes referred to as "dairy mold," since species of this genus impart flavor and aroma to many types of cheese. Little correlation was observed between aerobic plate counts and incidence of *Geotrichum* in frozen, blanched vegetables (Splittstoesser *et al.*, 1980).

The second fungus, *Byssochlamys* (*B. fulva* and *B. nivea*), has become a problem with thermally processed fruits and fruit products. Spoilage of canned fruit results in a breakdown of texture, while growth in liquid products results in off-flavors and production of mycelial masses. A low-level toxic metabolite may also be produced. The ascospores of this organism are relatively heat-resistant and can survive temperatures commonly used in the processing of acidic foods. Effective sanitation appears to be the best method for controlling this organism (Splittstoesser, 1976).

The occurrence and frequency of isolation of both *Geotrichum* and *Byssochalmys* from processing equipment may be related to the effectiveness of in-house sanitation programs. Therefore, these fungi may serve as indicator systems for evaluation of these programs in fruit and vegetable processing and bottling plants. Details regarding detection and enumeration of these organisms have been published (Pusch *et al.*, 1976; Splittstoesser, 1976).

F. Viruses

Goyal *et al.* (1979) found that no significant statistical relationship could be demonstrated between enterovirus concentration in oysters and the bacteriological and physicochemical quality of water and shellfish. Viruses in water were, however, moderately correlated with total coliforms in water and oysters and with fecal coliforms in oysters.

Richards (1985) reported that cases of shellfish-associated enteric virus illness are increasing, whereas bacterial illness from shellfish is decreasing. He proposed that enteroviruses be used as an interim indicator for the possible presence of human pathogenic viruses in seafoods.

Coliphage detection appears to be more promising as a simple, inexpensive indicator of fecal contamination. In a collaborative study of water quality, highly significant correlations were found between coliphages and coliforms in natural water systems (Wentsel *et al.*, 1982). The number of fecal or total coliforms present could be determined by enumeration of coliphages. Coliphages in these systems could be enumerated within 6 hr. This procedure was also evaluated for use with chicken, turkey, and pork sausage (Kennedy *et al.*, 1984). High coliphage levels generally reflected high fecal coliform counts, particularly for fresh meat samples. The procedure was rapid (16 hr), and the authors proposed that coliphages might offer a promising alternative as rapid indicators of fecal contamination in foods. However, many more studies are needed to substantiate their utility as indicators.

G. Other Organisms

Numerous other organisms have occasionally been cited and suggested as potential indicators for food quality and/or safety. They include psychrotrophs (Ayres, 1963), *Streptococcus salivarius*, staphylococci, clostridia (Thatcher and Clark, 1973), *Bifidibacterium*, *Bacteroides*, and lactobacilli (Jay, 1978). However, they have not enjoyed widespread utilization as indicators for various reasons.

APPENDIX: DETECTION, ISOLATION, AND ENUMERATION METHODOLOGIES— REVIEWS OF RECENT STUDIES

Suggested Standard Methods for Detection. Table A-1 outlines some suggested standard methods for detection, quantitation, and identification of microorganisms. Included in this outline are isolation media and confirmatory tests for indicator microorganisms as well as for known foodborne pathogens. As the title implies, these are only some suggested standard methods. Rapid and automated methods have not been included. For more detailed information regarding methodologies, the reader should consult the list of selected references (Table A-II).

Reviews and Collaborative Studies Involving Isolation, Characterization, and Enumeration. Several pertinent review articles have been published involving isolation, characterization, and enumeration of both microbial indicators and foodborne pathogens. These are listed in Table A-III. In an attempt to familiarize the reader with methodologies simultaneously evaluated by more than one laboratory, a list of collaborative studies has also been compiled and given in Table A-IV.

Modifications in Media and Techniques. Within the last 8 years, many modifications in both media and techniques have significantly improved the recovery of microorganisms from their environment. Table A-V lists some of the more pertinent modifications. These studies were included in this chapter because classical or standard methods sometimes fail to accurately detect the microbial population of a given product or sample. Advantages and limitations of media and methods are discussed in many of the references cited in this table.

TABLE A-I

Some Suggested Standard Detection Methods for Microbiological Quantitation and Identification[a]

Determination	Isolation media	Confirmatory tests
Total plate count[b]	Plate count agar; nutrient agar; TGE agar	Microscopic examination of isolates
Coliform count	Violet red bile agar; Lauryl sulfate tryptose broth; EMB agar; MacConkey agar	EMB agar; growth and gas evolution in brilliant green bile broth within 48 hr at 35°C
E. coli	EC broth; LST broth; EE broth; VRB agar; MacConkey agar	EMB agar; ImViC reaction; biochemical tests
Streptococcal count	KF streptococcal agar; selective enterococcus agar; azide blood agar; SF medium; bile esculin agar	Microscopic examination of isolates; determination of growth temperature ranges; tolerance to high pH and salt; catalase and biochemical tests
Salmonella	SC and TT broths; BG and BS agar; Hektoen enteric agar; EMB agar; MacConkey agar	TSI and LI agars; serology; biochemical tests
Enterobacteriaceae	EE broth	Violet red bile glucose agar; biochemical tests for specific genera
S. aureus	Baird-Parker medium; EY tellurite; S-110 medium	Coagulase tube test; DNAse test
Clostridial count[c]	Peptone yeast extract broth containing bromcresol purple and whole peas; cooked meat medium; Andersen's pork pea medium; sulfite polymyxin sulfadiazine agar; anaerobic agar; reinforced clostridial agar; iron milk medium; SFP agar	Growth on liver–veal and tryptose sulfite cycloserine agars; growth and gas evolution in fluid thioglycolate; nonmotility, nitrite, and gelatinase tests; microscopic examination of isolates; animal bioassays; biochemical tests; isolation of spores
Bacillus	TGE agar with bromcresol purple; trypticase glucose agar; B. cereus selective agar; phenol red agar	Nutrient agar with manganese; microscopic examination of isolates; biochemical tests; isolation of spores
Yersinia enterocolitica	Cold enrichment in saline or phosphate buffer; MacConkey's and EMB agars; Yersinia selective agar	Selective broth enrichment followed by plating on SS and MacConkey's agar; TSI, urea, and motility reactions; serology

4. Standard and Experimental Methods of Identification and Enumeration

TABLE A-I (*Continued*)

Determination	Isolation media	Confirmatory tests
Campylobacter jejuni	Filtration of suspensions through 0.65-μm membranes; plating of filtrates on Skirrow's medium with microaerobic atmosphere; *Campylobacter* agar with supplements	Microscopic examination of isolates; oxidase and catalase tests; H_2S production; growth at 45°C
Vibrio	Thiosulfate citrate bile-salts sucrose agar; Glucose salt teepol and bismuth sulfite salt broths	Microscopic examination of isolates; bacteriophage susceptibility; serology; biochemical tests such as oxidase reactions; growth in liquid media with added salt
Listeria	Blood agar with nalidixic acid; blood agar; tryptone agar; Brain–heart infusion agar	Growth at 4°C; motility, catalase, and hemolysis tests; biochemical tests; serology
Viruses	Slurry of food sample is prepared and clarified. Viruses are concentrated by polyethylene glycol and ultracentrifugation. They are enumerated by plaque count following inoculation and incubation in a susceptible tissue culture.	Electron microscopic examination of concentrated rinsings from plaques
Yeast and mold count	Potato dextrose agar; Sabouraud dextrose agar; malt agar	Microscopic examination of isolates

[a] These are some suggested methods for general use. Other methods (Table A-V) may be more appropriate, depending on the time needed for reporting of results, the type of food to be analyzed, and the type of organism to be detected. Whatever method(s) are chosen, the possible presence of injured cells must be considered, and steps must be included for their recovery. Similar methods are outlined, in the Difco Manual, Tenth Edition.

[b] Can be used to determine mesophilic and psychrotrophic counts by using appropriate incubation times and temperatures. Surface or spread plates should be used for psychrotrophs.

[c] Prepared and/or incubated under anaerobic conditions. If isolated colonies are to be subcultured, the use of anaerobic agar tubes (roll tubes) for initial isolation may be most convenient.

TABLE A-II

Some Selected References Containing Methods for Isolation and Enumeration of Microorganisms in Foods

Title	Source
Microbial Ecology of Foods	International Commission on Microbiological Specifications for Food
Meat Microbiology	Brown, M.H. (Ed.), Applied Science Publishers
Methods in Food and Dairy Microbiology	Diliello, L. R., Avi Publishing Co.
Isolation and Identification Methods for Food Poisoning Organisms	Corry, J. E. L., Roberts, D., and Skinner, F. A. (Eds.), Academic Press, Inc.
Identification Methods for Microbiologists, Second Edition	Skinner, F. A. (Ed.), Academic Press
Food-Borne Infections and Intoxications, Second Edition	Riemann, H., Davis, U. C., and Bryan, F. (Eds.), Academic Press
Laboratory Methods in Food and Dairy Microbiology, Revised Edition	Harrigan, W. F., and McGance, M. E., Academic Press
Food Microbiology: Public Health and Spoilage Aspects	deFigueiredo, M. P., and Splittstoesser, D. F., Avi Publishing Co.
Basic Food Microbiology	Banwart, G. J., Avi Publishing Co.
Food Mycology	Rhodes, M. E. (Ed.), G. K. Hall & Co.
Food and Beverage Mycology	Beuchat, L. R., (Ed.), Avi Publishing Co.
Manual of Methods for General Bacteriology	Gerhardt, P. (Ed.-in-Chief), American Society for Microbiology.
Compendium of Methods for the Microbiological Examination of Foods (First Edition, 1976; Second Edition, 1984)	Speck, M. L. (Ed.), American Public Health Association
Rapid Methods and Automation in Microbiology	Tilton, R. C. (Ed.), American Society for Microbiology
Bacteriological Analytical Manual, Fifth Edition	Food and Drug Administration, Bureau of Foods, Division of Microbiology, Association of Official Analytical Chemists (Publishers)
Microorganisms in Foods: Their Significance and Methods of Enumeration	Thatcher, F. S., and Clark, D. S. (Eds.), University of Toronto Press
Microbiological Methods, Fifth Edition	Collins, C. H., and Lyne, P. M. (Eds.), Butterworths
Standard Methods for the Examination of Dairy Products, Fifteenth Edition	Richardson, G. H. (Ed.), American Public Health Association
Inhibition and Inactivation of Vegetative Microbes	Skinner, F. A., and Hugo, W. B. (Eds.), Academic Press
Difco Manual, Tenth Edition	Difco Laboratories

TABLE A-II (*Continued*)

Title	Source
Repairable Lesions in Microorganisms	Hurst, A., and Nasim, A. (Eds.), Academic Press
The Revival of Injured Microbes	Andrew, M. H. E., and Russell, A. D. (Eds.), Academic Press
Modern Food Microbiology, Second Edition	Jay, J. M., D. Van Nostrand Co.
Laboratory Methods in Food and Dairy Microbiology	Harrigan, W. F., and McGance, M. E., Academic Press
Food Microbiology: Advances and Prospects	Roberts, T. A., and Skinner, R. A. (Eds.), Academic Press
Maintenance of Microorganisms: A Manual of Laboratory Methods	Kirsop, B. E., and Snell, J. J. S. (Eds.), Academic Press
CRC Handbook Series in Nutrition and Food. Section G, Volume III. Culture Media for Microorganisms and Plants.	Rechcigl, Jr., M. (Ed.-in-Chief), CRC Press
Collection of Methods for the Microbiological Examination of Foods (English)	Schmidt-Lorenz (Ed.), Verlag Chemie International
A Modern Introduction to Food Microbiology	Board, R. G., Blackwell Scientific Publications
Experimental Microbial Ecology	Burns, R. G., and Slater, J. H. (Eds.), Blackwell Scientific Publications
Manual of Clinical Microbiology, Third Edition	Lennette, E. H., Balows, A., Hausler, Jr., W. J., and Truant, J. P. (Eds.), American Society for Microbiology
Foodborne Microorganisms and Their Toxins. Developing Methodologies.	Pierson, M. D., and Stern, N. J. (Eds.), Marcel Dekker
Advances in Meat Research, Volume 2. Meat and Poultry Microbiology.	Pearson, A. M., and Dutson, T. R. (Eds.), Avi Publishing
Instrumental Methods for Rapid Microbiological Analysis	Nelson, W. H. (Ed.), VCH Publishers
Manual of Industrial Microbiology and Biotechnology	Demain, A. L., and Solomon, N. A. (Eds.), American Society for Microbiology
Toxicity Testing Using Microorganisms	Dutka, B. J., and Bitton, G. (Eds.), CRC Press

TABLE A-III

Some Review Articles Involving Microbial Isolation and Identification

Microorganism	Food product or method	References
Total counts	Butter	Joyce and Ould (1978)
Total counts	Evaluation of several rapid diagnostic tests	Szakal (1979)
Psychrotrophs	Sources and causes of contamination in milk	Thomas and Thomas (1978)
Psychrotrophs	Selective media	Brant et al. (1978)
Psychrotrophs	Role in milk spoilage	Richter (1979)
Psychrotrophs	Occurrence in foods	Kraft and Rey (1979)
Psychrotrophs	Detection in milk and dairy products	Cousin (1982)
Gram-positive bacteria	Rapid identification methodology	Kaplan (1981)
Gram-negative bacteria	Rapid identification methodology	Pezzlo (1981)
Salmonellae and enterobacteriaceae	Isolation methodology	Reusse (1978)
Enterobacteriaceae	Bacteriocin typing for identification	Gillies (1979)
Enterobacteriaceae	Serotyping	Orskov and Orskov (1979)
E. coli	Review of foodborne illness	Kornacki and Marth (1982)
Enteropathogenic E. coli	Methods for recovery from foods	Mehlman and Romero (1982a)
Enterobacteriaceae	Selecting miniaturized identification kits	Cox et al. (1984)
Coliforms	Recent advances for water analysis	LeChevallier and McFeters (1984)
Coliforms and Salmonella	Methods for recovery of injured cells from food.	Andrews (1986)
Salmonella	Detection by immunoassays	Ibrahim (1986)
Salmonella	Methods for isolation and identification	Anonymous (1982)
Salmonella	RV enrichment medium for isolation	Vassiliadis (1983)
Salmonella	Detection in foods	D'Aoust (1984a)
Salmonella	Improved methods for detection	LaHellec and Colin (1984)
Salmonella	Enrichment and plating methodologies	Fagerberg and Avens (1976)
Salmonella	Improved methods for detection	Hobbs (1977b)

TABLE A-III (Continued)

Microorganism	Food product or method	References
Salmonella	Phage typing	Guinee and Van Leeuwen (1979)
Salmonella	Factors affecting isolation	Harvey and Price (1979)
Salmonella	Ten-year review of incidence, monitoring, and epidemiology	Silliker (1980)
Salmonella	Review of preenrichment and enrichment techniques	D'Aoust (1981)
P. aeruginosa	Pyocin and phage typing; serological characterization	Govan (1978); Bergan (1978); Lanyi and Bergan (1978)
Y. enterocolitica	Characterization and role in food hygiene	Morris and Feeley (1976)
Y. enterocolitica	Isolation and identification	Winblad (1979)
Y. enterocolitica	Recovery from foods	Stern (1982d)
Y. enterocolitica	Pathogenicity and significance in foods	Zink et al. (1982)
V. cholerae	Procedures for typing	Mukerjee (1979)
V. parahaemolyticus	Enumeration in foods and on utensils	Sakazaki et al. (1979)
V. parahaemolyticus	Public health significance	Beuchat (1982)
V. cholerae	Incidence and detection in marine foods	DePaola (1981)
V. cholerae	Incidence in shellfish	Madden et al. (1982)
D. nigrificans	Identification and characterization in food spoilage	Speck (1981)
C. jejuni	Recovery from foods and sensitivity to isolation techniques	Palumbo (1986)
C. jejuni	Methods for recovery from foods	Stern (1982a)
C. jejuni	Recovery from food	Blaser (1982)
C. jejuni, C. coli	Characterization of toxins	Johnson and Lior (1984)
C. jejuni	Importance to meat industry	Kotula and Stern (1984)
Plesiomonas shigelloides	Detection in water and food	Miller and Koburger (1985)
Anaerobic spore-formers	Presence and characterization in dairy products	Donnelly and Busta (1981a)
Anaerobic bacteria	Headspace gas chromatography as a tool in identification	Maerdh et al. (1981)

(continued)

TABLE A-III *(Continued)*

Microorganism	Food product or method	References
Anaerobes	Rapid identification methodology	Moore (1981)
C. botulinum	Biochemistry of toxins	Sakaguchi (1982)
C. botulinum	Heat-processed seafoods	Licciardello (1983)
C. botulinum	Fishery products	Hobbs (1976)
C. perfringens	Role in food poisoning	Hobbs (1977a)
C. perfringens	Food poisoning	Frazer (1979)
C. perfringens	Growth and sporulation in foods	Craven (1980)
Bacillus stearothermophilus	Growth and control in food spoilage	Ito (1981)
Bacillus coagulans	Identification and characterization in food spoilage	Thompson (1981)
Bacillus cereus	Foodborne illness	Johnson (1984)
Staphylococcus aureus	Characterization of toxins	Freer and Arbuthnott (1982)
Staphylococcus	Identification and characterization	Sperber (1977)
S. aureus	Detection methodology	Terplan and Zaadhof (1978)
Staphylococcus	Isolation and identification	Oeding (1979)
Coagulase-negative Staphylococci	Epidemiological typing	Parisi (1985)
Enterococci	Identification and typing	Jelinkova (1979)
Food-poisoning bacteria	The role of bacteria in food poisoning	Turnbull (1980)
Lactic acid bacteria	Classification and differentiation	Stamer (1979)
Fungi	Various products	Camaschella (1977)
Fungi	Methodology for microscopic determination in tomato products	Olson (1980)
Fungi	Mycotoxin detection in cereals	Osborne (1982)
Fungi	Evaluation of biological methods for detecting mycotoxins	Watson and Lindsay (1982)
Fungi	Mycotoxin isolation and characterization	Franck (1984)
Fungi	Mycotoxin isolation and characterization	Egan et al. (1982)
Fungi	Review of mode of action of aflatoxins	Dashek and Llewellyn (1983)

4. Standard and Experimental Methods of Identification and Enumeration

TABLE A-III *(Continued)*

Microorganism	Food product or method	References
Fungi	Enumeration in foods and feeds	Jarvis et al. (1983)
Fungi	Immunoassays for mycotoxins	Chu (1984)
Fungi	Sampling methods for airborne species	Solomon (1984)
Aspergillus	Detection of aflatoxins in milk and milk products	Applebaum et al. (1982)
Fusarium	Detection and analysis of mycotoxins	Gilbert (1984)
Alternaria	Analysis and characteristics of toxins	Schade and King (1984)
Alternaria	Importance of toxins in foods	King and Schade (1984)
Alternaria	Food contamination by toxins	Watson (1984)
Alternaria	Biosynthesis of mycotoxins	Stinson (1985)
Yeasts and fungi	Enumeration in foods	Anderson (1977)
Yeasts	Occurrence and control of food-spoilage yeasts	Fleet (1978)
Yeasts	Role in food spoilage	Miller (1979)
Yeasts	Media and methods for growth and detection	Beech (1981)
Cold-tolerant yeasts	Growth and detection in chilled and frozen foods	Davenport (1981)
Xerotolerant yeasts	Detection and control of food spoilage	Tilbury (1981)
Viruses	Virus transmission and isolation from food	Chalapati (1976)
Enteroviruses	Shellfish	Gerba and Goyal (1978)
Viruses	Viral contamination of foodstuffs of animal origin	Labie (1979)
Viruses	Role of foodstuffs in transmission	Zavate et al. (1979)
Viruses	Rapid laboratory techniques for detection	World Health Organization (1981)
Viruses	Significance in the environment	Anonymous (1982)
Viruses	Importance of foodborne viruses	Blackwell et al. (1985)
Viruses	Potential use as indicators	Richards (1985)
Indicator organisms	Frozen, blanched vegetables	Splittstoesser (1983)

(continued)

TABLE A-III *(Continued)*

Microorganism	Food product or method	References
Indicator organisms	Meat and poultry products	Tompkin (1983)
Indicator organisms	Dairy products	Reinbold (1983)
Indicator organisms	Fish and shellfish	Matches and Abeyta (1983)
Indicator for viruses	Shellfish	Metcalf (1978)
Indicator for viruses	Foods preserved by heat	Oliver and Salo (1978)
Indicator for viruses	Foods preserved by ionizing radiation	Rowley et al. (1978b)
Various	Laboratory diagnosis of foodborne diseases	Kazal (1976)
Various	Enumeration methodology for organisms in dried foods	Mossel and Shennan (1976)
Various	Microbiology of dried foods	Harrewijn and Mossel (1977)
Various	Choice of indicator organisms to assess sanitary quality	Mossel and Harrewijn (1977)
Various	Review of techniques for Gram staining	Spengler et al. (1978)
Various	Use of bacteria to detect toxic metabolites	Berkowitz (1979)
Various	Bacteria associated with meat	Gill (1979)
Various	Serotyping of bacteria	Henriksen (1979)
Various	Review of bacterial metabolism	Kilbourn (1979)
Various	Microbial associations with foods of animal origin	Mossel (1979)
Various	Bacterial contamination of cured meat	Roberts (1979)
Various	Microbial association with shellfish	Van den Broek et al. (1979b)
Various	Bacterial agents of foodborne disease	Doyle and Foster (1981)
Various	Relevance of distribution to control of microbial hazards	Kilsby and Pugh (1981)
Various	Selective culture media	Bridson (1982)
Various	Review of microbiological aspects of poultry and poultry products	Cunningham (1982)
Various	Instrumentation in food microbiology	Cady (1977)
Various	Prospects of automation in microbiology	Leighton (1978)
Various	Microbial calorimetry as an	Monk (1978)

TABLE A-III (Continued)

Microorganism	Food product or method	References
	analytical method to measure growth	
Various	Review of automation and rapid methods	Newsom (1978)
Various	Automation for general and specific detection of bacteria	Bascomb (1981)
Various	Systems approach to microbial identification	D'Amato et al. (1981)
Various	Use of PGLC for identification	Gutteridge and Norris (1979)
Various	Immunological methods of detection	Anhalt (1981)
Various	Enumeration and estimation of microbial biomass	Atlas (1982)
Various	Microbiological examination of meat	Brown and Baird-Parker (1982)
Various	Three-class acceptance plan for microbiological sampling	Clark (1982)
Various	Rapid microbiological methodology	Stanndard et al. (1982)
Various	Direct and indirect methods for detecting bacteria in milk	O'Toole (1983b)
Various	Rapid identification using gas chromatography	Chopra (1983)
Various	Estimation of biomass	Harris and Kell (1985)
Various	Microbiology of hot-boned beef	Oblinger (1983)
Various	Seven-year evaluation of biochemical kits for rapid identification	Oberhofer (1983)
Various	Rapid methods for determining bacterial quality of red meat	Fung (1984)
Various	Computer program for teaching microbial identification	Tejedor and Chordi (1984)
Various	Comparative study of isolated bacterial enterotoxins	Gemmell (1984)
Various	Transport, culture, and identification of anaerobes	Citron (1984)
Various	Emerging foodborne pathogens	Bryant (1983)

TABLE A-IV

Collaborative Studies of Microbiological Analyses

Microorganism	Food product or method	Reference
Total count	Milk/spiral plate	Peeler et al. (1977)
Total count	Evaluation of plate loop technique for raw milk	Brodsky and Ciebin (1981)
Total count	Reproducibility of total and indicator counts in frozen cod	Slabyj et al. (1981)
Total count	Direct epifluorescent filter techniques for milk	Pettepher et al. (1983)
Enterobacteriaceae	Evaluation of semiautomated method (Repliscan)	Brown and Washington (1978)
Enterobacteriaceae	Performance of commercial identification systems	Marymont et al. (1978)
Enterobacteriaceae	Automated and computerized system (Automicrobic)	Smith et al. (1978)
Enterobacteriaceae	Evaluation of AutoMicrobic System	Isenberg et al. (1980)
Enterobacteriaceae	Identification with miniaturized systems	Cox et al. (1985)
Gram-negative bacilli	Collaborative identification of nonfermentative or oxidase-positive isolates	Jorgensen et al. (1984)
E. coli	Enumeration with membrane filter versus MPN methods	Sharpe et al. (1983)
E. coli	Comparison of MPN and Anderson-Baird-Parker direct plating in raw meats	Rayman et al. (1979)
E. coli	Recovery of injured cells by use of filter membranes and modified direct plating	Holbrook et al. (1980)
E. coli	Membrane filtration	Rycroft (1980)
Coliforms	Determination with MPN procedure	Silliker et al. (1979)
Coliforms	Rapid impedimetric procedures for meat	Firstenberg-Eden and Klein (1983)
Fecal coliforms	A-1 procedure as a screening mechanism	Andrews et al. (1981)
Fecal coliforms	Higher recovery by AOAC[a] A-1 method than by APHA[a] standard method	Hunt et al. (1981)
Fecal coliforms	Comparison of membrane filter methods for enumeration	Pagel et al. (1982)
Salmonella	Detection in raw meats	Erdman (1973)
Salmonella	Detection in dried foods	Silliker and Gabis (1973)

4. Standard and Experimental Methods of Identification and Enumeration

TABLE A-IV (*Continued*)

Microorganism	Food product or method	Reference
Salmonella	Detection in high moisture foods	Gabis and Silliker (1974a)
Salmonella	Detection in dried foods and feeds	Gabis and Silliker (1974b)
Salmonella	Detection in frozen meats	Silliker and Gabis (1974)
Salmonella	Indicators as substitutes for direct testing of dried foods and feeds	Silliker and Gabis (1976)
Salmonella	Detection in dried foods and feed ingredients	Gabis and Silliker (1977)
Salmonella	Minced meat; preenrichment and enrichment methods	Van Schothorst *et al.* (1978)
Salmonella	Enrichment-plating conditions for detection in foods	D'Aoust (1984b)
C. jejuni, Salmonella	Survey of raw meat	Turnbull and Rose (1982)
Enteric viruses	Evaluation of methods for recovery from shellfish	Larkin and Metcalf (1980)
B. cereus	Use of direct plating for high populations and MPN technique for low population	Lancette and Harmon (1980)
B. cereus	Method for differentiating members of group	Harmon (1982)
C. perfringens	Higher recovery from roast beef using tryptose sulfite cycloserine agar	Harmon (1976)
C. perfringens	Enumeration in foods	Hauschild *et al.* (1977)
C. perfringens	Maintaining viability with buffered glycerol–NaCl solution	Harmon and Placencia (1978)
C. perfringens	Enumeration in feces	Hauschild *et al.* (1979a)
S. aureus	Enumeration in foods	Rayman *et al.* (1978)
S. aureus	Enumeration in powdered milk	Chopin *et al.* (1985)
Fungi	Comparison of four media for enumeration in dairy products	Henson *et al.* (1982)
Fungi	Determination of aflatoxins using two LC (liquid chromatographic) methods	Campbell *et al.* (1984)
Fungi	Howard mold count	Cichowicz and Bandler (1982a)
Geotrichum	Detection in fruits and vegetables	Cichowicz and Bandler (1982b)
Geotrichum	Determination in cream-style corn	Cichowicz and Bandler (1980a)

(*continued*)

TABLE A-IV (*Continued*)

Microorganism	Food product or method	Reference
Geotrichum	Determination in citrus juices	Cichowicz and Bandler (1980b)
Yeasts	Evaluation of Abbott identification system	Cooper *et al.* (1984)
Various	Colony counting on hydrophobic grid membrane filters	Sharpe *et al.* (1983b)
Various	Evaluation of automated bacterial identification system	Stevens *et al.* (1984)
Various	Evaluation of viable counting methods	Greenwood *et al.* (1984)

[a] AOAC, Association of Official Analytical Chemists; APHA, American Public Health Association.

TABLE A-V

Some Modifications of Media and/or Methods for Quantitative Determination of Microorganisms in Foods

Microorganism	Food product	Method/medium of choice	References
Total counts and pathogens	Ready-cooked meals	Twenty-four-hour microbiological analysis	Catsaras and Dorso (1976)
Total counts	Casein	Prewarming diluent to 45°C and reconstituting with Stomacher	Cooke (1977)
Total counts	Various products	Spiral plate maker	Jarvis *et al.* (1977)
Total counts	Ready-to-eat cereals	Microbial recovery better in liquid MPN media than on agar pour plates	Koburger and Oblinger (1977)
Total counts	Milk	Standard methods agar, phosphate-buffered H_2O, 28°C for 72 hr	Langlois and Sanghirum (1977)
Total counts	Beef carcasses	Instrument for sampling; 90% of surface microflora can be recovered	Davidson *et al.* (1978)

TABLE A-V (*Continued*)

Microorganism	Food product	Method/medium of choice	References
Total counts	Beef and pork carcasses	A coring device for sampling meat surfaces; higher counts obtained with this method than with swabs	Emswiler et al. (1978)
Total counts	Tomato paste	Samples and media are incubated and analyzed for diacetyl or acetylmethylcarbinol in 4–12 hr	Luster (1978)
Total counts	Milk	Corrections which should be applied to geometric means to obtain unbiased and efficient estimate of the population	Moermans and Bossuyt (1978)
Total counts		Rapid method using semi-automatic micropipette and disposable glass capillary tubes	Slack and Wheldon (1978)
Total counts	Broiler carcasses	Excision of neck skin, combined with a miniaturized culturing system	Thomson and Bailey (1978)
Total counts		Platinum electodes to detect and enumerate gram-positive and gram-negative organisms	Wilkins (1978)
Total counts	Chicken skin	Limitation of peptone–iron agar to detect sulfide-producing strains	Daud et al. (1979)
Total counts		Two-electrode system; current differences were proportional to microbial population	Matsunaga et al. (1979)
Total counts	Milk, dairy products	Use of a miniaturized system with $CaCO_3$ saccharose lactose blood agar	Otte et al. (1979)
Total counts	Raw milk	Dispersion of bacterial clumps by treatment of samples with a mixer for 30 sec	Richard (1980)

(*continued*)

TABLE A-V (*Continued*)

Microorganism	Food product	Method/medium of choice	References
Total counts	Raw milk	Modification of the plate loop procedure	Olsen and Richardson (1980)
Total counts, indicator	Ground beef	Freezing and storage affect indicator groups differently; however, total counts are not significantly reduced	Williams et al. (1980)
Indicator organisms	Various products	Reduction of foaming by using Stomacher	Stersky et al. (1980)
Indicator microorganisms		Most abundant growth on triphenyl tetrazolium Cl agar with 2% lactose and 0.2% freeze-dried egg extract	Vassilev (1978)
Histame-producing bacteria	Fish	Histidine-containing agar medium for detection	Niven et al. (1981)
Nitrate reducers		Medium consisting of nitrate agar plus 1% starch plus 1% KI	Reisner (1978)
Proteolytic bacteria		Detection with standard methods caseinate agar	Espejo and Perez (1978)
Saccharolytic organisms		Improved detection with agar–starch plates	Dhawale et al. (1982)
Sulfide-producing bacteria	Poultry processing plants	Techniques to detect bacteria-producing volatile sulfides	McMeekin et al. (1978)
Gram-negative bacteria		Use of semisolid medium to detect motility	Bennett et al. (1979)
Gram-negative bacteria		New techniques in solid-phase radioimmunoassay for detection of bacterial lipopolysaccharide	Gutowski and Jacobs (1979)
Gram-negative bacteria		Rapid detection of oxidase from slants	Skillern and Overman (1983)
Gram-negative bacilli		Use of casein, tyrosine, and hypoxanthine for identification of nonfermentative bacteria	Kurup and Babcock (1979)
Gram-positive, gram-negative bacteria		Differentiation by adding fluorogenic compound to media	Ramsey et al. (1980)

TABLE A-V (Continued)

Microorganism	Food product	Method/medium of choice	References
Asporogenous, gram-positive bacteria		Use of selective isolation medium containing three antibiotics	Wakisaka et al. (1982)
Various	Meat	Rapid methods for enumeration	Bem et al. (1976)
Various	Shrimp	Use of 24-hr methods of analysis	Godwin et al. (1977)
Various		Rapid method for distinction of gram-negative from gram-positive bacteria by means of 3% KOH	Gregersen (1978)
Various		Use of fluorophores to aid visualization of colonies on agar	Lepp et al. (1979)
Various	Turkey eggs	Rapid monitoring technique using sterile adhesive tape	Arhienbuwa et al. (1980)
Various		Elimination of false-negative oxidase reactions due to medium acidification	Havelaar et al. (1980)
Various		Rapid test for indole formation	Hussain Qadri et al. (1980)
Various		Automated method for measuring bacterial growth	Marcelis et al. (1980)
Various		Automated recording incubator for tube cultures	Morimoto et al. (1980)
Various		Rapid test for esculin hydrolysis	Qadri et al. (1980)
Various	Canned vegetables	Use of thin-layer chromatography to detect microbial spoilage	Ackland et al. (1981)
Various		Use of carrageenan as a substitute for agar	Epifanio et al. (1981)
Various		Methods for distinguishing gram-negative from gram-positive bacteria	Carlone et al. (1982)
Various	Fish	Improved medium for detecting proteolytic activity	Karnop (1982)
Various		Growth delay analysis for	Takano and

(continued)

TABLE A-V (*Continued*)

Microorganism	Food product	Method/medium of choice	References
		evaluation of metabolic injury	Tsuchido (1982)
Various		Modified reagent (in DMS) for better detection of oxidase	Tarrand and Groeschel (1982)
Various		Determination of Gram type by reaction of polymyxin B and lipopolysaccharides	Wiegel and Quandt (1982)
Various	Prepared meals for reheating	Rapid detection with glucose utilization–nitrate reduction	Bomar (1983)
Various		Determination of biomass by rate of moisture uptake	O'Toole (1983b)
Various		Rapid method for selecting appropriate media for enumeration	Richard et al. (1983)
Various		Media gelled by Gelrite was equal to media gelled by agar	Shungu et al. (1983)
Various		Rapid method for isolating radiation-sensitive mutants	Tan et al. (1983)
Various		Determining Gram reaction by KOH and vancomycin	Von Graevenitz and Bucher (1983)
Various		Bacterial biomass determined by toluidine blue–membrane filtration	O'Toole (1983a)
Various		Rapid enzymatic characterization of anaerobes	Marler et al. (1984)
Various		Rapid identification of nonfermenters with kits	Lampe and van der Reijen (1984d)
Various		Rapid technique for counting cells directly on membrane filters	Bitton et al. (1983)
Various		Solidifying agent for culture media	Gardener and Jones (1984)
Various		Gelrite superior to agar in media	Lin and Casida (1984)

4. Standard and Experimental Methods of Identification and Enumeration

TABLE A-V (*Continued*)

Microorganism	Food product	Method/medium of choice	References
Various		Methanol fixation superior to heat fixation for Gram staining	Mangels et al. (1984)
Various		Improved technique for DNase detection	Waller et al. (1985)
Enterobacteriaceae	Dried milk	Recovery of injured cells by incubating samples in tryptone soya broth	Asperger and Winterer (1978)
Enterobacteriaceae		Rapid test for acetoin production	Hussain Qadri et al. (1978)
Enterobacteriaceae	Minced meats, frozen poultry	Development of violet red bile dextrose agar	Mossel et al. (1978)
Enterobacteriaceae	Ground meats	Use of violet red bile agar for reliable counts	Ng and Stiles (1978)
Enterobacteriaceae		Use of three-tube composite media for rapid identification	Ahuja et al. (1979)
Enterobacteriaceae		Identification using stored, frozen microdilution trays	Barry et al. (1979a)
Enterobacteriaceae		Detection of histidine decarboxylase	Enjalbert et al. (1979)
Enterobacteriaceae		Medium for rapid presumptive identification	Kaper et al. (1979)
Enterobacteriaceae		Aerobically incubated medium for decarboxylase testing	Maccani (1979)
Enterobacteriaceae		Identification using six biochemical reactions	Farid and Larsen (1980)
Enterobacteriaceae		Same-day identification with two commercial kits	Appelbaum et al. (1982)
Enterobacteriaceae		Use of single-tube medium for screening	Thompson and Borczyk (1984)
Enterobacteriaceae	Raw meat	Addition of catalase increased recovery of injured cells	Van Netten et al. (1984)
Coliforms	Milk and dairy products	Recovery of injured cells by surface-overlay with violet red bile agar on trypticase soy agar plates	Matsumoto et al. (1976)

(*continued*)

TABLE A-V (*Continued*)

Microorganism	Food product	Method/medium of choice	References
Coliforms	Milk and butter	Brilliant green bile broth	Cooke and Jorgenson (1977)
Coliforms	Butter	Modified pour plates	Loane et al. (1977)
Coliforms	Dairy products	Recovery of injured cells by plating on violet red bile-2 medium	Marshall et al. (1978)
Coliforms	Dairy products	Recovery of injured cells by pour plating on trypticase soy agar and overlaying with violet red bile agar	Ray and Speck (1978)
Coliforms	Various foods	Evaluation of the brilliant green lactose broth confirmation test	Seng and Jegathesan (1978)
Coliforms	Pasteurized milk	Development of rapid coliform test (8–10 hr)	Sudarsanam and Nambudripad (1978)
Coliforms		Rapid method for detection using tergitol pectin medium plus $CaCl_2$	Rapp and Ihle (1979)
Coliforms	Milk	Use of violet red bile agar for identification	Szita et al. (1979)
Coliforms	Milk	Fluid medium for determination	Takacs et al. (1979a)
Coliforms	Milk	Use of brilliant green lactose bile broth and MacConkey broth	Takacs et al. (1979b)
Coliforms	Various products	Need for confirmatory studies in brilliant green lactose bile broth	Tham and Danielsson (1980)
Coliforms	Cheese, meat	Detection with mineral-modified glutamate medium	Abbiss et al. (1981)
Coliforms	Various products	Recovery of injured cells by incubation for 1 hr at 37°C in a non-nutrient medium	Leela and Sankaran (1981)
Coliforms	Cheese	Use of total coliform count to monitor good manufacturing practice	Waite (1981)
Coliforms	Dairy products	Best detection with lactose glutamic acid broth	Bindschedler et al. (1981)

(*continued*)

TABLE A-V (*Continued*)

Microorganism	Food product	Method/medium of choice	References
Coliforms, E. coli	Dairy products	Recovery of injured cells produced higher counts	Cooke et al. (1982)
Coliforms	Dairy products	Better detection of injured cells with VRB-2 agar	Reber and Marshall (1982)
Coliforms	Water	Membrane filter medium for improved recovery of injured cells	LeChevallier et al. (1983a)
Coliforms	Ground beef	Recovery of freeze-injured cells	Gram et al. (1984)
Fecal coliforms	Frozen vegetables	Improved recovery of injured cells with TSA followed by overlay of VRBA	Splittstoesser et al. (1982)
Fecal coliforms	Water	Elimination of rosolic acid from M-FC medium	Presswood and Strong (1978)
Total and fecal coliforms	Oysters	Violet red bile agar and colicount samples (Millipore) procedures	Richards (1978)
Fecal coliforms, E. coli	Various products	Use of two rapid methods for enumeration	Andrews et al. (1979a)
Fecal coliforms	Shellfish	Use of automatic incubator with modified A-1 test for enumeration	Cook (1981)
Fecal coliforms	Shellfish	Incubation of presumptive media for less than 48 hr	Hastback (1981)
E. coli		Use of lauryl tryptose broth for confirmatory medium	Anderson et al. (1979)
E. coli		Recovery of injured cells by catalase addition to medium	Egan (1979)
E. coli	Snow crab	Rapid method for enumeration	Powell et al. (1979)
E. coli	Various products	Recovery of injured cells by use of filter membranes and modified direct plating	Holbrook et al. (1980)
E. coli	Frozen meat	Recovery of injured cells by incubation on membrane filters for 4 hr at 35°C on tryptone soya agar	Mackey et al. (1980)

(*continued*)

TABLE A-V (*Continued*)

Microorganism	Food product	Method/medium of choice	References
E. coli		Synergistic inhibition of growth and gas production	Meadows *et al.* (1980)
E. coli	Ground meat	Detection with violet red bile agar at 44°C	Stiles and Ng (1980)
E. coli	Water	Use of lauryl tryptose mannitol broth plus tryptophan for confirmation	Barrow (1981)
E. coli		Preincubation for 2 hr with tryptone glucose extract or plate count agars overlayed with violet red bile agar	Draughon and Nelson (1981)
E. coli	Frozen meats	More precision and higher counts using direct plating method	Rayman and Aris (1981)
E. coli		Detection of enterotoxins by tissue culture	Paul *et al.* (1981)
E. coli		Single-tube, rapid (4 hr) confirmatory test	Smith and Rockliff (1982)
E. coli		Rapid test for enterotoxin-producing colonies	Finkelstein and Yang (1983)
E. coli	Seafood	Fluorogenic assays for rapid, sensitive enumeration	Alvarez (1984)
E. coli		Identification from MacConkey agar and spot indole test	Bale *et al.* (1984)
E. coli	Frozen foods	Direct plating method for enumeration	Hall (1984)
E. coli		Detection of enterotoxin by ELISA	Ronnberg *et al.* (1984)
E. coli		Rapid isolation and identification by detecting β-glucuronidase	Trepeta and Edberg (1984)
E. coli, coliforms	Various	New selective and differential agar (lauryl sulfate aniline blue)	Wright (1984)
E. coli, *Shigella*		Rapid immunosorbent assay for screening	Pal *et al.* (1985)
Citrobacter freundii		Recovery media is critical for injured cells	Verrips and Kwast (1982)

4. Standard and Experimental Methods of Identification and Enumeration

TABLE A-V (*Continued*)

Microorganism	Food product	Method/medium of choice	References
Salmonella	Poultry	A system composed of six sugars for screening isolates	Cox and Williams (1976)
Salmonellae	Frozen poultry	Highest detection by analysis of enclosed organs	Reusse et al. (1976)
Salmonella	Various products	Lysine iron cystine neutral red broth	D'Aoust (1977)
Salmonellae	Ground meat	Best isolation at 43°C and secondary enrichment in tetrathionate broth	Kafel and Bryan (1977)
Salmonellae		Commercially available miniaturized system (API-50E) for 50 chemical reactions	Ottaviani and Aliani (1977)
Salmonellae	Frozen vegetables	Recovery of freeze-injured cells with buffered peptone–H_2O as preenrichment	Sadovski (1977)
Salmonellae	Foods, feeds	Timed release of selective agents into nonselective basal broth	Sveum and Hartman (1977)
Salmonella	Minced meat	Enrichment in modified Rappaport medium at 43°C	Vassiliadis et al. (1977)
Salmonella		Use of brilliant green MacConkey agar and deoxycholate citrate agar for maximum numbers	Chau and Leung (1978)
Salmonellae	Broiler carcasses	Recovery of heat-injured cells by direct enrichment in either selenite cystine or selenite brilliant green broth	Cox and Mercuri (1978)
Salmonella		Recovery of heat-injured cells on Levine eosin methylene blue salts agar with additives	D'Aoust (1978)
Salmonella, Enterobacteriaceae		Dulcitol malonate phenylalanine agar	Eskenazi and Littell (1978)
Salmonellae	Chicken, ground beef	Enrichment medium containing tryptic soy	Kinner and Moats (1978)

(*continued*)

TABLE A-V (*Continued*)

Microorganism	Food product	Method/medium of choice	References
Salmonellae		broth, sodium cholate, and $MgCl_2$ Effect of selective agars on growth of rare serotypes	Kroninger and Banwart (1978)
Salmonellae	Beef and poultry meat	Xylose lysine deoxy cholate agar plus novobiocin; tryptic soy xylose lysine agar	Moats (1978)
Salmonellae	Various products	Factors affecting recovery of injured cells in preenrichment media	Pietzsch and Mulindwa (1978)
Salmonellae	Pasteurized egg albumen, frozen poultry	Buffered peptone water or tryptone soya broth for preenrichment	Siems (1978)
Salmonellae	Meats and poultry	Control of nonspecific staining in fluorescent antibody technique	Swaminathan et al. (1978a)
Salmonellae	Raw meat and poultry	Direct enrichment in tetrathionate broth plus brilliant green dye; preenrichment in buffered peptone–H_2O followed by enrichment in tetrathionate broth	Thomason and Dodd (1978)
Salmonella		Brilliant green agar plus antibiotics incubated at 43°C	Watson and Walker (1978)
Salmonella	Various products	Combined use of brilliant green, bismuth sulfite, Hektoen enteric, and xylose lysine deoxycholate agars	Andrews et al. (1979b)
Salmonella		Influence of competitive microflora upon isolation	Mulindwa and Pietzsch (1979)
Salmonellae	Milk powder	Use of buffered peptone water for preenrichment	Muller (1979)
S. typhi		Typing according to 14 biochemical patterns	Paramasinan et al. (1979)

4. Standard and Experimental Methods of Identification and Enumeration

TABLE A-V *(Continued)*

Microorganism	Food product	Method/medium of choice	References
Salmonella	Various products	Modification of lysine iron agar for improved isolation	Rappold and Bolderdijk (1979)
Salmonellae	Various products	Use of a selective motility medium for secondary selective enrichment	Smeltzer and Duncalfe (1979)
S. heidelberg		Recovery of injured cells was better on simple medium than on complex medium	Tang and Jackson (1979)
S. enteritidis		Detection of deoxyribonuclease activity	Tomasulo and Braunstein (1979)
Salmonellae	Pork sausage	Use of modified Rappaport medium	Vassiliadis et al. (1979b)
Salmonella		Use of brilliant green agar with sodium deoxycholate	Vassiliadis et al. (1979a)
Salmonella	Broilers, frozen meat	Direct enrichment in selenite cystine broth	Cox et al. (1980)
Salmonella	High and low moisture foods	Refrigeration (4°C) of nonselective and selective enrichment broth cultures for 72 hr	D'Aoust et al. (1980)
Salmonellae	Broiler chickens	Increased isolation by delayed secondary enrichment	Rigby and Pettit (1980)
Salmonellae	Meats and poultry	Rapid detection with direct immunoenzyme method	Swaminathan and Ayres (1980)
Salmonella	Milk chocolate	Preenrichment in nonfat dry milk with brilliant green or buffered peptone water at 35°C	Wilson et al. (1980)
Salmonella	Chicken giblets	Use of Rappaport medium for isolation	Harvey and Price (1981)
Salmonella	Chicken carcasses	Greater detection by enrichment in modified Rappaport medium	Mavrommati et al. (1981)
Salmonella	Meat products	Addition of $CaCO_3$ as a buffer to enrichment media	Perutkova (1981)

(continued)

TABLE A-V (*Continued*)

Microorganism	Food product	Method/medium of choice	References
Salmonella	Casein	Lactose broth as preenrichment medium. One cell/gm of product could be detected	Poelma et al. (1981)
Salmonellae	Various products	Combining preenrichment and selective enrichment to detect both injured and uninjured cells	Sveum and Kraft (1981)
Salmonella	Pork sausage	Greater detection with modified Rappaport medium at 43°C	Vassiliadis et al. (1981)
Salmonella	Broiler carcasses	Highest detection by selective enrichment in tetrathionate brilliant green broth and plating on bismuth sulfite agar	D'Aoust et al. (1982b)
Salmonella	Water	Better recovery by adding novobiocin to modified Rappaport medium	Alcaide et al. (1982)
Salmonella	Poultry	Use of surfactants is not warranted	D'Aoust et al. (1982a)
Salmonella		Novobiocin addition to XLD and HE agars to improve isolation	Restaino et al. (1982)
Salmonella		Use of X-ray microprobe to detect cells	Richter and Banwart (1982)
Salmonella	Nonfat dry milk	Improved detection by soak method	Andrews et al. (1983)
Salmonella		Best enterotoxin detection by using cell sonicates	Baloda et al. (1983)
Salmonella		Direct enrichment in SC and TT broths for best recovery	Cox et al. (1983)
Salmonella	Chicken	Importance of BS agar and LIA for routine use	Dube (1983)
Salmonella	Various	Enhanced isolation with modified Rappaport medium	Van Schothorst and Renaud (1983)
Salmonella	Water	More reliable detection with glass microfiber filters	Block et al. (1983)
Salmonella	Broilers	Better isolation with TT	Truscott (1983)

TABLE A-V (*Continued*)

Microorganism	Food product	Method/medium of choice	References
Salmonella		broth and dulcitol BG agar Adverse effects of sulphacetamide and mandelate	Jones et al. (1984)
Salmonella	Dry products	Rapid (8 hr) detection method	Rappold et al. (1984)
Salmonella	Chicken	Aerobic incubation of enrichment and plating media	Bailey et al. (1984)
Salmonella	Various	Recovery from refrigerated media	Mavrommati et al. (1984)
Salmonella	Raw beef	Modified Rappaport medium	Tongpim et al. (1984)
Salmonella	Chicken liver	Isolation with Rappaport-Vassiliadis soya peptone medium	McGibbon et al. (1984)
Salmonella		Treatment of samples with protease to inhibit non specific reactions in ELISA	Rigby (1984)
Salmonella	Meat	Best isolation with RV medium	Vassiliadis et al. (1984)
Salmonella	Water	Modified agar for detection	Hussong et al. (1984)
Salmonella	Beef and chicken	Method for enumerating freeze-damaged cells in a mixed population	White and Hall (1984)
Salmonella	Egg	Best isolation with TT broth at 43°C	Yde and Ghysels (1984)
S. typhimurium		Liquid media better for recovery of injured cells than solid media	Mackey and Derrick (1982)
Shigella flexneri		Best isolation with Y medium	Magalhaes and Andrade (1983)
Campylobacter		Use of Skirrow's medium plus amphotericin B	Wang et al. (1978)
Campylobacter		A simple biological technique for reducing oxygen tension	Karmali and Fleming (1979)
Campylobacter	Broilers	Use of heart infusion broth plus blood	Grant et al. (1980)
C. jejuni	Various prod-	Isolation using thioglycol-	Oosterom et al.

(*continued*)

TABLE A-V (*Continued*)

Microorganism	Food product	Method/medium of choice	References
	ucts	late, lysed horse blood, antimicrobial compounds, and ox bile	(1981)
C. jejuni	Lamb carcasses	Isolation using tryptic soy agar, lysed horse blood, and four antibiotics; 32 cells/cm^2 could be detected	Stern (1981a)
C. jejuni	Turkey eggs and meat	Detection of low numbers with selective enrichment–plating procedure using brucella broth and agar	Acuff et al. (1982)
C. jejuni		Use of new selective medium containing four antibiotics	Bolton and Robertson (1982)
Campylobacter		Improved growth in a medium containing peptone, yeast extract, K_2HPO_4, $(NH_4)_2SO_4$, Na_2SO_3, soluble starch, and agar	Mehlman and Romero (1982b)
C. jejuni	Ground beef	Use of Cary-Blair diluent for refrigerated storage and glycerol or dimethyl sulfoxide for frozen storage	Stern and Kotula (1982)
C. jejuni	Ground beef	Campy-BAP medium most sensitive for recovery	Stern (1982d)
C. jejuni	Milk, turkey	Enrichment in brucella broth; detection on C. jejuni selective agar	Wyatt and Timm (1982)
C. jejuni		Enrichment broth better than direct plating	Bolton et al. (1983)
Campylobacter		Use of blood-free medium for detection	Bolton and Coates (1983)
C. jejuni	Milk	Method for detection of 1 cell/ml	Lovett et al. (1983)
C. jejuni	Chicken	Increased isolation in medium at 42°C, for 48 hr, at pH 7.0	Park et al. (1983)

4. Standard and Experimental Methods of Identification and Enumeration

TABLE A-V (*Continued*)

Microorganism	Food product	Method/medium of choice	References
C. jejuni		Optimum recovery by cold enrichment in Campy-Thio broth	Rubin and Woodard (1983)
C. jejuni		New selective enrichment broth	Martin et al. (1983)
C. jejuni	Chicken liver	Thioglycolate broth plus supplements best enrichment medium	Barot and Bokkenheuser (1984)
C. jejuni	Milk	Best recovery by adding $NaHSO_3$ and holding under N_2	Koidis and Doyle (1984)
C. jejuni		Improved preservation with egg-based medium	Nair et al. (1984)
C. jejuni		Detecting heat injury	Palumbo (1984)
C. jejuni	Various	Greater recovery with constant gas flow	Heisick et al. (1984)
C. jejuni		Injured cells adversely affected by incubation at 42°C and polymyxin B	Ray and Johnson (1984)
C. jejuni	Chicken	Enrichment broth better than direct plating	Rothenberg et al. (1984)
C. jejuni		Best recovery of injured cells with brucella broth	Abram and Potter (1985)
C. jejuni	Raw milk	Method for isolating less than 1 cell per gram	Hunt et al. (1985)
C. jejuni		Inhibition of growth on basal media	Ng et al. (1985a)
P. aeruginosa		Recovery of injured cells with plate count agar. Selective procedures decrease counts	Kukulinsky-Fuller and Nelson (1977)
Pseudomonads	Poultry meat	Selective medium composed of heart infusion agar plus antibiotics	Mead and Adams (1977)
P. aeruginosa	Water	mPA-C medium; enumeration after 24 hr at 41.5°C	Brodsky and Ciebin (1978)
Pseudomonads		Use of the API-20 enteric strip	Morris et al. (1978)
P. aeruginosa		Isolation using a new selective agent	Marold et al. (1981)

(*continued*)

TABLE A-V (Continued)

Microorganism	Food product	Method/medium of choice	References
Pseudomonas		Use of selective isolation medium containing three antibiotics	Wakisaka and Koizumi (1982)
Pseudomonas		Modified MacConkey agar for pigment production	Daly et al. (1984)
Pseudomonas		New selective media for recovery	Gould et al. (1985)
P. aeruginosa		Enhanced recovery with new selective agar	Robin and Janda (1984)
Yersinia	Various products	Enrichment with 1% sorbitol and 0.15% bile salts. Isolation on MacConkey agar	Mehlman et al. (1978)
Y. enterocolitica	Milk	Enrichment in Butterfield's phosphate buffer; enumeration in modified Rappaport broth	Schiemann and Toma (1978)
Y. enterocolitica		Identification by production of pyrrolidonyl peptidase	Zaremba (1978)
Y. enterocolitica		Cellobiose arginine lysine agar for isolation	Dudley and Shotts (1979)
Y. enterocolitica		Effect of dyes upon quantitative recovery	Schiemann (1979a)
Y. enterocolitica		Use of selective agar for isolation	Schiemann (1979b)
Y. enterocolitica		Cold enrichment in tryptone soya broth; isolation on DHL agar	Van Pee and Stragier (1979)
Y. enterocolitica	Various meats, shellfish, vegetables	Rapid recovery by use of KOH treatment	Aulisio et al. (1980)
Y. enterocolitica	Meat	Detection with two modified selenite broths	Lee et al. (1980)
Y. enterocolitica		Recovery of heat-injured cells was directly related to the heating menstrum used for injury	Restaino et al. (1980)
Y. enterocolitica		Formulation of a broth medium for optimum	Schiemann (1980)

4. Standard and Experimental Methods of Identification and Enumeration

TABLE A-V (*Continued*)

Microorganism	Food product	Method/medium of choice	References
		growth; effects of selective agents	
Y. enterocolitica		Isolation using a medium consisting of casein hydrolysate, peptone, sodium oxalate, and bile salts	Soltesz et al. (1980)
Y. enterocolitica		Recovery by cold enrichment using phosphate-buffered saline; identification using API 20 E system	Barclay (1981)
Y. enterocolitica	Vegetables	New sensitive and rapid technique for isolation	Park et al. (1981)
Y. enterocolitica	Raw milk	Enrichment in medium with sucrose, tris buffer, sodium azide, and ampicillin	Vidon and Delmas (1981)
Y. enterocolitica		Test for potential virulance by ability to autoagglutinate in tissue culture media at 35°C	Stern and Damare (1982)
Y. enterocolitica	Poultry	Best isolation by enrichment in PBS, sorbitol, bile salts, and KOH treatment	deBoer et al. (1982)
Y. enterocolitica		Selective media vary in ability to support growth	Head et al. (1982)
Y. enterocolitica	Meats	More rapid procedure for recovery	Doyle and Hugdahl (1983)
Y. enterocolitica		Virulence screening method with Congo red agar	Prpic et al. (1983)
Y. enterocolitica		Use of lysine arginine iron agar for five biochemical tests in one tube	Weagant (1983)
Y. enterocolitica	Water	New selective medium for isolation	Bercovier et al. (1984)
Vibrio parahaemolyticus		Recovery of injured cells with trypticase soy agar plus 0.25% NaCl	Emswiler et al. (1976)

(*continued*)

TABLE A-V (*Continued*)

Microorganism	Food product	Method/medium of choice	References
V. parahaemolyticus	Seafood	Recovery of injured cells in trypticase soy broth	Ray et al. (1978)
V. parahaemolyticus	Mussels	Limitations of current methodology	Van den Broek et al. (1979a)
V. parahaemolyticus	Seafood	Rapid, presumptive identification by testing for seven biochemical reactions in one medium	Kaper et al. (1980)
V. parahaemolyticus	Oysters	Recovery of injured cells with enrichment broths plus magnesium and iron salts	Ma-Lin and Beuchat (1980)
V. parahaemolyticus	Seafood	Loss of polymyxin B from enrichment broth	Blanchfield et al. (1982)
V. parahaemolyticus	Seafood	Enrichment for 8 hr at 37°C	Dupray and Cormier (1983)
V. parahaemolyticus	Fish	Alkaline enrichment broth plus NaCl plus dye for best detection	Hofer and Silva (1984)
V. fluvialis	Water	Alkaline peptone NaCl novobiocin medium for enrichment	Nishibuchi et al. (1983)
V. vulnificus	Shellfish	Isolation with thiosulfate citrate bile-salts sucrose agar	Tamplin et al. (1982)
V. vulnificus		Selective medium with salicin	Brayton et al. (1983)
V. cholerae		Rapid agglutination test involving the flagellar H-antigen complex	Sil and Bhattacharyya (1978)
V. cholerae		Hydrolysis of nucleic acids	Tsan (1978)
V. cholerae		Simple laboratory method for isolation and identification	Huq (1979)
V. cholerae		Improved detection with oxidase test	Rogol et al. (1981)
Vibrio		Development of two new media for isolation	Ozsan and Mercangoz (1980)
Desulfotomaculum nigrificans	Soy protein–based products	Enumeration with modified infant soy formula broth	Donnelly and Busta (1981b)
Klebsiella		Isolation with MacConkey	Bagley and

TABLE A-V (Continued)

Microorganism	Food product	Method/medium of choice	References
		inositol carbenicillin agar	Seidler (1978)
Proteus		Rapid method based on detection of urease	Senior (1981)
Chromobacterium	Various	Best detection with Bennett agar at 25°C	Koburger and May (1982)
Serratia marcescens		Rapid identification by coagglutinatian	Roig et al. (1983)
Bacillus	Various products	A key of biochemical tests to aid in rapid identification	Seenappa and Kempton (1981)
B. cereus		Differentiation by appearance of filamentous appendages on spores	Hachisuka and Kozuka (1981)
B. cereus	Various products	Detection with polymyxin pyruvate egg-yolk mannitol bromothymol-blue agar	Holbrook and Anderson (1980)
B. cereus	Beans, wheat	Best differentiation with MYP agar	Harmon et al. (1984)
Anaerobes		Improved cooked meat medium for growth	Wren (1979)
Anaerobic bacteria		Rapid method for detecting β-lactamase	Lee and Rosenblatt (1983)
Clostridia		Use of medium for sulfite reduction test	Kawabata (1980)
C. botulinum	Honey	Dialysis–enrichment culture method for detecting spores	Sugiyama et al. (1978)
C. botulinum		Development of a defined medium which supports growth and sporulation of type E	Hawirko et al. (1979)
C. perfringens	Ground beef	Best isolation of small numbers of cells with tryptose sulfite cycloserine agar	Emswiler et al. (1977a)
C. perfringens	Various products	Medium consisting of litmus milk plus thioglycolate medium with additives and antibiotics	Erickson and Deibel (1978)
C. perfringens	Meats and sea-	Strains typed by mouse	Fruin (1978)

(continued)

TABLE A-V (*Continued*)

Microorganism	Food product	Method/medium of choice	References
	foods	serum neutralization method	
C. perfringens	Various products	Confirmation by growth in motility-NO$_3$ medium, lactose gelatin medium and petone yeast extract medium with salicin and raffinose	Harmon and Kautter (1978)
C. perfringens		Medium for sporulation	Sacks and Thompson (1978)
C. perfringens		Autoclaving of tryptose sulfite cycloserine agar to permit growth	Brodsky and Ciebin (1979)
C. perfringens	Various	Method for isolation and enumeration	Debevere (1979)
C. perfringens		Development of a defined medium for studies of spore coat and enterotoxin synthesis	Sacks and Thomas (1979)
C. perfringens	Various products	Recommendations for the bacteriological examination of foods	Schau (1979)
C. perfringens	Poultry meat	Detection with three plating media. Need for confirmatory tests when examining natural samples	Adams and Mead (1980)
C. perfringens		Recovery of injured cells affected by selective agents	Apiluktivongsa and Walker (1980)
C. perfringens	Milk and dairy products	Optimum detection with tryptone sulfite neomycin agar	El-Bassiony (1980)
C. perfringens		New test for presumptive identification	Hansen and Elliott (1980)
C. perfringens	Various products	Application of serological typing in identification	Stringer et al. (1980)
C. perfringens		Rapid and simplified confirmatory tests	Mead et al. (1981)
C. perfringens		Incubating plates at 37°C for best recovery of heat-injured spores	Labbe and Norris (1982)

4. Standard and Experimental Methods of Identification and Enumeration

TABLE A-IV (*Continued*)

Microorganism	Food product	Method/medium of choice	References
C. perfringens		Use of iron milk medium	St. John et al. (1982)
C. perfringens	Water	Enumeration by membrane filtration	Burger et al. (1984)
C. perfringens		Sporulation medium	Tortora (1984)
C. sporogenes		Modified PA agar for better recovery of spores	Grischy et al. (1983)
C. sporogenes		Egg yolk TSA for recovery of injured spores	Michels and Kagei (1983)
C. sporogenes		Isolation using a semidefined basal medium plus amino acids	Fryer and Mead (1979)
C. sporogenes		Highest detection with Andersen's pork pea infusion. Pour plate procedure may not be adequate for recovery of heat-stressed spores	Polvino and Bernard (1982)
C. thermosaccharolyticum		Best detection and enumeration in medium consisting of peas in 2% peptone	Ashton (1981)
Listeria		A selective blood agar for pathogenic strains	Skalka and Smola (1983)
Gram-positive cocci		Rapid differentiation between micrococci and staphylococci with peptone agar plus furazolidone	Von Rheinbaben and Hadlok (1981)
Staphylococci		Recovery of injured cells with tryptone soya broth plus 10% NaCl	Ibrahim (1976)
Staphylococci	Dairy	Diagnostic paper strips	Ibrahim (1977)
Staphylococci		Recommendation against use of lysostaphin susceptibility test	Heddaeus et al. (1978)
Staphylococci	Various products	Selective cultivation with Kranep agar	Sinell (1978)
Staphylococci	Animal	Use of standardized plate culture DNase tests	Devriese and Van de Kerckhove (1979)
Staphylococci		Identification by acid production from carbohy-	Devriese and Van de Kerck-

(*continued*)

TABLE A-V (*Continued*)

Microorganism	Food product	Method/medium of choice	References
		drates using phenol red broth	hove (1980)
Staphylococcus, Micrococcus		Differenctiation based on sensitivity to lysostaphin	Colwell et al. (1981)
Staphylococcus	Various products	Addition of nonfat dry milk to food samples to enhance the recovery of thermonuclease	Park et al. (1979)
S. aureus		Pork plasma medium for direct enumeration	Devoyod et al. (1976)
S. aureus		Rapid identification using simplified thermonuclease test	Lachica (1976)
S. aureus		Recovery of injured cells by adding catalase or pyruvate to media	Brewer et al. (1977)
S. aureus	Various products	Recovery of injured cells with modified Vogel and Johnson agar	Andrews and Martin (1978)
S. aureus	Various products	Rapid method for detection of heat-stable staphylococcal nuclease	Koupal and Diebel (1978)
S. aureus	Minced meat	Isolation on Baird-Parker medium	Niskanen and Aalto (1978)
S. aureus	Beef sausage	Recovery of injured cells by plating on tryptic soy agar	Smith and Palumbo (1978)
S. aureus		Assay of catalase activity during injury and recovery	Andrews and Martin (1979)
S. aureus	Various products	Use of modified pork plasma agar for enumeration	Hauschild et al. (1979b)
S. aureus	Various products	Assaying pork plasma before incorporation into agar medium	Mossel and Eelderink (1979)
S. aureus	Various products	Preincubation in glucose-free nutrient broth; detection with Baird-Parker agar	Takacs and Domjan (1979a)
S. aureus	Sausage	Rapid detection of thermonuclease by direct	Emswiler-Rose et al. (1980)

TABLE A-V (Continued)

Microorganism	Food product	Method/medium of choice	References
S. aureus		plating of sausage casing disks on assay agar Rapid (2 hr) identification using protein A hemagglutination	Essers and Radebold (1980)
S. aureus	Various products	Rapid procedure for enumeration and identification	Lachica (1980)
S. aureus		Selective isolation with Baird-Parker medium containing acriflavine plus antibiotics	Devriese (1981)
S. aureus	Cheese	Simple method for detection of thermonuclease activity	Ibrahim (1981)
S. aureus		Injured cells can give false-negative coagulase results	Korkeala and Hirvelae (1981)
S. aureus		Identification based upon demonstration of both strongly positive thermonuclease and coagulase reactions	Sukroongreung et al. (1981)
S. aureus		Use of thermonuclease test as a primary procedure for identification	Shanholtzer and Peterson (1982)
S. aureus		Liquid modification of BP medium to detect low numbers	Van Doorne et al. (1981)
S. aureus		Improved detection of enterotoxins by modified ELISA techniques	Fey et al. (1982)
S. aureus		Isolation and identification in a single step	Mintzer-Morgenstern and Katzenelson (1982)
S. aureus		Detection of low numbers by enrichment in liquid BP agar at 43°C, and confirmation by TNase test	Van Doorne et al. (1982)
S. aureus		Four-hour microtube coagulase test	Goldstein and Roberts (1982)

(continued)

TABLE A-V (*Continued*)

Microorganism	Food product	Method/medium of choice	References
S. aureus		Enumeration of low numbers with modified MPN	Patterson and Damoglou (1983)
S. aureus		Baird-Parker with pig plasma (if supply of plasma is available)	Becker et al. (1983)
S. aureus		Use of neutralizing medium to recover injured cells	Dey and Engley (1983)
S. aureus		Rapid identification with a commercial latex kit	Dibb et al. (1983)
S. aureus	Milk	Use of thermostable nuclease for rapid identification	Hill (1983)
S. aureus	Various	Better detection with rabbit plasma fibrinogen agar than with BP agar	Beckers et al. (1984)
S. aureus	Various	Replacement of egg yolk with Tween 80, $MgCl_2$, and pyruvate in BP medium	Lachica (1984)
S. aureus	Water	Modified medium with membrane filter	Stengren and Starzyk (1984)
Streptococci	Dried and frozen foods	Limitations of KF and bile esculin agars	Mundt (1976)
Streptococci	Various products and water	Medium containing esculin, kanamycin, and azide	Mossel et al. (1978)
Streptococci		Presumptive identification on agar media	Facklam et al. (1979)
Streptococci		Rapid method for identification	Waitkins et al. (1980)
Streptococci		Four-hour identification method	Bosley et al. (1983)
Streptococci		Ten-minute detection of hippurate hydrolysis	Mugg (1983)
Enterococci		Presumptive identification by naphthylamide hydrolysis	Facklam et al. (1982)
Enterococci		Identified in 4 hr with bile esculin agar plus Autobac system	Brown et al. (1983)
Enterococci		Selective and differential	Littel and

TABLE A-V (Continued)

Microorganism	Food product	Method/medium of choice	References
		medium with fluorogenic substates	Hartman (1983)
Enterococci		Selective enterococcus medium	Herrmann et al. (1978)
Enterococci	Meat and meat products	Cultivation on kanamycin esculin azide agar or citrate azide Tween carbonate agar	Reuter (1978)
Enterococci		Medium that differentiates enterococci from other group D streptococci within 4–24 hr	Qadri et al. (1979)
Enterococci		Best isolation with m-*Enterococcus* agar	Pagel and Hardy (1980)
Lactic streptococci	Dairy products	Improved enumeration on Elliker agar plus $(NH_4)_2HPO_4$	Barach (1979)
Lactic streptococci		Use of agar medium for differentiation of strains	Mullan and Walker (1979)
Lactic streptococci	Skim milk	Measurement of growth by direct measurement of cell dry weight	Ross et al. (1980)
Lactobacillus, Pediococcus	Brewery products	Sucrose agar medium	Woodward (1978)
Lactobacillus		Modificaton of minitek system for rapid identification	Benno and Mitsuoka (1983)
Propionibacteria	Cheese	Selection with trypticase-lactate or $(NH_4)_2SO_4$ lactate agars	Peberdy and Fryer (1976)
Yeasts and fungi	Various products	Use of modified rose bengal medium with two antibiotics	Baggerman (1981)
Yeasts		Miniaturized urea broth test	Roberts et al. (1978)
Yeasts	Poultry	Rapid identification by serological and urease tests	Taguchi et al. (1979)
Yeasts		Modified dye pour plate auxanographic method for identification	Weymann et al. (1979)
Yeasts	Orange juice	Recovery of injured cells on plate count agar	Graumlich (1981)

(*continued*)

TABLE A-V (*Continued*)

Microorganism	Food product	Method/medium of choice	References
Yeasts	Fruit	Molybdate agar plus calcium propionate	Rale and Vakil (1984)
S. cerevisiae		Repair of injured cells after storage in H$_2$O at 22°C. Cells plated on yeast nitrogen base glucose (filter-sterilized) medium	Graumlich and Stevenson (1978)
Fungi	Tomato products	Chemical method to assess level of contamination	Jarvis (1977)
Fungi	Beans and soybeans	Higher detection rate with direct plating compared to serial dilution	Mislivec and Bruce (1977)
Fungi	Various products	Agar medium with yeast extract and oxytetracycline	Stec (1977)
Fungi	Various products	Malt extract agar with antibiotics	Blaser (1978b)
Fungi	Nonfluid foods	Direct microscopic detection	Blaser (1978a)
Fungi		Fluorescein diacetate ethidium bromide method to determine viability	Calich *et al.* (1979)
Fungi	Various products	Use of dichloran rose bengal medium to control spreading	King *et al.* (1979)
Fungi	Tomato products	Detection of contamination by chemical assay for chitin	Bishop *et al.* (1982)
Fungi	Tomato products	Converting Howard mold count into number of mold filaments per unit volume	Aldred (1983)
Fungi		Simple screening method for intracellular mycotoxins	Filtenborg *et al.* (1983)
Fungi	Airborne sampling	Rose bengal streptomycin agar	Morring *et al.* (1983)
Fungi	Tomato, fruit products	Chemical detection of glucosamine	Cousin *et al.* (1984)
Fungi		Detection of growth by spectrophotometry	Granade *et al.* (1985)

TABLE A-V (*Continued*)

Microorganism	Food product	Method/medium of choice	References
Aspergillus	Peanuts	Selective medium for *A. flavus* and *A. parasiticus*	Pitt et al. (1983)
Xerophilic fungi	Dried and semi-dried foods	Detection with low-water-activity medium containing glycerol and dichloran	Hocking and Pitt (1980)
A. flavus	Dried, processed foods	Use of *Aspergillus* differential medium plus antibiotics	Beuchat (1979)
Viruses	Oysters	Procedure for recovery	Vaughn et al. (1979)
Foodborne viruses	Various products	Use of polyelectrolytes to enhance recovery	Kostenbader and Cliver (1981)
Viruses	Oysters	Extraction with NF milk salt freon solution at pH 9	Sullivan and Peeler (1982)
Viruses	Vegetables	Eluted with alkaline solutions and concentrated by organic flocculation	Ward et al. (1982)
Viruses	Water	Concentration with a filter aid	Singh et al. (1983)
Viruses		Recovery with chaotropic agents	Wait and Sobsey (1983)
Viruses	Water	Simple field method for concentration	Torangos et al. (1984)
Viruses	Vegetables	Best extraction with beef extract, pH 8.0	Badawy et al. (1985)
Enteroviruses		Rapid detection by DNA hybridization	Hyypiae et al. (1984)
Enteroviruses		ELISA technique to detect several types	Deng and Cliver (1984)
Enteroviruses	Meats	Method for concentrations. 2–4 plaque-forming units per gram of meat could be determined	Kalitina (1978)
Enteroviruses	Oysters	Improved method for concentration by adjusting pH and NaCl concentration	Sobsey et al. (1978)
Enteroviruses		Identification by enzyme-linked immunoassay	Herrmann et al. (1979)

(*continued*)

TABLE A-V (*Continued*)

Microorganism	Food product	Method/medium of choice	References
Enteroviruses	Mussels	An improved method for recovery	Soncini et al. (1979)
Enteroviruses	Shellfish	Improved recovery method	Metcalf et al. (1980a)
Enteroviruses	Shellfish	Method for recovery of small numbers (3 plaque-forming units per 100 gram of shellfish)	Metcalf et al. (1980b)
Enteroviruses	Shellfish	Best detection with glass wool hydroextraction method	Tierney et al. (1980)
Poliovirus	Oysters	Extraction with polyelectrolyte and modified acid precipitation	Landry et al. (1980)

ACKNOWLEDGMENTS

The author wishes to express sincere gratitude to Jane Voss for typing this manuscript and to Miriam Chipley for compiling the bibliography. Without their help, this review would not have been possible.

REFERENCES

Abbiss, J. S., Wilson, J. M., Blood, R. M., and Jarvis, B. 1981. A comparison of minerals-modified glutamate medium with other media for the enumeration of coliforms in delicatessen foods. J. Appl. Bacteriol. 51, 121–127.

Abeyta, Jr., C., Michalovskis, A., and Wekell, M. M. 1985. Differentiation of *Clostridium perfringens* from related clostridia in iron milk medium. J. Food Protection 48, 130–134.

Abram, D. D., and Potter, N. M. 1985. A research note: diluents and the enumeration of stressed *Campylobacter jejuni*. J. Food Protection 48, 135–137.

Ackland, M. R., Trewhella, E. R., Reeder, J., and Bean, P. G. 1981. The detection of microbial spoilage in canned foods using thin-layer chromatography. J. Appl. Bacteriol. 51, 277–281.

Acuff, G. R., Vanderzant, C., Gardner, F. A., and Golam, F. A. 1982. Evaluation of an

4. Standard and Experimental Methods of Identification and Enumeration

enrichment-plating procedure for the recovery of *Campybacter jejuni* from turkey eggs and meat. J. Food Prot. 45, 1276–1278.

Adams, B. W., and Mead, G. C. 1980. Comparison of media and methods for counting *Clostridium perfringens* in poultry meat and further-processed products. J. Hyg. 84, 151–158.

Adams, B. W., Mead, G. C., and Pennington, D. E. 1980. A scrape-sampling method for the microbiological examination of poultry carcasses. Br. Poult. Sci. 21, 71–75.

Ahuja, S., Mago, M. L, and Saxena, S. N. 1979. Evaluation of the efficacy of composite media for the rapid identification of members of the *Enterobacteriaceae*. Indian J. Pathol. Microbiol. 22, 249–253.

Ajl, S. J., Kadis, S., and Montie, T. C. (Editors). 1970. "Microbial Toxins," Vol. 1, Academic Press, New York, N.Y.

Alcaide, E., Martinez, J. P., Martinex-Germes, P., and Garay, E. 1982. Improved *Salmonella* recovery from moderate to highly polluted waters. J. Appl. Bacteriol. 53, 143–146.

Aldred, J. B. 1983. The relationship between the Howard Mould count and the enumeration of mould filaments in tomato products. J. Assoc. Public Anal. 21, 73–76.

Aleixo, J. A. G., Swaminathan, B., and Minnich, S. A. 1984. *Salmonella* detection in foods and feeds in 27 hours by an enzyme immunoassay. J. Microbiol. Methods 2, 135–145.

Alexande, R. N., and Marshall, R. J. 1982. Moisture loss from agar plates during incubation. J. Food Prot. 45, 162–163.

Alouf, J. E., Fehrenbach, F. J., Freer, J. H., and Jeljaszewicz, J. (Editors). 1984. "Bacterial Protein Toxins," Academic Press, New York, N.Y., 473 pp.

Alvarez, R. J. 1984. Use of fluorogenic assays for the enumeration of *Escherichia coli* from selected seafoods. J. Food Sci. 49, 1186–1187.

American Public Health Association. 1972. "Standard Methods for the Examination of Dairy Products," 13th Edition. W. J. Hausler (Editor). American Public Health Assoc., Washington, D.C.

Ames, B., Durstan, W., Yamasaki, E., and Lee, F. 1973. Carcinogens are mutagens: a simple test system combining liver homogenate for activation and bacteria for detection. Proc. Natl. Acad. Sci. U.S.A. 70, 2281–2285.

Anderson, A. W. 1977. The significance of yeasts and molds in foods. Food Technol. 31, 45–50.

Anderson, D., and Fugate, J. 1977. Use of gas-liquid chromatography for the confirmation of *Clostridium perfringens*. Abs. Ann. Meet. Amer. Soc. Microbiol. p. 246.

Anderson, J. E., Adams, D. M., and Walter, Jr., W. M. 1983. Conditions under which bacterial amylases survive ultrahigh temperature sterilization. J. Food Sci. 48, 1622–1625.

Anderson, J. G., Meadows, P. S., Mullins, B. W., and Patel, K. 1979. Inconsistent results with the *Escherichia coli* confirmatory medium lactose ricinoleate broth. FEMS Microbiol. Lett. 5, 53–56.

Anderson, M. E., Sebaugh, J. L., Marshall, R. T., and Stringer, W. C. 1980. A method for decreasing sampling variance in bacteriological analyses of meat surfaces. J. Food Prot. 43, 21–22.

Andrews, G. P., and Martin, S. E. 1978. Modified Vogel and Johnson Agar for *Staphylococcus aureus*. J. Food Prot. 41, 530–532.

Andrews, G. P., and Martin, S. E. 1979. Catalase activity during the recovery of heat-stressed *Staphylococcus aureus* MF-31. Appl. Environ. Microbiol. 38, 390–394.

Andrew, M. H. E., and Russell, A. D. 1984. (Editors). "The Revival of Injured Microbes." Soc. Appl. Bacteriol. Sym. Series No. 12. Academic Press, New York, N.Y.

Andrews, W. H. 1986. Resusicitation of injured *Salmonella* spp. and coliforms from foods. J. Food Protection 49, 62–75.
Andrews, W. H., Wilson, C. R., Poelma, P. L., Romero, A., McClure, F. D., and Gentile, D. E. 1978a. Enumeration of coliforms and *Salmonella* in food prepared by blending and stomaching. J. Assoc. Off. Anal. Chem. 61, 1324–1327.
Andrews, W. H., Wilson, C. R., Poelma, P. L., Romero, A., Rude, R. A., Duran, A. P., McClure, F. D., and Gentile, D. E. 1978b. Usefulness of the stomacher in a microbiological regulatory laboratory. Appl. Environ. Microbiol. 35, 89–93.
Andrews, W. H., Duran, A. P., McClure, F. D., and Gentile, D. E. 1979a. Use of two rapid A-1 methods for the recovery of fecal coliforms and *Escherichia coli* from selected food types. J. Food Sci. 44, 289–291.
Andrews, W. H., Wilson, C. R., Poelma, P. L., and Romero, A. 1979b. Relative productivity of five selective plating agars for the recovery of *Salmonella* from selected food types. J. Assoc. Off. Anal. Chem. 62, 320–326.
Andrews, W. H. Wilson, C. R., Poelma, P. L., Bullock, L. K., McClure, F. D., and Gentile, D. E. 1981. Interlaboratory evaluation of the AOAC method and the A-1 procedure for recovery of fecal coliforms from foods. J. Assoc. Off. Anal. Chem. 64, 1116–1121.
Andrews, W. H., Wilson, C. R., and Poelma, P. L. 1983. Improved *Salmonella* species recovery from nonfat dry milk pre-enriched under reduced rehydration. J. Food Sci. 48, 1162–1165.
Angelotti, R., Wilson, J. L., Litsky, W., and Walter, W. G. 1964. Comparative evaluation of the cotton swab and RODAC methods for the recovery of *Bacillus subtilis* spore contamination from stainless steel surfaces. Health Lab. Sci. 1, 289–296.
Anhalt, J. P. 1981. Fluorescent Antibody Procedures and Counterimmunoelectrophoresis. *In* "Laboratory Procedures in Clinical Microbiology." J. A. Washington, (ed.), Springer-Verlag, New York, pp. 249–277.
Anonymous, 1978. Quantitative determination of airborne microorganisms. Ind. Aliment. 17, 955–958.
Anonymous, 1982. Significance of viruses in the environment. Environ. Int. 7, 1–51.
Anonymous, 1983. Draft guideline for validation of the *Limulus* amebocyte lysate test as an end-product endotoxin test for human and animal parenteral drugs, biological products, and medical devices. Pharmacop. Forum 9, 3012–3020.
Apiluktivongsa, P., and Walker, H. W. 1980. Influence of selective agents on recovery of cold-stressed cells of *Clostridium perfringens*. J. Food Sci. 45, 574–576.
Appelbaum, P. C., and Leathers, D. J. 1984. Evaluation of the rapid NFT system for identification of gram negative, non-fermenting rods. J. Clin. Microbiol. 20, 730–734.
Appelbaum, P. C., Arthur, R. R., and Parker, M. E. 1982. Comparison of two methods for same-day identification of *Enterobacteriaceae*. Am. J. Clin. Pathol. 78, 351–355.
Applebaum, R. S., Brackett, R. E., Wiseman, D. W., and Marth, E. H. 1982. Aflatoxin: Toxicity to dairy cattle and occurrence in milk and milk products—a review. J. Food Prot. 45, 725–777.
Arhienbuwa, F. E., Adler, H. E., and Wiggins, A. D. 1980. A method of survillance for bacteria on the shells of turkey eggs. Poult. Sci. 59, 28–33.
Asada, S., Takano, M., and Shibasaki, I. 1979. Deoxyribonucleic acid strand breaks during drying of *Escherichia coli* on a hydrophobic filter membrane. Appl. Environ. Microbiol. 37, 266–273.
Ashton, D. H. 1981. Thermophilic organisms involved in food spoilage: thermophilic anaerobes not producing hydrogen sulfide. J. Food Prot. 44, 146–148.

Asperger, H., and Brandl, E. 1979. The coliform count as a microbiological criterion for refrigerated market milk. Wien. Tierarztl. Monatsschr. 66, 7–13.
Asperger, H., and Winterer, H. 1978. Recovery of injured *Enterobacteriaceae* in dried milk. Wien. Tierarztl. Monatsschr. 65, 241–246.
Asselineau, J. 1966. "The Bacterial Lipids." Holden-Day Inc., 728 Montgomery Street, San Francisco, California.
Atlas, R. M. 1982. Enumeration and estimation of microbial biomass. *In* "Experimental Microbial Ecology," R. G. Burns and J. H. Slater (eds.), Blackwell Scientific Publications, Oxford, England, pp. 84–102.
Aulisio, C. C. G., Mehlman, I. J., and Sanders, A. C. 1980. Alkali method for rapid recovery of *Yersinia enterocolitica* and *Yersinia pseudotuberculosis* from foods. Appl. Environ. Microbiol. 39, 135–140.
Ayres, J. C. 1963. Low temperature organisms as indexes of quality of fresh meat. *In* "Microbiological Quality of Foods." L. W. Slanetz, C. O. Chichester, A. R. Gaufin and Z. J. Ordal (Editors). Academic Press, New York, N.Y.
Ayres, J. C. 1967. Use of fluorescent antibody for the rapid detection of enteric organisms in egg, poultry, and meat products. Food Technol. 21, 145–154.
Bachrach, U., and Bachrach, Z. 1974. Radiometric method for the detection of coliform organisms in water. Appl. Microbiol. 28, 169–171.
Badawy, A. S., Gerba, C. P., and Kelley, L. M. 1985. Development of a method for recovery of rotavirus from the surface of vegetables. J. Food Protection. 48, 261–264.
Baggerman, W. I. 1981. A modified Rose Bengal medium for the enumeration of yeasts and molds from foods. Eur. J. Appl. Microbiol. Biotechnol. 12, 242–247.
Bagley, S. T., and Seidler, R. J. 1978. Primary *Klebsiella* identification with MacConkey-inositol-carbenicillin agar. Appl. Environ. Microbiol. 36, 536–538.
Bailey, J. S., Reagan, J. O., Cox, N. A., and Thomson, J. E. 1984. Comparison of aerobic and anaerobic incubation conditions for optimal recovery of *Salmonella*. J. Food Protection. 47, 615–617.
Bailey, J. S., Cox, N. A., Thomson, J. E., and Fung, D. Y. C. 1985. Identification of *Enterobacteriaceae* in foods with the AutoMicrobic system. J. Food Protection 48, 147–149.
Baldock, J. D. 1974. Microbiological monitoring of the food plant: methods to assess bacterial contamination on surfaces. J. Milk Food Technol. 37, 361–368.
Bale, M. J., McLaws, S. M., Fenn, J. P., and Matsen, J. M. 1984. Use of and cost savings with morphologic criteria and the spot indole test as a routine means of identification of *Escherichia coli*. Diagn. Microbiol. Infect. Dis. 2, 187–191.
Baloda, S. B., Faris, A., Krovacek, K., and Wadstroem, T. 1983. Rapid detection of a new cell-associated *Salmonella* cytotoxic enterotoxin in cell sonicates by Chinese hamster ovary cells. FEMS Microbiol. Lett. 20, 201–206.
Bang, F. B. 1956. A bacterial disease of *Limulus polyphemus*. Bull. Johns Hopkins Hosp. 115, 337.
Barach, J. T. 1979. Improved enumeration of lactic acid streptococci on Elliker agar containing phosphate. Appl. Environ. Microbiol. 38, 173–174.
Barak, M., and Ulitzur, S. 1980. Bacterial bioluminescence as an early indicator of marine fish spoilage. Eur. J. Appl. Microbiol. 10. 155–165.
Barclay, G. 1981. The laboratory isolation and identification of *Y. enterocolitica* and *Y. pseudotuberculosis*. Can. J. Med. Technol. 43, 34–41.
Barot, M. S., and Bokkenheuser, V. D. 1984. Systematic investigation of enrichment media for wild-type *C. jejuni* strains. J. Clin. Microbiol. 20, 77–80.

Barrell, R. A. E., and Paton, A. M. 1979. A semi-automatic method for the detection of salmonellas in food products. J. Appl. Bacteriol. 46, 153-159.
Barrow, G. I. 1981. A comparison of confirmatory media for coliform organisms and *Escherichia coli* in water. J. Hyg. 87, 369-375.
Barry, A. L., Badal, R. E., and Effinger, L. J. 1979a. Identification of *Enterobacteriaceae* in frozen microdilution trays prepared by micro-media systems. J. Clin. Microbiol. 10, 492-496.
Barry, A. L., Badal, R. E., and Effinger, L. J. 1979b. Reproducibility of three microdilution systems for identification of *Enterobacteriaceae*, compared with API 20E and Micro-ID test systems. Curr. Microbiol. 3, 21-25.
Barton, B. W. 1980. The ice cream methylene blue test. Environ. Health 88, 247-249.
Bascomb, S. 1976. Rapid identification of bacteria from clinical specimens by continuous flow analysis. *In* "Rapid Methods and Automation in Microbiology." M. M. Johnston and S. W. B. Newsom (Editors). Learned Information Ltd., Oxford, N.Y.
Bascomb, S. 1981. Application of automation to the general and specific detection of bacteria. Lab. Pract. 30, 461-464.
Bascomb, S., and Spencer, R. C. 1980. Automated methods for identification of bacteria from clinical specimens. J. Clin. Pathol. 33, 36-46.
Basel, R. M., Richter, E. R., and Banwart, G. J. 1983. Monitoring microbial numbers in food by density centrifugation. Appl. Environ. Microbiol. 45, 1156-1159.
Batish, V. K., Chandler, H., and Ranganathan, B. 1982. Characterization of deoxyribonuclease-positive enterococci isolated from milk and milk products. J. Food Prot. 45, 348-352.
Batty, I., and Walker, P. D. 1966. Colonial morphology and fluorescent labelled antibody staining in the identification of species of the genus *Clostridium*. *In* "Identification Methods for Microbiologists" Part A. B. M. Gibbs and F. A. Skinner (Editors). Academic Press. New York, New York.
Baynes, N. C., Comrie, J., and Prain, J. H. 1983. Detection of bacterial growth by the Malthus conductance meter. Med. Lab. Sci. 40, 149-158.
Becker, H., Terplan, G., and Zaadhof, K. J. 1983. Suitability of newer selective media for the detection of *S. aureus* in foods. Zentralbe. Bakteriol. Mikrobiol. Hyg. I Abt. B, 177, 113-126.
Beckers, H. J., Van Leusden, F. M., Bindschedler, O., and Guerraz, D. 1984. Evaluation of a pour-plate system with a rabbit plasma-bovine fibrinogen agar for the enumeration of *Staphylococcus aureus* in food. Can. J. Microbiol. 30, 470-474.
Beech, F. W. 1981. Media and methods for growing yeasts: proceedings of discussion meeting. *In* "Biology and Activities of Yeasts." F. A. Skinner, S. M. Passmore and R. R. Davenport (Editors). Academic Press for the Society of Applied Bacteriology.
Beezer, A. E., Newell, R. D., and Tyrrell, H. J. V. 1979. Characterisation and metabolic studies of *Saccharomyces cerevisiae* and *Kluyveromyces fragilis* by flow microcalorimetry. Antonie Van Leeuwenhoek 45, 55-63.
Bem, Z., Leistner, L., and Dressel, J. 1976. Rapid methods for counting bacteriological cells in meat. Fleischwirtschaft 56, 1803-1804.
Benborugh, J. E., and Martin, K. L. 1976. An indirect radiolabeled antibody staining technique for the rapid detection and identification of bacteria. J. Appl. Bacteriol. 41, 47-51.
Bennett, M., Nadler, H., and George, H. 1979. New motility medium for nonfermenting bacilli. Am. J. Med. Technol. 45, 143-145.
Bennett, V., and Cuatrecasas, P. 1975. Mechanism of action of *Vibrio cholerae* enterotoxin. J. Membrane Biol. 22, 1-28.

Benno, Y., and Mitsuoka, T. 1983. Modification of the Minitek system for identification of lactobacilli. Syst. Appl. Microbiol. 4, 123–131.

Bercovier, H., Brault, J., Cohen, S., Melis, R., Lambert, T., and Mollaret, H. H. 1984. A new isolation medium for the recovery of *Y. enterocolitica* from environmental sources. Curr. Microbiol. 10, 121–124.

Bergan, T. 1978. Phage typing of *Pseudomonas aeruginosa*. In "Methods in Microbiology." Vol. 10. T. Bergan and J. R. Norris (Editors). Academic Press Inc. Ltd., Oval Road, London, NW1 7DX, UK.

Berkman, R. M., and Wyatt, P. J. 1970. Differential light scattering measurements of heat-treated bacteria. Appl. Microbiol. 20, 510–512.

Berkowitz, D. 1979. Potential uses of bacteria in toxicology. Vet. Hum. Toxicol. 21, 422–426.

Bernheimer, A. W. 1976. (Editor). "Mechanisms in Bacterial Toxinology." John Wiley and Sons, New York, N.Y.

Beuchat, L. R. 1979. Survival of conidia of *Aspergillus flavus* in dried foods. J. Stored Prod. Res. 15, 25–31.

Beuchat, L. R. 1982. *Vibrio parahaemolyticus:* Public health significance. Food Technol. 36 (3) 80–83.

Bindschedler, O., deMan, J. C., and Curiat, G. 1981. Comparative study of several culture media to determine coliforms and *E. coli* in dairy and cocoa products. Zentralbe. Bakteriol. Parasitenkd. Infektionskv. Hyg. II. 136, 146–151.

Bishop, R. H., Duncan, C. L., Evancho, G. M., and Young, H. 1982. Estimation of fungal contamination in tomato products by a chemical assay for chitin. J. Food Sci. 47, 437.

Bissonnette, N., Lachance, R. A., Goulet, J., Landgraf, M., and Park, C. E. 1980. Evidence of thermonuclease production by *Bacillus* spp. and enterococci in naturally contaminated cheese. Can. J. Microbiol. 26, 722–725.

Bitton, G., Dutton, R. J., and Foran, J. A. 1983. New rapid technique for counting microorganisms directly on membrane filters. Stain Technol. 58, 343–346.

Blackwell, J. H., Cliver, D. O., Callis, J. J., Heidelbaugh, N. D., Larkin, E. P., McKercher, P. D., and Thayer, D. W. 1985. Foodborne viruses: their importance and need for research. J. Food Protection 48, 717–723.

Blanchfield, B., Stavrie, S., Jean, A., and Rivnick, H. 1982. Loss of polymyxin B from enrichment broth for *Vibrio parahaemolyticus*. J. Food Prot. 45, 744–746.

Blaser, M. J. 1982. *Campylobacter jejuni* and food. Food Technol. 36 (3) 89–92.

Blaser, P. 1978a. Comparison of methods for the quantitative detection of molds in foods. I. Selective staining procedures for the direct microscopic mold count. Zentralbl. Bakteriol. Parasitenkd. Infektionskr. Hyg., I. 166, 45–62.

Blaser, P. 1978b. Comparison of methods for the quantitative detection of molds in foods. III. Communication: comparison of different culture media for the mold plate count. Zentralbl. Bakteriol. Parasitenkd. Infektionskr. Hyg., I Abt. B 167, 146–164.

Blazevic, D. J., and Schierl, E. A. 1981. Rapid identification of Enterococci by reduction of Litmus milk. J. Clin. Microbiol. 14, 227–228.

Block, J. C., Dollard, M. A., Schwartzbrod, J., and Merle, G. 1983. Comparative study of two *Salmonella* concentration methods from surface waters. Environ. Technol. Lett. 4, 47–52.

Blomquist, G., Johansson, E., Soderstrom, B., and Wold, S. 1979. Classification of fungi by means of pyrolysis—gas chromatography—pattern recognition. J. Chromatogr. 173, 19–32.

Boileau, C. R., D'Hauteville, H. M., and Sansonetti, P. J. 1984. DNA hybridization technique to detect *Shigella* species and enteroinvasive *E. coli*. J. Clin. Microbiol. 20, 959–961.

Boling, E. A., Blanchard, G. C., and Russell, W. J. 1973. Bacterial identification by microcalorimetry. Nature 241, 472.

Bolton, F. J., and Coates, D. 1983. Development of a blood-free *Campylobacter* medium: Screening tests on basal media and supplements, and the ability of selected supplements to facilitate aerotolerance. J. Appl. Bacteriol. 54, 127–130.

Bolton, F. J., and Robertson, L. 1982. A selective medium for isolating *Campylobacter jejuni*. J. Clin. Pathol. 35, 462–467.

Bolton, F. J., Hinchliffe, P. M., Coates, D., and Robertson, L. 1982. A most probable number method for estimating small numbers of campylobacters in water. J. Hyg. 89, 185–190.

Bolton, F. J., Coates, D., Hinchliffe, P. M., and Robertson, L. 1983. Comparison of selective media for isolation of *Campylobacter jejuni/coli*. J. Clin. Pathol. 36, 78–83.

Bomar, M. T. 1983. Rapid detemination of critical bacterial counts in foods. Alimenta 22, 189–194.

Bonora, A., and Mares, D. 1982. A simple colorimetric method for detecting cell viability in cultures of eukaryotic microorganisms. Current Microbiol. 7, 217–222.

Bosley, G. S., Facklam, R. R., and Grossman, D. 1983. Rapid identification of enterococci. J. Clin. Microbiol. 18, 1275–1277.

Bostick, W. D., and Ausmus, B. S. 1978. Methodologies for the determination of adenosine phosphates. Anal. Biochem. 88, 78–92.

Bouhet, J. C., Van Chuong, P. P., Toma, F., Kirszenbaum, M., and Fromageot, P. 1976. Isolation and characterization of luteoskyrin and rugulosin, two hepatotoxin anthraquinonoids from *Penicillium islandicum* Sopp. and *Penicillum rugulosum* Thom. J. Agric. Food Chem. 24, 964–972.

Boye, E., Steen, H. B., and Skarstad, K. 1983. Flow cytometry of bacteria: A promising tool in experimental and clinical microbiology. J. Gen. Microbiol. 129, 973–980.

Brant, A. W., Mulder, R. W. A. W., Dorresteyn, I. W. J., and Pelgrom, R. F. M. 1978. A review and testing of selective media for psychrotrophic bacteria. Poult. Sci. 57, 1272–1278.

Braude, A. I. 1964. Bacterial Endotoxins. Scientific American. 1–12.

Brayton, P. R., West, P. A., Russek, E., and Colwell, R. R. 1983. New selective plating medium for isolation of *Vibrio vulnificus* biogroup. I. J. Clin. Microbiol. 17, 1039–1044.

Brettel, R., Lamprecht, I., and Schaarschmidt, B. 1980. Microcalorimetric investigations of the metabolism of yeasts. VII. Flow-calorimetry of aerobic batch cultures. Radiat. Environ. Biophys. 18, 301–309.

Brewer, D. G., Martin, S. E., and Ordal, Z. J. 1977. Beneficial effects of catalase or pyruvate in a most-probable-number technique for the detection of *Staphylococcus aureus*. Appl. Environ. Microbiol. 34, 797–800.

Bridson, E. Y., 1982. Selective culture media—a "Renewed" activity. Med. Lab. Sci. 39, 1–2.

Britz, T. J., and Steyn, P. L. 1979. Volatile fatty acid production by the dairy and clinical propionibacteria and related coryneforms. Phytophylactia 11, 111–115.

Brodsky, M. H. 1982. Enumeration of indicator organisms in food with the ISO-GRID hydrophobic grid-membrane filter method. *In* "Rapid Methods and Automation in Microbiology." R. C. Tilton (Editor). Amer. Soc. Microbiol. Washington, D.C.

Brodsky, M. H., and Ciebin, B. W. 1978. Improved medium for recovery and enumeration

4. Standard and Experimental Methods of Identification and Enumeration 145

of *Pseudomonas aeruginosa* from water using membrane filters. Appl. Environ. Microbiol. 36, 36–42.

Brodsky, M. H., and Ciebin, B. W. 1979. Effect of heat treatment on the performance of tryptose-sulfite-cycloserine agar for enumeration of *Clostridium perfringens*. Appl. Environ. Microbiol. 37, 1038–1040.

Brodsky, M. H., and Ciebin, B. W. 1981. Collaborative evaluation of the plate loop technique for determining viable bacterial counts in raw milk. J. Food Prot. 43, 287–291.

Brodsky, M. H., Entis, P., Sharpe, A. N., and Jarvis, G. A. 1982. Enumeration of indicator organisms in foods using the automated hydrophobic grid-membrane filter technique. J. Food Prot. 45, 292–296.

Brown, L. H., Peterson, E. M., and de la Maza, L. M. 1983. Rapid identification of enterococci. J. Clin. Microbiol. 17, 369–370.

Brown, M. H., and Baird-Parker, A. C. 1982. The microbiological examination of meat. In "Meat Microbiology," M. H. Brown (ed.), Applied Science, Barking, Essex, England, pp. 423–520.

Brown, S. D., and Washington, J. A., II. 1978. Evaluation of the Repliscan system for identification of *Enterobacteriaceae*. J. Clin. Microbiol. 8, 695–699.

Brun, Y., Fleurette, J., and Forey, F. 1978. Micromethod for biochemical identification of coagulase-negative staphylococci. J. Clin. Microbiol. 8, 503–508.

Brune, H. E., and Cunningham, F. E. 1971. A review of microbiological aspects of poultry processing. World's Poult. Sci. J. 27, 223–240.

Bryan, F. L. 1974. Microbiological food hazards today—based on epidemiological information. Food Technol. 28, 52–59.

Bryant, R. G. 1983. Food microbiology update: emerging foodborne pathogens. Appl. Biochem. Biotechnol. 8, 437–454.

Buckwold, F. J., Ronald, A. R., Harding, G. K. M., Marrie, T. J., Fox, L., and Cates, C. 1979. Biotyping of *Escherichia coli* by a simple multiple-inoculation agar plate technique. J. Clin. Microbiol. 10, 275–278.

Buddemeyer, E. U., Wells, G. M., Hutchinson, R., Cooper, M. D., and Johnston, G. S. (1978). Radiometric estimation of the replication time of bacteria in culture: An objective and precise approach to quantitative microbiology. J. Nucl. Med. 19, 619–625.

Burdash, N. M., West, M. E., Bannister, E. R., and Manos, J. P. 1978. An identification scheme for gram-negative nonfermentative bacilli. Health Lab. Sci. 15, 95–103.

Burger, J. S., Nupen, E. M., and Brabow, W. O. K. 1984. Evaluation of four growth media for membrane filtration counting of *Clostridium perfringens* in water. Water S.A. 10, 185–188.

Burman, L. G., and Ostensson, R. J. 1978. Time—and media—saving testing identification of microorganisms by multipoint inoculation on undivided agar plates. J. Clin. Microbiol. 8, 219–227.

Busta, F. F. 1976. Practical implications of injured microorganisms in food. J. Milk Food Technol. 39, 138–145.

Busta, F. F. 1978. Introduction to injury and repair of microbial cells. Adv. Appl. Microbiol. 23, 195–200.

Buttiaux, R., and Mossel, D. A. A. 1961. The significance of various organisms of fecal origin in foods and drinking water. J. Appl. Bacteriol. 24, 353–364.

Butzler, J. P. 1984. (Editor). "*Campylobacter* Infection in Man and Animals." CRC Press, Inc., Boca Raton, Fla.

Cady, P. 1977. Instrumentation in food microbiology. Food Prod. Dev. 11, 80, 82–85.

Cady, P., Hardy, D., Martins, S., Dufour, S. W., and Kraeger, S. J. 1978. Automated

impedance measurements for rapid screening of milk microbial count. J. Food Protection. 41, 277–280.
Calich, V. L. G., Purchio, A., and Paula, C. R. 1979. A new fluorescent viability test for fungal cells. Mycopathologia 66, 175–177.
Callahan, L. T., III. 1974. Purification and characterization of *Pseudomonas aeruginosa* exotoxin. Infection and Immunity, 9, 113–118.
Camaschella, P. 1977. Detection techniques for molds in foods. Ind. Aliment. 16, 68–74.
Campbell, A. D., Francis, Jr., O. J., Beebe, R. A., and Stoloff, L. 1984. Determination of aflatoxins in peanut butter, using two liquid chromatographic methods: Collaborative study. J. Assoc. Off. Anal. Chem. 67, 312–317.
Carlone, G. M., Valadez, M. J., and Pickett, M. J. 1982. Methods for distinguishing gram-positive from gram-negative bacteria. J. Clin. Microbiol. 16, 1157–1159.
Catasara, M., and Dorso, Y. 1976. Method for a 24-hour microbiological analysis of foodstuffs. I. Principles and application to ready-cooked meals. Bull. Acad. Fra. 49, 237–241.
Cattaneo, P., and Cantoni, C. 1978. Identification and rapid determination of histamine in fish samples. Ind. Aliment. 17, 303–307.
Cerny, G. 1978. Studies on the aminopeptidase test for the distinction of gram-negative from gram-positive bacteria. Eur. J. Appl. Microbiol. 5, 113–122.
Chakrabarti, A. G. 1984. Fluorometric assay for aflatoxins. Bull. Environ. Contam. Toxicol. 33, 515–518.
Chalapati, R. V. 1976. Virus transmission through foods. J. Food Sci. Technol. (Mysore) 13, 287–293.
Chambon, P., Dano, S. D., Chambon, R., and Geahchan, A. 1983. Rapid determination of aflatoxin M1 in milk and dairy products by HPLC. J. Chromatogr. 259, 372–374.
Chappelle, E. W., and Levin, G. V. 1968. Use of firefly bioluminescent reaction for rapid detection and counting of bacteria. Biochem. Med. 2, 41–51.
Chau, P. Y., and Leung, Y. K. 1978. Inhibitory action of various plating media on the growth of certain *Salmonella* serotypes. J. Appl. Bacteriol. 45, 321–332.
Chester, B. 1979. Semiquantitative catalase test as an aid in identification of oxidative and nonsaccharolytic gram-negative bacteria. J. Clin. Microbiol. 10, 525–528.
Chipley, J. R. 1972. Comparison of the antigenicity of phenol and ethylenediaminetetraacetate complexes isolated from cell walls of *Salmonella enteritidis*. Appl. Microbiol. 23, 651–653.
Chipley, J. R. 1974. Release of lipopolysaccharide, phospholipids and enzymes from *Salmonella enteritidis* by ethylenediaminetetra—acetic acid. Microbios 10, 139–150.
Chipley, J. R. 1981. The effects of high-energy irradiation on mycotoxin production by *Aspergillus*. *In* "GIAM VI. Global Impacts of Applied Microbiology. Sixth International Conference." S. O. Emejuaiwe. O. Ogunbic, and S. O. Sanni (Editors). Academic Press, New York.
Chipley, J. R., and Cremer, M. L. 1980. Microbiological problems in the food service industry. Food Technol. 34: 59–68, 84.
Chipley, J. R., Mabee, M. S., Applegate, K. L., and Dreyfuss, M. S. 1974. Further characterization of tissue distribution and metabolism of ^{14}C aflatoxin B_1 in chickens. Appl. Microbiol. 28, 1027–1029.
Chomarat, M., Coullioud, D., Carret, G., and Flandrois, J. P. 1982. Evaluation of *Salmonella* and *Shigella* identification with the MS-2-BID automatic system. Zentralbl. Bakteriol. Mikrobiol. Hyg. I, ABt.A, 251, 43–51.
Chopin, A., *et al.* 1985. ICMSF Methods Studies. *XV.* Comparison of four media and

methods for enumerating *Staphylococcus aureus* in powdered milk. J. Food Protection. 48, 21–27.
Chopra, I. 1983. Rapid diagnosis of microbial infections by gas chromatography. Lab. Pract. 32, 25–28.
Christopher, F. M., Smith, G. C., and Vanderzant, C. 1982. Examination of poultry giblets, raw milk, and meat for *Campylobacter jejuni*. J. Food Prot. 45, 160–162.
Chu, F. S., 1984. Immunoassays for analysis of mycotoxins. J. Food Protection 47, 562–569.
Cichowicz, S. M., and Bandler, R. 1980a. Determination of *Geotrichum* in cream style corn: collaborative study. J. Assoc. Off. Anal. Chem. 63, 481–482.
Cichowicz, S. M., and Bandler, R. 1980b. Determination of *Geotrichum* mold in citrus juices: collaborative study. J. Assoc. Off. Anal. Chem. 63, 483–484.
Cichowicz, S. M., and Bandler, R. 1982a. Howard mold count of fruit nectars, purees, and pastes: Collaborative study. J. Assoc. Off. Anal. Chem. 65, 1093–1094.
Cichowicz, S. M., and Bandler, R. 1982b. Determination of *Geotrichum* mold in comminuted fruits and vegetables: Collaborative study. J. Assoc. Off. Anal. Chem. 65, 1095–1096.
Citron, D. M., 1984. Specimen collection and transport, anaerobic culture techniques, and identification of anaerobes. Rev. Infect. Dis. 6, S51–S58.
Clark, D. S. 1971. Studies on the surface plate method of counting bacteria. Can. J. Microbiol. 17, 943–946.
Clark, D. S. 1967. Comparison of pour and surface plate methods for determination of bacterial counts. Can. J. Microbiol. 13, 1409–1414.
Clark, D. S. 1982. International perspectives for microbiological sampling and testing of foods. J. Food Prot. 45, 667–671.
Clarke, K. R., and Owens, N. J. P. 1983. A simple and versatile micro-computer program for the determination of MPN. J. Microbiol. Methods 1, 133–138.
Clarke, P. H., and Steel, K. J. 1966. Rapid and simple biochemical tests for bacterial identification. *In* "Identification Methods for Microbiologists" Part A. B. M. Gibbs and F. A. Skinner (Editors). Academic Press, New York, New York.
Clarke, P. M., and Payton, M. A. 1983. An enzymatic assay for acetate in spent bacterial culture supernatants. Anal. Biochem. 130, 402–405.
Cliffe, A. J. McKinnon, C. H., and Berridge, N. J. 1973. Microcalorimetric estimation of bacteria in milk. J. Soc. Dairy Technol. 26, 209–210.
Colwell, R. R., Gunn, B. A., Singleton, F. L., Peele, E. R., Keiser, J. K., and Kapfer, C. O. 1981. Comparison of methods for identifying *Staphylococcus* and *Micrococcus* spp. J. Clin. Microbiol. 15, 195–200.
Cook, D. W. 1981. Automatic incubator for use with Modified A-1 test for enumerating fecal coliform bacteria in shellfish growing waters. J. Assoc. Off. Anal. Chem. 64, 771–773.
Cooke, B. C. 1977. Influence of temperature and method of reconstitution on standard plate counts of casein and caseinates. J. Appl. Bacteriol. 43, 299–302.
Cooke, B. C. and Jorgenson, M. 1977. An evaluation of modified minerals glutamate medium for use in the presumptive coliform test on dairy products. N. Z. J. Dairy Sci. Technol. 12, 272–273.
Cooke, B. C., Hill, B. M., Pitcher, P. A., and Thomson, A. E. 1977. A comparison of methods for the confirmation of *Escherichia coli* isolated from dairy products. N. Z. J. Dairy Sci. Technol. 12, 270–271.
Cooke, B. C., Josephson, B. R., and Norris, L. M. 1982. An evaluation of resuscitation

modifications to the standard presumptive coliform analysis for the recovery of coliforms and *E. coli* from dairy products. Aust. J. Dairy Technol. 37, 149–150.

Cooper, B. H., Johnson, J. B., and Thaxton, E. S. 1978. Clinical evaluation of the Uni-Yeast-Tek system for rapid presumptive identification of medically important yeasts. J. Clin. Microbiol. 7, 349–355.

Cooper, B. H., Prowant, S., Alexander, B., and Brunson, D. H. 1984. Collaborative evaluation of the Abbott yeast identification system. J. Clin. Microbiol. 19, 853–856.

Cordray, J. C., and Huffman, D. L. 1985. Comparison of three methods for estimating surface bacteria on pork carcasses. J. Food Protection 48, 582–584.

Costilow, R. N. 1981. Biophysical factors in growth. *In* "Manual of Methods for General Bacteriology." P. Gerhardt, R. G. E. Murray, R. N. Costilow, E. W. Nester, W. A. Wood, N. R. Krieg, and G. B. Phillips (Editors). Am. Soc. Microbiol., Washington, D.C.

Couse, N. L., and King, J. W. 1982. Quantitation of the spiral plating technique for use with the *Salmonella* Mammalian microsome assay. Environ. Mutagenesis 4, 445–455.

Cousin, M. A. 1982. Presence and activity of psychotrophic microorganisms in milk and dairy products: a review. J. Food Prot. 45, 172–207.

Cousin, M. A., Zeidler, C. S., and Nelson, P. E. 1984. Chemical detection of mold in processed foods. J. Food Sci. 49, 439–445.

Cousins, C. M., Pettipher, G. L., McKinnon, C. H., and Mansell, R. 1979. A rapid method for counting bacteria in raw milk. Dairy Ind. Int. 44 (4), 27–31, 39.

Cowell, N. D., and Morisetti, M. D. 1969. Microbiological techniques—some statistical aspects. J. Sci. Food Agric. 20, 573–578.

Cox, N. A., and Mercuri, A. J. 1978. Recovery of salmonellae from broiler carcasses by direct enrichment. J. Food Prot. 41, 521–524.

Cox, N. A., and Mercuri, A. J. 1979a. Rapid biochemical testing procedures for *Enterobacteriaceae* in foods. Food Technol. 33, 57–62.

Cox, N. A., and Mercuri, A. J. 1979b. Recovering low levels of various *Salmonella* serotypes from deep-frozen broiler carcasses by direct enrichment. J. Food Prot. 42, 660–661.

Cox, N. A., and Williams, J. E. 1976. A simplified biochemical system to screen *Salmonella* isolates from poultry for sterotyping. Poult Sci. 55, 1968–1971.

Cox, N. A., Mercuri, A. J., Thomson, J. E., and Chew, V. 1976. Swab and excised tissue sampling for total and *Enterobacteriaceae* counts of fresh and surface-frozen broiler skin. Poult. Sci. 55, 2405–2408.

Cox, N. A., Bailey, J. S., Thomson, J. E., and Carson, M. O. 1980. Lactose preenrichment versus direct enrichment for recovering *Salmonella* from deep-chilled broilers and frozen meat products. Poult. Sci. 59, 2431–2436.

Cox, N. A., Thomson, J. E., and Bailey, J. S. 1981. Sampling of broiler carcasses for *Salmonella* with low volume water rinse. Poult. Sci. 60, 768–770.

Cox, N. A., Bailey, J. S., Thomson, J. E., and Fung, D. Y. C. 1982. Adaptation of commercial systems for rapid identification of bacteria in foods. *In* "Rapid Methods and Automation in Microbiology." R. C. Tilton (Editor). Amer. Soc. Microbiol., Washington, D.C.

Cox, N. A., Bailey, J. S., and Thomson, J. E. 1983. Comparison of preenrichment to direct enrichment and the effect of pyruvate in media for recovery of salmonellae in feed. Poult. Sci. 62, 947–951.

Cox, N. A., Fung, D. Y. C., Goldschmidt, M. C., Bailey, J. S., and Thompson, J. E. 1984. Selecting a miniaturized system for identification of *Enterobacteriaceae*. J. Food Protection 47, 74–77.

Cox, N. A., Van Wart, M., Bailey, J. S., and Thomson, J. E. 1985. Identification of *Enterobacteriaceae* from foods with the Spectrum-10. J. Food Protection 48, 76–79.
Coykendall, A. L., and Lizotte, P. A. 1978. *Streptococcus mutans* isolates identified by biochemical tests and DNA base contents. Arch. Oral Biol. 23, 427–428.
Craven, S. E. 1980. Growth and sporulation of *Clostridium perfringens* in foods. Food Technol. 34, 80–87.
Craythorn, J. M., Barbour, A. G., Matsen, J. M., Britt, M. R., and Garibaldi, R. A. 1980. Membrane filter contact technique for bacteriological sampling of moist surfaces. J. Clin. Microbiol. 12, 250–255.
Cunningham, F. E. 1982. Microbiological aspects of poultry and poultry products—an update. J. Food Prot. 45, 1149–1164.
Curbey, W. A., and Gall, L. S. 1977. Instrumental approach to microbiological analysis of body fluids. Pub. Health Lab. 35, 118–124.
Dahlen, G., and Ericsson, I. 1983. Differentiation between gram-negative anaerobic bacteria by pyrolysis gas chromatography of lipopolysaccharides. J. Gen. Microbiol. 129, 557–563.
Daly, J. A., Boshard, R., and Matsen, J. M. 1984. Differential primary plating medium for enhancement of pigment production by *Pseudomonas aeruginosa*. J. Clin. Microbiol. 19, 742–743.
Dam, R., Froning, G. W., and Skala, J. H. 1970. Recommended methods for the analysis of eggs and poultry meat. North Central Regional Research Bull. No. 205.
D'Amato, R. F., Holmes, B., and Bottone, E. J. 1981. The systems approach to diagnostic microbiology. CRC Crit. Rev. Microbiol. 9, 1–44.
D'Aoust, J. Y. 1977. Limitations of lysine-iron-cystine-neutral red broth in the presumptive identification of *Salmonella*. Appl. Environ. Microbiol. 34, 595–596.
D'Aoust, J. Y. 1978. Recovery of sublethally heat-injured *Salmonella typhimurium* on supplemental plating media. Appl. Environ. Microbiol. 35, 483–486.
D'Aoust, J. Y. 1981. Update on preenrichment and selective enrichment conditions for detection of *Salmonella* in foods. J. Food Prot. 44, 369–374.
D'Aoust, J. Y. 1984a. *Salmonella* detection in foods: present status and research needs for the future. J. Food Protection. 47, 78–81.
D'Aoust, J. Y. 1984b. Effective enrichment-plating conditions for detection of *Salmonella* in foods. J. Food Protection 47, 588–590.
D'Aoust, J. Y., Maishment, C., Burgener, D. M., Conley, D. R., Loit, A., Milling, M., and Purvis, U. 1980. Detection of *Salmonella* in refrigerated preenrichment and enrichment broth cultures. J. Food Prot. 43, 343–345.
D'Aoust, J. H., Maishment, C., Stotland, P., and Boville, A. 1982a. Surfactants for the effective recovery of *Salmonella* in fatty foods. J. Food Prot. 45, 249–252.
D'Aoust, J. Y., Stotland, P., and Boville, A. 1982b. Sampling methods for detection of *Salmonella* in raw chicken carcasses. J. Food Sci. 47, 1643–1644.
Darveau, R. P., and Hancock, R. E. W. 1983. Procedure for isolation of bacterial lipopolysaccharides from both smooth and rough *Pseudomonas aeruginosa* and *Salmonella typhimurium* strains. J. Bacteriol. 155, 831–838.
Dashek, W. V., and Llewellyn, G. C. 1983. Mode of action of the hepatocarcinogens, aflatoxins, in plant systems: A review. Mycopathologia 81, 83–94.
Daud, H. B., McMeekin, T. A., and Thomas, C. J. 1979. Spoilage association of chicken skin. Appl. Environ. Microbiol. 37, 399–401.
Davenport, R. R. 1981. Cold-tolerant yeasts and yeast-like organisms. *In* "Biology and Activities of Yeasts." F. A. Skinner, S. M. Passmore and R. R. Davenport (Editors). Academic Press for the Society of Applied Bacteriology.

Davidson, C. M., Taylor, M., and Zellerman, G. G. 1978. Method for sampling beef carcasses. Appl. Environ. Microbiol. 35, 811–812.

Davis, C. E., Freedman, S. D., Douglas, H., and Barude, A. I. 1969. Analysis of sugars in bacterial endotoxins by gas-liquid chromatography. Anal. Biochem. 28, 243–256.

Dawson, L. E., and Stadelman, W. J. 1960. Microorganisms and their control on fresh poultry meat. Michigan State Univ. Tech. Bull. 278.

Dawson, L. E., Chipley, J. R., Cunningham, F. E., and Kraft, A. A. 1979. Incidence and control of microorganisms on poultry products. Michigan State Univ. Agri. Exper. Station Res. Rep. 383.

Debevere, J. M. 1979. A simple method for the isolation and determination of *Clostridium perfringens*. Eur. J. Appl. Microbiol. Biotechnol. 6, 409–414.

deBoer, E., Hartog, B. J., and Osterom, J. 1982. Occurrence of *Yersinia enterocolitica* in poultry products. J. Food Prot. 45, 322–325.

DeCastro, F., and Widomski, M. J. 1978. Rapid detection of bacteria by means of an electronic particle counter. Abst. Ann. Meet. Am. Soc. Microbiol., 302.

Dees, S. B., and Moss, C. W. 1978. Identification of *Achromobacter* species by cellular fatty acids and by production of keto acids. J. Clin. Microbiol. 8, 61–66.

deFigueiredo, M. P., and Jay, J. M. 1976. Coliforms, Enterococci, and other microbial indicators. *In* "Food Microbiology: Public Health and Spoilage Aspects." M. P. de-Figueiredo and D. F. Splittstoesser (Editors). AVI Publishing Co., Westport, CT.

deFigueiredo, M. P., and Splittstoesser, D. F. 1976. (Editors). "Food Microbiology: Public Health and Spoilage Aspects." AVI Publishing Co. Westport, CT.

Delarras, C., Lapan, P., and Gayral, J. P. 1979. *Micrococcaceae* isolated from meat and dairy products. (Taxonomic study). Zentralbl. Bakteriol. Parasitenkd. Infektionskr. Hyg., I Abt. B 168, 377–385.

Delmore, R. P., Corosa, Jr., P. A., Piscia, P., and Thompson, W. N. 1981. Evaluation of the microbiological sampling efficiency of a centrifugal air-sampling device. Dev. Ind. Microbiol. 22, 617–626.

Deng, M., and Cliver, D. O. 1984. A broad-spectrum enzyme-linked immunosorbent assay for the detection of human enteric viruses. J. Virol. Methods 8, 87–98.

DePaola, A. 1981. *Vibrio cholerae* in marine foods and environmental waters: a literature review. J. Food Sci. 46, 66–70.

DeScrilli, A., Leali, L., and Quaroni, E. 1976. Statistical evaluation of automatic plate counts using the Artek counter. Arch. Vet. Ital. 27, 106–111.

Devoyod, J. J., Millet, L., and Mocquot, G. 1976. An agar medium for the direct enumeration of *Staphylococcus aureus:* pork plasma medium for *Staphylococcus aureus* (PPSA). Can. J. Microbiol. 22, 1603–1611.

Devriese, L. A. 1981. Baird-Parker medium supplemented with acriflavine, polymynins and sulphonamide for the selective isolation of *Staphylococcus aureus* from heavily contaminated materials. J. Appl. Bacteriol. 50, 351–357.

Devriese, L. A., and Van de Kerckhove, A. 1979. A comparison of methods and the validity of deoxyribonuclease tests for the characterization of staphylococci isolated from animals. J. Appl. Bacteriol. 46, 385–393.

Devriese, L. A., and Van de Kerckhove, A. 1980. Carbohydrate dissimilation tests in the identification of staphylococci. Antonie Van Leeuwenhoek, 46, 65–72.

Dey, B. P., and Engley, Jr., F. B. 1983. Methodology for recovery of chemically treated *S. aureus* with neutralizing medium. Appl. Environ. Microbiol. 45, 1533–1537.

Dezeure-Wallays, B., and Van Hoof, J. 1980. Study on the occurrence of salmonellae on air-chilled broiler carcasses. Vlaams Diergeneeskd Tÿdschr. 49, 285–298.

Dezfulian, M., and Bartlett, J. G. 1985. Selective isolation and rapid identification of *Clostridium botulinum* types A and B by toxin detection. J. Clin. Microbiol. 21, 231–233.
deZutter, L., Abrams, R., and Van Hoff, J. 1982. Bacteriological survey of beef carcasses: Correlation between swab and maceration methods. Arch. Lebensmittelhyg. 33, 36–37.
Dhawale, M. R., Wilson, J. J., Khachatourians, G. G., and Ingledew, W. M. 1982. Improved method for detection of starch hydrolysis. Appl. Environ. Microbiol. 44, 747–750.
Dhople, A. M., and Hanks, M. J. 1973. Quantitative extraction of adenosine triphosphate from cultivable and host-grown microbes. Calculation of adenosine triphosphate pools. Appl. Microbiol. 26, 399–403.
Dibb, W. L., Hellum, K. B., Oesterwold, B., and Oeding, P. 1983. Comparison of four methods for differentiation of *Staphylococcus aureus* from other Micrococcaceae in the routine laboratory. Acta Pathol. Microbiol. Immunol. Scand., Ser. B. 91 B, 307–310.
Dobrogosz, W. J. 1981. Enzymatic activity. *In* "Manual of Methods for General Bacteriology." P. Gerhardt, R. G. E. Murray, R. N. Costilow, E. W. Nester, W. A. Wood, N. R. Krieg and G. B. Phillips (Editors). Am. Soc. Microbiol., Washington, D.C.
Dodds, K. L., Holley, R. A., and Kempton, A. G. 1983. Evaluation of the catalase and *Limulus* amoebocyte lysate tests for rapid determination of the microbial quality of vacuum-packed cooked turkey. Can. Inst. Food Sci. Technol. 16, 167–172.
Donkersloot, J. A., Robrish, S. A., and Krichevsky, M. I. 1972. Fluorometric determination of deoxyribonucleic acid in bacteria with ethidium bromide. Appl. Microbiol. 24, 179–183.
Donnelly, C. B., Gilchrist, J. E., Peeler, J. T., and Campbell, J. E. 1976. Spiral plate count method for the examination of raw and pasteurized milk. Appl. Environ. Microbiol. 32, 21–27.
Donnelly, L. S., and Busta, F. F. 1981a. Anaerobic sporeforming microorganisms in dairy products. J. Dairy Sci. 64, 161–166.
Donnelly, L. S., and Busta, F. F. 1981b. Alternative procedures for enumeration of *Desulfotomaculum nigrificans* spores in raw ingredients of soy protein-based products. J. Food Sci. 46, 1527–1531.
Dow, C. S., France, A. D., Khan, M. S., and Johnson, T. 1979. Particle size distribution analysis for the rapid detection of microbial infection of urine. J. Clin. Pathol. 32, 386–390.
Doyle, M. P. 1979. Microbiological methods and product quality. Cereal Foods World 24, 534–536.
Doyle, M. P., and Foster, E. M. 1981. Bacterial agents of food-borne disease. Wis. Med. J. 80 (10), 24–27.
Doyle, M. P., and Hugdahl, M. B. 1983. Improved procedure for recovery of *Yersinia enterocolitica* from meats. Appl. Environ. Microbiol. 45, 127–135.
Draughon, F. A., and Nelson, P. J. 1981. Comparison of modified direct-plating procedures for recovery of injured *Escherichia coli*. J. Food Sci. 46, 1188–1191.
Dreyfuss, M. S., and Chipley, J. R. 1980. Comparison of effects of sublethal microwave radiation and conventional heating on the metabolic activity of *Staphylococcus aureus*. Appl. Environ. Microbiol. 39, 13–16.
Drion, E. F., and Mossel, D. A. A. 1972. Mathematical–ecological aspects of the examination of Enterobacteriaceae of foods processed for safety. J. Appl. Bacteriol. 35, 233–239.
Drucker, D. B., and Gibson, L. F. 1982. Pyrolysis gas chromatography of *Streptococcus faecalis*; effect of cultural conditions on pyrochromatograms. Microbios. 32, 93–100.

Drucker, D. B., Hillier, V. F., and Lee, S. M. 1982. Comparison of computer methods for taxonomy of some streptococci using gas chromatographic chemotaxonomic data. Microbios 35, 139–150.

Dube, S. D. 1983. Outbreak of food poisoning caused by lactose-fermenting *Salmonella tuebingen*. J. Clin. Microbiol. 17, 698–699.

Dudley, M. V., and Shotts, F. B., Jr., 1979. Medium isolation of *Yersinia enterocolitica*. J. Clin. Microbiol. 10, 180–183.

Dudley, M., Nauschuctz, W. F., and Juchau, S. V. 1983. Clinical evaluation of new AMS GSC plus card. Diagn. Microbiol. Infect. Dis. 1, 139–143.

Duke, M. 1980. An appraisal of the practical value of identikits for rapid identification of microorganisms in hospital and quality control laboratories. Lab. Pract. 29, 377–380.

Dupray, E., and Cormies, M. 1983. Optimal enrichment time for isolation of *Vibrio parahaemolyticus* from seafood. Appl. Environ. Microbiol. 46, 1234–1235.

Edwin, C., Tatini, S. R., Strobel, R. S., and Maheswaran, S. K. 1984. Production of monoclonal antibodies to staphylococcal enterotoxin A. Appl. Environ. Microbiol. 48, 1171–1175.

Edwin, E. E., Shreeve, J. E. and Jackman, R. 1978. A rapid colony test for thiaminase activity. J. Appl. Bacteriol. 44, 305–312.

Egan, A. F. 1979. Enumeration of stressed cells of *Escherichia coli*. Can. J. Microbiol. 25, 116–118.

Egan, H., Stoloff, L., Castegnaro, M., Scott, P., O'Neill, I. K., and Bartsch, H. 1982. (Editors). "Environmental Carcinogens. Selected Methods of Analysis. Volume 5—Some Mycotoxins." Int. Agency for Res. on Cancer, Lyon, France. 502 pp.

El-Bassiony, T. A. 1980. Occurrence of *Clostridium perfringens* in milk and dairy products. J. Food Prot. 43, 536–537.

Ellender, R. D., Cook, D. W., Sheladia, V. L., and Johnson, R. A. 1980. Enterovirus and bacterial evaluation of Mississippi oysters. Gulf Res. Rep. 6, 371–376.

Ellner, P. D., and Myers, D. A. 1981. Preliminary evaluation of the AutoSCAN-3, an instrument for automated reading and interpretation of microdilution trays: identification of aerobic gram-negative bacilli. J. Clin. Microbiol. 14, 326–328.

Emswiler, B. S., and Kotula, A. W. 1978. Differentiation of *Salmonella* serotypes by pyrolysis-gas-liquid chromatography of cell fragments. Appl. Environ. Microbiol. 35, 97–104.

Emswiler, B. S., Pierson, M. D., and Shoemaker, S. P. 1976. Sublethal heat stress of *Vibrio parahaemolyticus*. Appl. Environ. Microbiol. 32, 792–798.

Emswiler, B. S., Pierson, C., and Kotula, A. W. 1977a. Comparative study of two methods for detection of *Clostridium perfringens* in ground beef. Appl. Environ. Microbiol. 33, 735–737.

Emswiler, B. S., Pierson, C. J., and Kotula, A. W. 1977b. Stomaching vs. blending. Food Tech. 31, 40–42.

Emswiler, B. S., Nichols, J. E., Kotula, A. W., and Rough, D. K. 1978. Device for microbiological sampling of meat surfaces. J. Food Prot. 41, 546–548.

Emswiler-Rose, B. S., Johnston, R. W., Harris, M. E., and Lee, W. H. 1980. Rapid detection of staphylococcal thermonuclease on casings of naturally contaminated fermented sausages. Appl. Environ. Microbiol. 40, 13–18.

Emswiler-Rose, B. S., Gehle, W. D., Johnston, R. W., Okrend, A., Moran, A., and Bennett, B. 1984. An enzyme immunoassay for detection of salmonellae in meat and poultry products. J. Food Sci. 49, 1018–1020.

Enjalbert, F., Richard, C., Attisso, M., and Cremieux, A. 1979. Value of the detection of histidine decarboxylase (HDC) in species of the *Klebsiella-Enterobacter-Serratia* group. Ann. Microbiol. 130A, 385–388.

Epifanio, E. C., Veroy, L. R., Uyenco, F., Cajipe, G. J. B., and Laserna, E. C. 1981. Carrageenan from *Eucheuma striatum* (Schmitz) in bacteriological media. Appl. Environ. Microbiol. 41, 155–158.

Erdman, I. E. 1973. International collaborative assay for the detection of *Salmonella* in raw meats. Can. J. Microbiol. 19, 715–720.

Erickson, J. E., and Deibel, R. H. 1978. New medium for rapid screening and enumeration of *Clostridium perfringens* in foods. Appl. Environ. Microbiol. 36, 567–571.

Eskenazi, S., and Littell, A. M. 1978. Dulcitol-malonate-phenylalanine agar for the identification of *Salmonella* and other *Enterobacteriaceae*. Appl. Environ. Microbiol. 35, 199–201.

Espejo, S. J. 1977. Comparison of three methods for grading the bacteriological quality of raw milk in warm countries. J. Arch. Zootec. 26, 271–279.

Espejo, S. J., and Perez, F. D. 1978. Comparative study between the standard method and the improved technic with SMCA, for the detection and count of proteolytic bacteria. J. Arch. Zootec. 27, 291–299.

Essers, L., and Radebold, K. 1980. Protein A-hemagglutination test, a reliable method for rapid identification of *S. aureus*. Zentralbl. Bakteriol. Mikrobiol. Hyg. I Abt. A 247, 170–176.

Evancho, G. M., Ashton, D. H., and Zwarun, A. A. 1974. Use of a radiometric technique for the rapid detection of growth of clostridial species. J. Food Sci. 39, 77–79.

Ewetz, L., and Thore, A. 1978. Correlation between hemin content and the chemiluminescent luminol reaction with bacteria. Appl. Environ. Microbiol. 36, 790–793.

Ewing, W. H. 1970. Differentiation of *Enterobacteriaceae* by biochemical reactions. U.S. Dept. Health, Educ., and Welfare, Natl. Communicable Disease Center, Atlanta, Ga.

Ewing, W. H., and Davis, B. R. 1970. Media and tests for differentiation of *Enterobacteriaceae*. U.S. Dept. of Health, Educ., and Welfare, Natl. Communicable Disease Center, Atlanta, Ga.

Ewing, W. H., Ball, M. M., Bartes, S. F., and McWhorter, A. C. 1970. The biochemical reactions of certain species and bioserotypes of *Salmonella*. J. Infectious Diseases 121, 288–294.

Facklam, R. R., Padula, J. F., Wortham, E. C., Cooksey, R. C., and Rountree, H. A. 1979. Presumptive identification of group A, B, and D streptococci on agar plate media. J. Clin. Microbiol. 9, 665–672.

Facklam, R. R., Thacker, L. G., Fox, B., and Eriquez, L. 1982. Presumptive identification of streptococci with a new test system. J. Clin. Microbiol. 15, 987–990.

Fagerberg, D. J., and AVens, J. S. 1976. Enrichment and plating methodology for *Salmonella* detection in food. A review. J. Milk Food Technol. 39, 628–646.

Farber, J. M., and Idziak, E. S. 1982. Detection of glucose oxidation products in chilled fresh beef undergoing spoilage. Appl. Environ. Microbiol. 44, 521–524.

Farber, J. M., and Sharpe, A. N. 1984. Improved bacterial recovery by membrane filters in the presence of food debris. Appl. Environ. Microbiol. 48, 441–443.

Farid, A. F., and Larsen, J. L. 1980. A simple scheme for identification of aerobic gram-negative bacteria with special reference to *Enterobacteriaceae*. Zentralbl. Veterinarmed., B27, 567–575.

Fey, H., Stiffler-Rosenberg, G., Wartenweiler-Burkhard, G., Mueller, C., and Rueegg, O. 1982. Detection of staphylococcal enterotoxins (SET). Schweiz. Arch. Tierheilkd. 124, 297–306.

Filtenborg, O., Frisvad, J. C., and Svendsen, J. A. 1983. Simple screening method for molds producing intracellular mycotoxins in pure cultures. Appl. Environ. Microbiol. 45, 581–585.

Fincher, E. L. 1965. Surface sampling application, methods, recommendations. Proceedings of an institute on the control of infections in hospitals. Univ. of Michigan, Natl. Communicable Disease Center, Atlanta, Ga. 189–199. (Reprint Number 66).

Fincher, E. L., and Mallison, G. F. 1967. Intramural sampling of airborne microorganisms. In "Air Sampling Instruments Manual," Third Edition. Amer. Conf. of Governmental Industrial Hyg., 1014 Broadway, Cincinnati, Ohio.

Finkelstein, R. A., and Yang, Z. 1983. Rapid test for identification of heat-labile enterotoxin-producing *Escherichia coli* colonies. J. Clin. Microbiol. 18, 23–28.

Firstenberg-Eden, R. 1983. Rapid estimation of the number of microorganisms in raw meat by impedance measurement. Food Technol. 37, 64–65.

Firstenberg-Eden, R., and Eden, G. 1984. Impedance Microbiology. J. Wiley and Sons, Somerset, New Jersey.

Firstenberg-Eden, R., and Klein, C. S. 1983. Evaluation of a rapid impedimetric procedure for the quantitative estimation of coliforms. J. Food Sci. 48, 1307–1311.

Firstenberg-Eden, R., and Tricarico, M. K. 1983. Impedimetric determination of total, mesophilic and psychotrophic counts in raw milk. J. Food Sci. 48, 1750–1754.

Fitts, R. 1985. Development of a DNA-DNA hybridization test for the presence of *Salmonella* in foods. Food Technol. 39, 95–102.

Fitts, R., Diamond, M., Hamilton, C., and Neri, M. 1983. DNA-DNA hybridization assay for detection of *Salmonella* spp. in foods. Appl. Environ. Microbiol. 46, 1146–1151.

Fleet, G. H. 1978. Yeasts as food spoilage agents. Food Technol. Aust. 30, 420–423.

Fleischer, M., Shapton, N., and Cooper, P. J. 1984. Estimation of yeast numbers in fruit mix for yogurt. Comparison of impedance and its components as measurements using the Bactometers 32 and M120B. J. Soc. Dairy Technol. 37, 63–65.

Flossdorf, J. 1983. A rapid method for the determination of the base composition of bacterial DNA. J. Microbiol. Methods. 1, 305–311.

Foegeding, P. M., and Busta, F. F. 1980. *Clostridium perfringens* cells and phospholipase C activity at constant and linearly rising temperatures. J. Food Sci. 45, 918–924.

Forrest, W. W. 1972. Microcalorimetry. In "Methods in Microbiology." J. R. Norris and D. W. Ribbons (Editors). Academic Press, New York.

Franck, B. 1984. Mycotoxins from mold fungi. Weapons of uninvited fellow-boarders of man and animal: Structures, biological activity, biosynthesis, and precautions. Angew. Chem. (Engl. Ed.) 23, 493–505.

Frazer, A. G. 1979. Food poisoning caused by *Clostridium perfringens*. Papua New Guinea Med. J. 22, 87–97.

Freed, R. C., Evenson, M. L., Reiser, R. F., and Bergdoll, M. S. 1982. Enzyme-linked immunosorbent assay for detection of staphylococcal enterotoxins in foods. Appl. Environ. Microbiol. 44, 1349–1355.

Freer, J. H., and Arbuthnott, J. P. 1982. Toxins of *Staphylococcus aureus*. Pharmacol. Ther. 19, 55–106.

Friun, J. T. 1978. Types of *Clostridium perfringens* isolated from selected foods. J. Food Prot. 41, 768–769.

Fromm, D. 1959. An evaluation of techniques commonly used to quantitatively determine the bacterial population on chicken carcasses. Poultry Sci. 38, 887–893.

Fryer, T. F., and Mead, G. C. 1979. Development of a selective medium for the isolation of *Clostridium sporogenes* and related organisms. J. Appl. Bacteriol. 47, 425–431.

4. Standard and Experimental Methods of Identification and Enumeration 155

Fujita, T., Monk, P. R., and Wadso, I. 1978. Calorimetric identification of several strains of lactic acid bacteria. J. Dairy Res. 45, 457–463.

Fung, D. Y. C. 1980. Teaching of automation and rapid methods in food microbiology. J. Food Prot. 43, 733–735.

Fung, D. Y. C. 1982. Overview of development in miniaturized microbiological techniques. In "Rapid Methods and Automation in Microbiology." R. C. Tilton (Editor). Amer. Soc. Microbiol., Washington, D.C.

Fung, D. Y. C. 1984. Rapid methods for determining the bacterial quality of red meats. J. Environ. Health 46, 226–228.

Fung, D. Y. C., Lee, C. Y., and Kastner, C. L. 1980. Adhesive tape method for estimating microbial loads on meat surfaces. J. Food Prot. 43, 295–297.

Fung, D. Y. C., Goldschmidt, M. C., and Cox, N. A. 1984. Evaluation of bacterial diagnostics kits and systems at an instructional workshop. J. Food Protection 47, 68–73.

Gabis, D. A., and Silliker, J. H. 1974a. Comparison of analytical schemes for detection of *Salmonella* in high moisture foods. Can. J. Microbiol. 20, 663–669.

Gabis, D. A., and Silliker, J. H. 1974b. The influence of selective enrichment media and incubation temperatures on the detection of salmonellae in dried foods and feeds. Can. J. Microbiol. 20, 1509–1511.

Gabis, D. A., and Silliker, J. H. 1977. The influence of selective enrichment broths, differential plating media, and incubation temperatures on the detection of *Salmonella* in dried foods and feed ingredients. Can. J. Microbiol. 23, 1225–1231.

Gabis, D. A., Brazis, R., and Midura, T. F. 1976a. Sample collection, shipment and preparation for analysis. In "Compendium of Methods for the Microbiological Examination of Foods." M. L. Speck (Editor). Am. Public Health Assoc., Washington, D.C.

Gabis, D. A., Vesley, D., and Favero, M. S. 1976b. Sampling, equipment, supplies, and environment. In "Compendium of Methods for the Microbiological Examination of Foods." M. L. Speck (Editor). Am. Public Health Assoc., Washington, D.C.

Gall, L. S., Baum, J. M., and Curby, W. A. 1978. Serological techniques for bacterial identification using a particle counter. Absts. Ann. Meet. Am. Soc. Microbiol. p. 302.

Gardener, S., and Jones, J. G. 1984. A new solidifying agent for culture media which liquefies on cooling. J. Gen. Microbiol. 130, 731–733.

Gemmell, C. G. 1984. Comparative study of the nature and biological activities of bacterial enterotoxins. J. Med. Microbiol. 17, 217–235.

Gendloff, E. H., Pestka, J. J., Swanson, S. P., and Hart, L. P. 1984. Detection of T-2 toxin in *Fusarium sporotrichiodes*–infected corn by enzyme-linked immunosorbent assay. Appl. Environ. Microbiol. 47, 1161–1163.

Gerats, G. E., and Snijders, J. M. A. 1978. Assessment of bacterial counts in the meat producing industry. III. The stomacher method and the spiral plate method. Arch. Lebensmittelhyg. 29, 57–61.

Gerba, C. P., and Goyal, S. M. 1978. Detection and occurrence of enteric viruses in shellfish: a review. J. Food Prot. 41, 743–754.

Gerba, C. P., Goyal, S. M. Cech, I., and Bogdan, G. F. 1980. Bacterial indicators and environmental factors as related to contamination of oysters by enteroviruses. J. Food Prot. 43, 99–101.

Gerhardt, P., Murray, R. G. E., Costilow, R. N., Nester, E. W., Wood, W. A., Krieg, N. R., and Phillips, G. B. 1981. (Editors). "Manual of Methods for General Bacteriology." Am. Soc. Microbiol., Washington, D.C.

Giammanco, G., Pignato, S., Agodi, A., Toucas, M., and D'Hauteville, H. 1982. Taxonomic value of a chromogenic test for the detection of amino-peptidases in the genus *Shigella*. Ann. Microbiol. B133, 343–346.

Gibbs, P. A., Patterson, J. T., and Early, J. 1979. A comparison of the fluorescent antibody method and a standardized cultural method for the detection of salmonellas. J. Appl. Bacteriol. 46, 501–505.

Gibson, D. M., Ogden, I. D., and Hobbs, G. 1984. Estimation of the bacteriologial quality of fish by automated conductance measurements. Int. J. Food Microbiol. 1, 127–134.

Gibson, J. B., Crull, S. L., and Borchardt, K. A. 1978. Micromethod for rapid identification of gram-negative, nonfermentative bacteria. Health Lab. Sci. 15, 9–14.

Gilbert, J. 1984. The detection and analysis of *Fusarium* mycotoxins. In "The Applied Mycology of *Fusarium*," M. O. Moss and J. E. Smith (eds.), Cambridge University Press, Cambridge England, pp. 175–193.

Gilbert, M., Penel, A., Kosikowski, F. V., Henion, J. D., Maylin, G. A., and Lisk, D. J. 1977. Electron affinity gas chromatographic determination of beta-nitropropionic acid as its pentafluorbenzyl derivatives in cheese and mold filtrates. J. Food Sci. 42, 1650–1653.

Gilchrist, J. E., Campbell, J. E., Donnelly, C. B., Peeler, J. T., and Delaney, J. M. 1973. Spiral plate method for bacterial determination. Appl. Microbiol. 25, 244–252.

Gill, C. O. 1979. Intrinsic bacteria in meat. J. Appl. Bacteriol. 47, 367–378.

Gille, Y., and Guinet, R. 1978. Evaluation of the micromethod API 20C for yeast identification. Rev. Inst. Pasteur, Lyon. 11, 55–63.

Gillies, R. R. 1979. Bacteriocin typing of *Enterobacteriaceae*. In "Methods in Microbiology." Vol II. T. Bergan and J. R. Norris (Editors). Academic Press Inc., Ltd., Oval Road, London, NW1 7DX, U.K.

Gilliland, S. E., and Speck, M. L. 1977. Use of the Minitek system for characterizing lactobacilli. Appl. Environ. Microbiol. 33, 1289–1292.

Gilliland, S. E., Busta, F. F., Brinda, J. J., and Campbell, J. E. 1976. Aerobic plate count. In "Compendium of Methods for the Microbiological Examination of Foods." M. L. Speck (Editor). Am. Public Health Assoc., Washington, D.C.

Godwin, G. J., Grodner, R. M., and Novak, A. F. 1977. Twenty-four hour methods for bacteriological analyses in frozen raw breaded shrimp processing. J. Food Sci. 42, 750–754.

Goldschmidt, M. C., and Fung, D. Y. C. 1978. New methods for microbiological analysis of food. J. Food Prot. 41, 201–219.

Goldschmidt, M. C., and Fung, D. Y. C. 1979. Automated instrumentation for microbiological analysis. Food Technol. 33, 63–70.

Goldstein, J., and Roberts, J. W. 1982. Microtube coagulase test for detection of coagulase-positive staphylococci. J. Clin. Microbiol. 15, 848–851.

Goodman, T. G., and Hoffman, P. S. 1983. Hydrogenase activity in catalase-positive strains of *Campylobacter* spp. J. Clin. Microbiol. 18, 825–829.

Gould, W. D., Hagedorn, C., Bardinelli, T. R., and Zablotoqicz, R. M. 1985. New selective media for enumeration and recovery of fluorescent pseudomonads from various habitats. Appl. Environ. Microbiol. 49, 28–32.

Goulet, J., Levesque, G., Moreau, J. R., and Roth, L. A. 1983. A new simple method for microbiological sampling of meat surfaces. Can. J. Microbiol. 29, 631–636.

Govan, J. R. W. 1978. Pyocin typing of *Pseudomonas aeruginosa*. In "Methods in Microbiology." Vol. 10. T. Bergan and J. R. Norris (Editors). Academic Press Inc., Ltd., Oval Road, London NW1 7DX, U.K.

Goyal, S. M., Gerba, C. P., and Melnick, J. L. 1979. Human enteroviruses in oysters and their overlying waters. Appl. Environ. Microbiol. 37, 572–581.

Gram, L., and Sogaard, H. 1985. Microcalorimetry as a rapid method for estimation of bacterial levels in ground meat. J. Food Protection 48, 341–345.

Gram, L., Pedersen, P., and Soegaard, H. 1984. An evaluation of the effect of catalase and 3,3-thiodipropionic acid on the recovery of freeze-injured coliform bacteria. Int. J. Food Microbiol. 1, 155–162.

Granade, T. C., Hehmann, M. F., and Artis, W. M. 1985. Monitoring of filamentous fungal growth by in situ microspectrophotometry, fragmented mycelium absorbance density, and ^{14}C incorporation: Alternatives to mycelial dry weight. Appl. Environ. Microbiol. 49, 101–108.

Grant, I. H., Richardson, N. J., and Bokkenheuser, V. D. 1980. Broiler chickens as potential source of *Campylobacter* infections in humans. J. Clin. Microbiol. 11, 508–510.

Graumlich, T. R. 1981. Survival and recovery of thermally stressed yeasts in orange juice. J. Food Sci. 46, 1410–1411.

Graumlich, T. R., and Stevenson, K. E. 1978. Recovery of thermally injured *Saccharomyces cerevisiae*: effects of media and storage conditions. J. Food Sci. 43, 1865–1870.

Gray, L. D., and Kreger, A. S. 1985. Purification and characterization of an extracellular cytolysin produced by *Vibrio vulnificus*. Infect. Immun. 48, 62–72.

Green, B. L., Clausen, E., and Litsky, W. 1975. Comparison of the new millipore HC with conventional membrane filters for the enumeration of fecal coliform bacteria. Appl. Microbiol. 30, 697–699.

Greenwood, M. H., et al. 1984. The microbiology of selected retail food products with an evaluation of viable counting methods. J. Hyg. 92, 67–72.

Gregersen, T. 1978. Rapid method for distinction of gram-negative from gram-positive bacteria. Eur. J. Appl. Microbiol. 5, 123–127.

Griffiths, M. W., Phillips, J. D., and Muir, D. D. 1984. Methods for rapid detection of postpasteurization contamination in cream. J. Soc. Dairy Technol. 37, 22–27.

Gripenberg, M., Nissinen, A., Vaisanen, E., and Linder, E. 1979. Demonstration of antibodies against *Yersinia enterocolitica* lipopolysaccharide in human sera by enzyme-linked immunosorbent assay. J. Clin. Microbiol. 10, 279–284.

Grischy, R. O., Speck, R. V., and Adams, D. M. 1983. New media for enumeration and detection of *Clostridium sporogenes* (PA3679) spores. J. Food Sci. 48, 1466–1469.

Guillou, J. P., and Chevrier, L. 1979. Study and differentiation of some hydrolases active on triglycerides and esters, in anaerobic bacteria, using gas-liquid chromatography. Ann. Microbiol. 130B, 399–406.

Guinee, P. A. M., and Van Leeuwen, W. J. 1979. Phage typing of *Salmonella*. *In* "Methods in Microbiology." Vol. II. T. Bergan and J. R. Norris (Editors). Academic Press Inc., Ltd., Oval Road, London, NW1 7DX, U.K.

Guinee, P. A. M., Jansen, W. H., and Haas, H. M. E. 1983. Mechanized procedures for the serology of *Salmonella*. Zentralbl. Bakteriol. Mikrobiol. Hyg. I Abt. A. 255, 258–264.

Guthertz, L. S., and Okoluk, R. L. 1978. Comparison of miniaturized multitest systems with conventional methodology for identification of *Enterobacteriaceae* from foods. Appl. Environ. Microbiol. 35, 109–112.

Gutowski, J. A., and Jacobs, D. M. 1979. A solid-phase radioimmunoassay for bacterial lipopolysaccharide. Immunol. Commun. 8, 347–364.

Gutteridge, C. S., and Norris, J. R. 1979. The application of pyrolysis techniques to the identification of microorganisms. J. Appl. Bacteriol. 47, 5–43.

Gutteridge, C. S., and Norris, J. R. 1980. Effect of different growth conditions on the discrimination of three bacteria by pyrolysis gas-liquid chromatography. Appl. Environ. Microbiol. 40, 462–465.

Hachisuka, Y., and Kozuka, S. 1981. A new test of differentiation of *Bacillus cereus* and *Bacillus anthracis* based on the existence of spore appendages. Microbiol. Immunol. 25, 1201–1207.

Hall, L. P. 1984. A new direct plate method for the enumeration of *Escherichia coli* in frozen foods. J. Appl. Bacteriol. 56, 227–235.

Halmann, M., Velan, B., and Sery, T. 1977. Rapid identification and quantitation of small numbers of microorganisms by a chemiluminescent immunoreaction. Appl. Environ. Microbiol. 34, 473–477.

Hansen, K., Mikkelsen, T., and Moeller-Madsen, A. 1982. Use of the Limulus test to determine the hygienic status of milk products as characterized by levels of Gram-negative bacterial lipopolysaccharide present. J. Dairy Res. 49, 323–328.

Hansen, M. V., and Elliott, L. P. 1980. New presumptive identification test for *Clostridium perfringens:* reverse CAMP test. J. Clin. Microbiol. 12, 617–619.

Hanson, C. W., Marso, E., and Martin, W. J. 1978. Comparison of the Minitek test system with a conventional screening procedure for identification of *Enterobacteriaceae*. Health Lao. Sci. 15, 3–8.

Hardy, D., Kraeger, S. J., Dufour, S. W., and Cady, P. 1977. Rapid detection of microbial contamination in frozen vegetables by automated impedance measurements. Appl. Environ. Microbiol. 34, 14–17.

Harmon, S. M. 1976. Collaborative study of an improved method for the enumeration and confirmation of *Clostridium perfringens* in foods. J. Assoc. Off. Anal. Chem. 59, 606–612.

Harmon, S. M. 1982. New method for differentiating members of the *B. cereus* group: Collaborative study. J. Assoc. Off. Anal. Chem. 65, 1134–1139.

Harmon, S. M., and Kautter, D. A. 1978. Media for confirming *Clostridium perfringens* from food and feces. J. Food Prot. 41, 626–630.

Harmon, S. M., and Placencia, A. M. 1978. Method for maintaining viability of *Clostridium perfringens* in foods during shipment and storage: collaborative study. J. Assoc. Off. Anal. Chem. 61, 785–788.

Harmon, S. M., Kautter, D. A., and McClure, F. D. 1984. Comparison of selective plating media for enumeration of *Bacillus cereus* in foods. J. Food Protection 47, 65–67.

Harrewijn, G. A., and Mossel, D. A. A. 1977. Microbiology of dried foods. In "Global Impacts of Applied Microbiology. IV International Conference." J. S. Furato (Editor). Sociedade Brasileira de Microbiologia, Revista de Microbiologia, Sao Paulo, Brazil.

Harris, C. M., and Kell, D. B. 1985. The estimation of microbial biomass. Biosensors 1, 17–34.

Hartman, P. A. 1960. Enterococcus: Coliform ratios in frozen chicken pies. Appl. Microbiol. 8, 114–116.

Hartman, P. A. 1968. Miniaturized Microbiological Methods. Academic Press, Inc., New York, N.Y.

Hartman, P. A., Reinhold, G. W., and Saraswat, D. S. 1966. Indicator organisms—A review. I. Taxonomy of the fecal streptococci. Intern. J. System. Bacteriol. 16, 197–221.

Harvey, R. W. S., and Price, T. H. 1979. Principles of *Salmonella* isolation. J. Appl. Bacteriol. 46, 27–56.

Harvey, R. W. S., and Price, T. H. 1981. Comparison of Selenite F, Muller—Kauffmann Tetrathionate and Rappaport's Medium for *Salmonella* isolated from chicken giblets after pre-enrichment in buffered peptone water. J. Hyg. 87, 219–224.

Hastback, W. G. 1981. Short incubation of presumptive media for detection of fecal coliforms in shellfish. Appl. Environ. Microbiol. 42, 1125–1127.

Hatcher, W. S. DiBenedetto, S., Taylor, L. E., and Murdock, D. I. 1977. Radiometric analysis of frozen concentrated orange juice for total viable microorganisms. J. Food Sci. 42, 636–639.

Hauschild, A., and Hilsheimer, R. 1983. Detection of *C. botulinum* in honey by a procedure involving membrane filtration. Can. Inst. Food Sci. Technol. J. 16, 256–258.

Hauschild, A. H. W., Gilbert, R. J., Harmon, S. M., O'Keeffe, M. F. O., and Vahlefeld, R. 1977. Collaborative study for the enumeration of *Clostridium perfringens* in foods. Can. J. Microbiol. 23, 884–892.

Hauschild, A. H. W., Desmarchelier, P., Gilbert, R. J., Harmon, S. M., and Vahlefeld, R. 1979a. Comparative study for the enumeration of *Clostridium perfringens* in feces. Can. J. Microbiol. 25, 953–963.

Hauschild, A. H. W., Park, C. E., and Hilshemer, R. 1979b. A modified pork plasma agar for the enumeration of *Staphylococcus aureus* in foods. Can. J. Microbiol. 25, 1052–1057.

Havelaar, A. H., Hoogendorp, C. J., Wesdorp, A. J., and Scheffers, W. A. 1980. False-negative oxidase reaction as a result of medium acidification. Antonie Van Leeuwenhoek 46, 301–312.

Hawirko, R. Z., Naccarato, C. A., Lee, R. P. W., and Malba, P. Y. 1979. Outgrowth and sporulation studies on *Clostridium botulinum* type E; influence of isoleucine. Can. J. Microbiol. 25, 522–534.

Head, C. B., Whitty, D. A., and Ratnam, S. 1982. Comparative study of selective media for recovery of *Yersinia enterocolitica*. J. Clin. Mcirobiol. 16, 615–621.

Heddaeus, H., Heczko, P. B., and Pulverer, G. 1978. Evaluation of the lysostaphin-susceptibility test for the classification of staphylococci. J. Med. Microbiol. 12, 9–15.

Hedges, A. J., Shannon, R., and Hobbs, R. P. 1978. Comparison of the precision obtained in counting viable bacteria by the Spiral Plate Maker, the Droplette and the Miles and Misra methods. J. Appl. Bacteriol. 45, 57–65.

Heisick, J., Lanier, J., and Peeler, J. T. 1984. Comparison of enrichment methods and atmosphere modification procedures for isolating *C. jejuni* from foods. Appl. Environ. Microbiol. 48, 1254–1255.

Henningson, E., Roffey, R., and Bovallius, A. 1982. A comparative study of apparatus for sampling airborne microorganisms. Gnana 20, 155–159.

Henriksen, S. D. 1979. Serotyping of Bacteria. *In* Methods in Microbiology, Vol. 12, Bergan, T., and Norris, J.R. (Eds.), Academic Press, Inc. p. 1–13.

Henson, O. E., Hall, P. A., Arends, R. E., Arnold, Jr., E. A., Knecht, R. M., Johnson, C. A., Pusch, D. J., and Johnson, M. G. 1982. Comparison of four media for the enumeration of fungi in dairy products—a collaborative study. J. Food Sci. 47, 930–932.

Herman, J. P. M., Jakubcza, E., Izard, D., and LeClerc, H. 1980. Role of microcalorimetry in taxonomy of some *Enterobacteriaceae*. Can. J. Microbiol. 26, 413–419.

Herrmann, J. E., Hendry, R. M., and Collins, M. F. 1979. Factors involved in enzyme-linked immunoassay of viruses and evaluation of the method for identification of enteroviruses. J. Clin. Microbiol. 10, 210–217.

Herrmann, M., Morenz, J., Hofmann, H., and Hofmann, G. 1978. Investigations of the use of the selective enterococcus (SE) medium for routine diagnostic procedures for enterococci. Z. Gesamte Hyg. Grenzgeb. 23, 374–377.

Hildebrandt, G., Weiss, H., and Siems, H. 1977. Random sampling in judging the occurrence of salmonellae in frozen chicken. Fleischwirtschaft 57, 255–258.

Hill, B. M. 1983. The thermo-nuclease test as a method for identifying *Staphylococcus aureus*. Aust. J. Dairy Technol. 38, 95–96.

Hill, W. E., 1981. DNA hybridization method for detecting enterotoxigenic *Escherichia coli* in human isolates and its possible application to food samples. J. Food Saf. 3, 233–247.

Hill, W. E., Madden, J. M., McCordell, B. S., Shak, D. B., Jagow, J. A., Payne, W. L., and Boutin, B. K. 1983a. Foodborne enterotoxigenic *E. coli*: Detection and enumeration by DNA colony hybridization. Appl. Environ. Microbiol. 45, 1324–1330.

Hill, W. E., Payne, W. L., and Aulisio, C. C. G. 1983b. Detection and enumeration of virulent *Yersinia enterocolitica* in food by DNA colony hybridization. Appl. Environ. Microbiol. 46, 641–643.

Hirsh, D. C., and Martin, L. D. 1983a. Rapid detection of *Salmonella* spp. by using Felix-01 bacteriophage and HPLC. Appl. Environ. Microbiol. 45, 260–264.

Hirsh, D. C., and Martin, L. D. 1983b. Detection of *Salmonella* spp. in milk using Felix-01 bacteriophage and high-pressure liquid chromatography. Appl. Environ. Microbiol. 46, 1243–1245.

Ho, B. 1983. An improved *Limulus* gellation assay. Microbios Lett. 24, 81–84.

Hobbs, B. G. 1976. *C. botulinum* and its importance in fishery products. Adv. Food Res. 22, 135–185.

Hobbs, B. C. 1977a. *Clostridium perfringens* food poisoning. In "Global Impacts of Applied Microbiology. IVth International Conference." J. S. Furato (Editor). Sociedade Brasileira de Microbiologia, Revista de Microbiologia, Sao Paulo, Brazil.

Hobbs, B. C. 1977b. Improved Methods for *Salmonella* detection in foods. In "Global Impacts of Applied Microbiology. IVth International Conference." J. S. Furate (Editor). Sociedade Brasileira de Microbiologia, Revista de Microbiologia, Sao Paulo, Brazil.

Hocking, A. D., and Pitt, J. I. 1980. Dichloranglycerol medium for enumeration of xerophilic fungi from low-moisture foods. Appl. Environ. Microbiol. 39, 488–492.

Hofer, E., and Silva, C. H. D. 1984. An evaluation of the efficiency of enrichment media in the isolation process for *Vibrio parahaemolyticus*. Zentralbl. Bakteriol. Mikrobiol. Hyg. I Abt. A, 256, 456–465.

Hofherr, L., and Lund, M. E. 1979. Characterization of staphylococci using the API 20E system. Am. J. Med. Technol. 45, 127–129.

Hofherr, L., Votava, H., and Blazevic, D. J. 1978. Comparison of three methods for identifying nonfermenting gram-negative rods. Can. J. Microbiol. 24, 1140–1144.

Holbrook, R., and Anderson, J. M. 1980. An improved selection and diagnostic medium for the isolation and enumeration of *Bacillus cereus* in foods. Can. J. Microbiol. 26, 753–759.

Holbrook, R., Anderson, J. M., and Baird-Parker, A. C. 1980. Modified direct plate method for counting *Escherichia coli* in foods. Food Technol. Aust. 32, 78–83.

Holmes, B., Wilcox, W. R., Lapage, S. P., and Malnick, H. 1977. Test reproducibility of the API (20E), Enterotube, and Pathotec systems. J. Clin. Pathol. 30, 381–387.

Holmes, B., Dowling, J., and Lapage, S. P. 1979. Identification of gram-negative nonfermenters and oxidase-positive fermenters by the Oxi/Ferm Tube. J. Clin. Pathol. 32, 78–85.

Hood, M. A., Ness, G. E., and Blake, N. J. 1983a. Relationship among fecal coliforms, *Escherichia coli*, and *Salmonella* spp. in shellfish. Appl. Environ. Microbiol. 45, 122–126.

Hood, M. A., Ness, G. E., Rodrick, G. E., and Blake, N. J. 1983b. Effects of storage on microbial loads of two commercially important shellfish species. Appl. Environ. Microbiol. 45, 1221–1228.

Hsu, S. C., and Williams, T. J. 1982. Evaluation of factors affecting the membrane filter technique for testing drinking water. Appl. Environ. Microbiol. 44, 453–460.

Hunt, D. A., Lucas, J. P., McClure, F. D., Springer, J., and Newell, R. 1981. Comparison of modified A-1 method with Standard EC Test for recovery of fecal coliform bacteria from shellfish. J. Assoc. Off. Anal. Chem. 64, 607–610.

Hunt, J. M., Francis, D. W., Peeler, J. T., and Lovett, J. 1985. Comparison of methods for isolating *Campylobacter jejuni* from raw milk. Appl. Environ. Microbiol. 50, 535–536.

Hunter, K. W., Brimfield, A. A., Miller, M., Finkelman, F. D., and Chu, S. F. 1985.

Preparation and characterization of monoclonal antibodies to the trichothecene mycotoxin T-2. Appl. Environ. Microbiol. 49, 168–172.

Huq, M. I. 1979. A simple laboratory method for the diagnosis of *V. cholerae.* Trans. R. Soci. Trop. Med. Hyg. 73, 553–556.

Hurley, M. A., and Roscoe, M. E. 1983. Automated statistical analysis of microbial enumeration by dilution series. J. Appl. Bacteriol. 55, 159–164.

Hurst, A., and Nasim. A. 1984. (Editors). Repairable Lesions in Microorganisms. Academic Press, New York, New York.

Hussain Qadri, S. M., Nichols, C. W., Qadri, S. G. M., and Villarreal, A. 1978. Rapid test for acetyl-methyl-carbinol formation by *Enterobacteriaceae.* J. Clin. Microbiol. 8, 463–464.

Hussain Qadri, S. M., Peddecord, M., and Zubairi, S. 1980. Rapid test for indole formation with non-proliferating bacteria. Antonie Van Leeuwenhoek 46, 419–423.

Hussong, D., Enkiri, N. K., and Burge, W. D. 1984. Modified agar medium for detecting environmental salmonellae by the MPN method. Appl. Environ. Microbiol. 48, 1026–1030.

Hutter, K. L., and Eipel, H. E. 1978. Flow cytometric determination of cellular substances in algae, bacteria, moulds, and yeasts. Antonie Van Leeuwenhoek 44, 269–282.

Hutter, K. J., and Eipel, H. E. 1979. Microbial determinations by flow cytometry. J. Gen. Microbiol. 113, 369–375.

Hysert, D. W., Kovecses, F., and Morrison, N. M. 1976. A firefly bioluminescence ATP assay method for rapid detection and enumeration of brewery microorganisms. J. Am. Soc. Brewery Chemists 34, 351–360.

Hyypiae, T., Staalhandske, P., Vainionpaeae, R., and Pettersson, U. 1984. Detection of enteroviruses by spot hybridization. J. Clin. Microbiol. 19, 436–438.

Ibrahim, G. F. 1976. Detection and enumeration of coagulase-positive staphylococci in dairy products: 2. Recovery of heat-stressed cells. Aust. J. Dairy Technol. 31, 138–140.

Ibrahim, G. F. 1977. Detection and enumeration of coagulase-positive staphylococci in dairy products: 3. Evaluation of the performance of the strip method. Aust. J. Dairy Technol. 32, 35–38.

Ibrahim, G. F. 1981. A simple sensitive method for determining staphylococcal thermonuclease in cheese. J. Appl. Bacteriol. 51, 307–312.

Ibrahim, G. F. 1986. A review of immunoassays and their application to salmonellae detection in foods. J. Food Protection 49, 299–310.

Isenberg, H. D., and Sampson-Scherer, J. 1977. Clinical laboratory evaluation of a system approach to the recognition of nonfermentative or oxidase-producing gram-negative, rod-shaped bacteria. J. Clin. Microbiol. 5, 336–340.

Isenberg, H. D., Gavan, T. L. Sonnenwirth, A., Taylor, W. I., and Washington, J. A. 1979. Clinical laboratory evaluation of automated microbial detection/identification system in analysis of clinical urine specimens. J. Clin. Microbiol. 10, 226–230.

Isenberg, H. D., Gavan, T. L., Smith, P. B., Sonnenwirth, A., Taylor, W., Martin, W. J., Rhoden, D., and Balows, A. 1980. Collaborative investigation of the AutoMicrobic System *Enterobacteriaceae* biochemical card. J. Clin. Microbiol. 11, 694–702.

Ito, K. A. 1981. Thermophilic organisms in food spoilage: flat-sour aerobes. J. Food Prot. 44, 157–163.

Izard, D., Savage, C., Enoyeh, E., Leclerc, H., and Troonen, H. 1982. Rapid and automated identification of *Enterobacteriaceae* with the Abbott MS-2 system and API-20E versus conventional methods. Zentralbl. Bakteriol. Mikrobiol. Hyg. I, Abt. A, 251, 26–34.

Izard, D., Husson, M. O., Vincent, P., Leclerc, H., Monget, D., and Boeufgras, J. M. 1984.

Evaluation of the four-hour rapid 20E system for identification of members of the family *Enterobacteriaceae*. J. Clin. Microbiol. 20, 51–54.

Jacobs, C. J., Prior, B. A., and deKock, M. J. 1983. A rapid screening method to detect ethanol production by microorganisms. J. Microbiol. Methods 1, 339–342.

Jakubczak, E., and LeClerc, H. 1980. Firefly assay of bacterial ATP: comparative study of extraction methods. Ann. Biol. Clin. 38, 297–304.

Jarvis, B. 1977. A chemical method for the estimation of mold in tomato products. J. Food Technol. 12, 581–591.

Jarvis, B. 1982. Rapid methods in food microbiology. A practical approach. Food Technol. Aust, 34, 518–523.

Jarvis, B., Lach, V. H., and Wood, J. M. 1977. Evaluation of the Spiral Plate Maker for the enumeration of microorganisms in foods. J. Appl. Bacteriol. 43, 149–157.

Jarvis, B., Seiler, D. A. L., Ould, A. J. L., and Williams, A. P. 1983. Observations on the enumeration of moulds in food and feedingstuffs. J. Appl. Bacteriol. 55, 325–336.

Jay, J. M. 1977. The *Limulus* lysate endotoxin assay as a test of microbial quality of ground beef. J. Appl. Bacteriol. 43, 99–109.

Jay, J. M. 1978. "Modern Food Microbiology." D. Van Nostrand Company, New York, N.Y.

Jelinkova, J. 1979. Identification and typing of enterococci. *In* Methods in Microbiology. Vol. 12. T. Bergan and J. R. Norris (Editors). Academic Press Inc., Ltd., Oval Road, London, NW1 7DX, U.K.

Jenkins, R. D., Hale, D. C., and Matsen, J. M. 1980. Rapid semiautomated screening and processing of urine specimens. J. Clin. Microbiol. 11, 220–225.

Johnson, K. M. 1984. *Bacillus cerus* foodborne illness—an update. J. Food Protection 47, 145–153.

Johnson, W. M., and Lior, H. 1984. Toxins produced by *Campylobacter jejuni* and *Campylobacter coli*. Lancet 1, 229–230.

Johnston, R. J., Butchart, A. M., and Kgamphe, S. J. 1978. A comparison of sampling methods for airborne bacteria. Enrivon. Res. 16, 279–284.

Jones, P. W., Collins, P., and Hayle, A. J. 1984. The effect of sodium sulphacetamide and sodium mandelate in brilliant green agar on the growth of salmonellas. J. Appl. Bacteriol. 57, 423–428.

Jorgensen, J. H., and Alexander, G. A. 1981. Automation of the Limulus Amoebocyte Lysate Test by using the Abbott MS-2 Microbiology System. Appl. Environ. Microbiol. 41, 1316–1320.

Jorgensen, J. H., and Smith, R. F. 1973. Preparation, sensitivity, and specificity of Limulus lysate for endotoxin assay. Appl. Microbiol. 26, 43–48.

Jorgensen, J. H., Lee, J. C., and Pahren, H. R. 1976. Rapid detection of bacterial endotoxin in drinking water and renovated wastewater. Appl. Environ. Microbiol. 32, 347–351.

Jorgensen, J. H., Dyke, J. W., Helgeson, N. G. P., Cooper, B. H., Redding, J. S., Crawford, S. A., Andruszewski, M. T., and Prowant, S. A. 1984. Collaborative evaluation of the Abbott Avantage System for identification of frequently isolated non-fermentative or oxidase-positive gram-negative bacilli. J. Clin. Microbiol. 20, 899–904.

Joyce, D. A., and Ould, A. J. L. 1978. Observations on some techniques for the microbiological testing of butter. J. Soc. Dairy Technol. 31, 227–230.

Kafel, S., and Bryan, F. L. 1977. Effects of enrichment media and incubation conditions on isolating salmonellae from ground-meat-filtrate. Appl. Environ. Microbiol. 34, 285–291.

Kalitina, T. A. 1978. A method for enterovirus concentration in virological examinations of meat. Vopr. Virusol. 5, 621–625.

Kaneko, T., Yokoyama, H., and Takahashi, T. 1984. A rapid determination of microbial counts in raw milk and yogart by an ATP assay technique. J. Food Hyg. Soc. Japan 25, 193–197.

Kaper, J., Seidler, R. J., Lockman, H., and Colwell, R. R. 1979. Medium for the presumptive identification of *Aeromonas hydrophila* and *Enterobacteriaceae*. Appl. Environ. Microbiol. 38, 1023–1026.

Kaper, J. B., Remmers, E. F., and Colwell, R. R. 1980. A medium for presumptive identification of *Vibrio parahaemolyticus*. J. Food Prot. 43, 936–938.

Kaplan, R. L. 1981. Rapid methods in microbiology: II. Rapid identification of gram-positive organisms. Am. J. Med. Technol. 47, 687–690.

Karachewski, N. O., Busch, E., and Wells, C. L. 1985. Comparison of PRASII, RapID ANA, and APl 20A systems for identification of anaerobic bacteria. J. Clin. Microbiol. 21, 122–126.

Karl, D. M. 1978. A rapid sensitive method for the measurement of quanine ribonucleotides in bacterial and environmental extracts. Anal. Biochem. 89, 581–595.

Karmali, M. A., and Fleming, P. C. 1979. Application of the Fortner principle to isolation of *Campylobacter* from stools. J. Clin. Microbiol. 10, 245–247.

Karnop, G. 1982. The role of proteolytic microorganisms in fish spoilage. I. Improvement of the technique for the detection of proteolytic bacteria. Arch. Lebensmittelhyg. 33, 57–61.

Karolus, J. J., Leblanc, D. H., Marsh, A. J., Mshar, R., and Furgalack, T. H. 1985. A research note: Presence of histamine in the bluefish, *Pomatomus saltatrix*. J. Food Protection. 48, 166–168.

Kass, E. H., and Wolff, S. M. (Editors). 1973. "Bacterial Lipopolysaccharides." Univ. of Chicago Press, Chicago, Ill.

Kawabata, N. 1980. Studies on the sulfite reduction test for clostridia. Microbiol. Immunol. 24, 271–279.

Kazal, H. L. 1976. Laboratory diagnosis of foodborne diseases. Ann. Clin. Lab. Sci. 6, 381–399.

Kelleher, S. D., and Zall, R. R. 1983. Ethanol accumulations in muscle tissue as a chemical indicator of fish spoilage. J. Food Biochem. 7, 87–94.

Kelley, R. W., and Kellogg, S. T. 1978. Computer-assisted identification of anaerobic bacteria. Appl. Environ. Microbiol. 35, 507–511.

Kellogg, S. T. 1979. MICRID: A computer-assisted microbiol identification system. Appl. Environ. Microbiol. 38, 559–563.

Kennedy, J. E., Jr., Oblinger, J. L., and Bitton, G. 1984. Recovery of coliphages from chicken, pork sausage, and delicatessen meats. J. Food Protection 47, 623–626.

Kennedy, J. E., Jr., and Oblinger, J. L. 1985. Application of bioluminescence to rapid determination of microbial levels in ground beef. J. Food Protection. 48, 334–340.

Kielwein, G. 1978. Occurrence and significance of enterococci in milk and dairy products. Arch. Lebensmittelhyg. 29, 127–128.

Kilbourn, J. P. 1979. Review of bacterial metabolism. J. Am. Med. Technol. 41, 91–102.

Kilian, M., and Bullow, P. 1979. Rapid identification of *Enterobacteriaceae*. II. Use of a β-glucuronidase detecting agar medium (PGUA agar) for the identification of *E. coli* in primary cultures of urine samples. Acta. Pathol. Microbiol. Scand., Ser. B 87, 271–276.

Kilsby, D. C. 1982. Sampling schemes and limits. *In* "Meat Microbiology." M. H. Brown (Editor). Applied Science Publishers, Ltd., New York, N.Y.

Kilsby, D. C., and Pugh, M. E. 1981. The reliance of the distribution of microorganisms within batches of food to the control of microbiological hazards from foods. J. Appl. Bacteriol. 51, 345–354.

Kim, T. K., Hammond, J. B., and Chipley, J. R. 1979. Chemical and electron microscopic studies of factors associated with the release of penicillinase from *Staphylococcus aureus*. Antonie Van Leeuwenhoek 45, 581–593.

King, A. D., Jr., and Schade, J. E. 1984. *Alternaria* toxins and their importance in food. J. Food Protection 47, 886–901.

King, A. D., Jr., Hocking, A. D., and Pitt, J. I. 1979. Dichloran-rose bengal medium for enumeration and isolation of molds from foods. Appl. Environ. Microbiol. 37, 959–964.

Kinner, J. A., and Moats, W. A. 1978. Selective action of sodium cholate-$MgCl_2$ broth and its possible use in isolation of salmonellae and other enteric pathogens. J. Food Prot. 41, 638–642.

Kirchman, D., Sigda, J., Kapuscinski, R., and Mitchell, R. 1982. Statistical analysis of the direct count method for enumerating bacteria. Appl. Environ. Microbiol. 44, 376–382.

Kirschner, L. E., and Puleo, J. R. 1979. Wipe-rinse technique for quantitating microbial contamination on large surfaces. Appl. Environ. Microbiol. 38, 466–470.

Klein, D. A., and Wu, S. 1974. Stress: a factor to be considered in heterotrophic microorganism enumeration from aquatic environments. Appl. Microbiol. 27, 429–431.

Klinger, I., Basker, D., and Juwen, B. J. 1981. Sampling for precise microbiological plate counts on broiler chicken carcasses. Poult. Sci. 60, 575–578.

Kloos, W. E., and Wolfshohl, J. R. 1982. Identification of *Staphylococcus* species with the API Staph-Ident system. J. Clin. Microbiol. 16, 509–516.

Koburger, J. A., and Dargan, R. A. 1985. The "Phoenix Phenomenon" during resuscitation of fungi in foods. J. Food Protection 48, 556–557.

Koburger, J. A., and May, S. O. 1982. Isolation of *Chromobacterium* spp. from foods, soil, and water. Appl. Environ. Microbiol. 44, 1463–1465.

Koburger, J. A., and Miller, M. L. 1985. A research note: evaluation of a fluorogenic MPN procedure for determining *Escherichia coli* in oysters. J. Food Protection 48, 244–245.

Koburger, J. A., and Oblinger, J. L. 1977. Organisms from positive MPN tubes inoculated with samples that yielded no growth on pour plates. J. Food Prot. 40, 484–485.

Koch, A. L. 1981. Growth measurement. In "Manual of Methods for General Bacteriology." P. Gerhardt, R. G. E. Murray, R. N. Costilow, E. W. Nester, W. A. Wood, N. R. Krieg and G. B. Phillips (Editors). Am. Soc. Microbiol., Washington, D.C.

Koch, K., Bohm, R., and Strauch, D. 1978. The differentiation of clostridial microcultures on nucleopore-filters by fluorescent antibodies as a rapid screening test for spores. II. Application of fluorescent antibodies. Zentralbl. Bakteriol. Parasitenkd. Infektionskr. Hyg., I Abt. A 241, 463–472.

Kodaka, H., Lombard, G. L., and Dowell, Jr., V. R. 1982. Gas-liquid chromatography technique for detection of hippurate hydrolysis and conversion of fumarate to succinate by microorganisms. J. Clin. Microbiol. 16, 962–964.

Koeppen, B., and Dalgaard, L. 1984. Assay for beta-glucuronidase utilizing headspace gas chromatography. Anal. Biochem. 136, 272–275.

Koidis, P., and Doyle, M. P. 1984. Procedure for increased recovery of *Campylobacter jejuni* from inoculated unpasteurized milk. Appl. Environ. Microbiol. 47, 455–460.

Konuma, H., Suzuki, A., and Kurata, H. 1982. Improved stomacher 400 bag application to the spiral plate system for counting bacteria. Appl. Environ. Microbiol. 44, 765–769.

Korkeala, H., and Hirvelae, V. 1981. The effect of various stress treatments on the coagulase test in *Staphylococcus aureus*. Nord. Vet. Med. 33, 434–440.

Kornacki, J. L., and Marth, E. H. 1982. Foodborne illness caused by *Escherichia coli*: A review. J. Food Prot. 45, 1051–1067.

Kostenbader, K. D., Jr., and Cliver, D. O. 1981. Flocculants for recovery of food-borne viruses. Appl. Environ. Microbiol. 41, 318–320.

Kotula, A. W., and Stern, N. J. 1984. The importance of *C. jejuni* to the meat industry: A review. J. Anim. Sci. 58, 1561–1566.
Koupal, A., and Deibel, R. H. 1978. Rapid qualitative method for detecting staphylococcal nuclease in foods. Appl. Environ. Microbiol. 35, 1193–1197.
Kovacs, S., and Takacs, J. 1979. Interrelationships of *Escherichia coli* and *Salmonella* contamination in sausages processed by salting, smoking and dehydration. Acta. Vet. Acad. Sci. Hung. 27, 337–342.
Kraft, A. A., and Rey, C. R. 1979. Psychrotrophic bacteria in foods: An update. Food Technol. 33, 66–71.
Kramer, J. 1977. A rapid microdilution technique for counting viable bacteria in food. Lab. Pract. 26, 657–676.
Kramer, J. M., and Gilbert, R. J. 1978. Enumeration of microorganisms in food: A comparative study of five methods. J. Hyg. 81, 151–159.
Kramer, J. M., Kendall, M., and Gilbert, R. J. 1979. Evaluation of the spiral plate and laser colony counting techniques for the enumeration of bacteria in foods. Eur. J. Appl. Microbiol. Biotechnol. 6, 289–299.
Kreiger, R. A., Snyder, O. P., and Pflug, I. J. 1984. A comparison of techniques for recovering *Bacillus subtilis* spores from inoculated meat substrates. J. Food Sci. 49, 366–369.
Kroninger, D. L., and Banwart, G. J. 1978. Effects of various selective agars on the growth of rare and common salmonellae. J. Food Sci. 43, 1328–1329.
Krysinski, E. P., and Heimsch, R. C. 1977. Use of enzyme-labelled antibodies to detect *Salmonella* in foods. Appl. Environ. Microbiol. 33, 947–954.
Kukulinsky-Fuller, J. C., and Nelson, F. E. 1977. Enumeration of temperature-stressed *Pseudomonas aeruginosa* utilizing selective procedures. J. Food Sci. 42, 415–420.
Kurup, V. P., and Babcock, J. B. 1979. Use of casein, tyrosine, and hypoxanthine in the identification of nonfermentative gram-negative bacilli. Med. Microbiol. Immunol. 167, 71–75.
Labadie, J., and Doumbia, M. 1984. Rapid counting of bacterial flora isolated from carcasses of beef, pork, and sheep with a resazurin test. Zentralbl. Bakteriol. Mikrobiol. Hyg. I. Abt. B. 179, 217–224.
Labbe, R. G., and Norris, K. E. 1982. Evaluation of plating media for recovery of heated *Clostridium perfringens* spores. J. Food Prot. 45, 686–688.
Labie, C. 1979. Viruses and food products of animal origin. Rev. Med. Vet. 130, 1427–1458.
Lachica, R. V. F. 1976. Simplified thermonuclease test for rapid identification of *Staphylococcus aureus* recovered on agar media. Appl. Environ. Microbiol. 32, 633–634.
Lachica, R. V. F. 1980. Accelerated procedure for the enumeration and identification of food-borne *Staphylococcus aureus*. Appl. Environ. Microbiol. 39, 17–19.
Lachica, R. V. F. 1984. Egg yolk-free Baird-Parker medium for the accelerated enumeration of foodborne *Staphylococcus aureus*. Appl. Environ. Microbiol. 48, 870–871.
LaHellec, C., and Colin, P. 1982. Miniaturized methods in poultry microbiology. *In* "Rapid Methods and Automation in Microbiology." R. C. Tilton (Editor). Am. Soc. Microbiol., Washington, D.C.
LaHellec, C., and Colin, P. 1984. Improved methods in *Salmonella* diagnostics. *In* Priority Aspects of Salmonellosis Research. H. E. Larsen (ed.), CEC, Luxemborg. p. 97–100.
Lame, H. 1978. Comparative study of three methods of estimating the surface bacterial flora. Rev. Med. Vet. 129, 615–624.
Lampe, A. S., and van der Reijen, T. J. K. 1984. Evaluation of commercial test systems for the identification of nonfermenters. Eur. J. Clin. Microbiol. 3, 301–305.

Lampen, J. O. 1966. Secretion of enzymes by microorganisms. *In* Function and Structure in Microorganisms. M. R. Pollock and M. H. Richmond (Editors). 15th Symp. of the Society for General Microbiol. Cambridge Univ. Press, London, England.

Lampi, R. A., Mikelson, D. A., Rowley, D. B., Prewite, J. J., and Wells, R. E. 1974. Radiometry and microcalorimetry techniques for rapid detection of food-borne microorganisms. Food Technol. 28, 52–57.

Lancette, G. A., and Harmon, S. M. 1980. Enumeration and confirmation of *Bacillus cereus* in foods: collaborative study. J. Assoc. Off. Anal. Chem. 63, 581–586.

Land, G. A., Harrison, B. A., Hulme, K. L., Cooper, B. H., and Byrd, J. C. 1979. Evaluation of the new API 20C strip for yeast identification against a conventional method. J. Clin. Microbiol. 10, 357–364.

Landry, E. F., Vaughn, J. M., and Vicale, T. J. 1980. Modified procedure for extraction of poliovirus from naturally-infected oysters using cat-floc and beef extract. J. Food Prot. 43, 91–94.

Langlois, B. E., and Sanghirum, A. 1977. Effect of various combinations of medium, diluent and incubation conditions on recovery of bacteria from manufacturing grade and grade A raw milk. Can. J. Food. Prot. 40, 222–227.

Langone, J. J., and Van Vunakis, H. 1982. (Editors). "Immunochemical Techniques. Part D. Selected Immunoassays." Academic Press, New York, N.Y. Methods Enzymol., Vol. 84.

Lanyi, B., and Bergan, T. 1978. Serological characterization of *Pseudomonas. In* "Methods in Microbiology." Vol. 10. T. Bergan and J. R. Norris (Editors). Academic Press Inc., Ltd., Oval Road, London NW1 7DX, U.K.

Larkin, E. P., and Metcalf, T. G. 1980. Cooperative study of methods for the recovery of enteric viruses from shellfish. J. Food Prot. 43, 84–86.

Larsson, L., and Holst, E. 1982. Feasibility of automated head-space gas chromatography in identification of anaerobic bacteria. Acta Pathol. Microbiol. Immunol. Scand. Ser. B 90B, 125–130.

Larsson, L., Marco, P. A., and Odham, G. 1978a. Analysis of amines and other bacterial products by head-space gas chromatography. Acta Pathol. Microbiol. Scand., Ser. B 86, 207–213.

Larsson, L., Marco, P. A., and Odham, G. 1978b. Detection of alcohols and volatile fatty acids by head-space gas chromatography in identification of anaerobic bacteria. J. Clin. Microbiol. 7, 23–27.

Larsson, L., Saellstroem-Baum, S., and Cochetiere-Collinet, M. 1984. A two-step extraction procedure for concentrating acidic organic volatiles in aqueous solution prior to gas chromatographic head-space analysis, as exemplified by short-chain fatty acids produced by *Bacillus cereus*. J. Microbiol. Methods 2, 9–14.

Lazarus, C. R., Abu-Bakar, A., West, R. L., and Oblinger, J. L. 1977. Comparison of microbial counts on beef carcasses by using the moist-swab contact method and secondary tissue removal technique. Appl. Environ. Microbiol. 33, 217–218.

LeChevallier, M. W., Cameron, S. C., and McFeters, G. A. 1983a. New medium for improved recovery of coliform bacteria from drinking water. Appl. Environ. Microbiol. 45, 484–492.

LeChevallier, M. W., Cameron, S. C., and McFeters, G. A. 1983b. Comparison of verification procedures for the membrane filter total coliform technique. Appl. Environ. Microbiol. 45, 1126–1128.

LeChevallier, M. W., and McFeters, G. A. 1984. Recent advances in coliform methodology for water analysis. J. Environ. Health 47, 5–9.

Lee, C. Y., Fung, D. Y. C., and Kastner, C. L. 1982. Computer-assisted identification of bacteria on hot-boned and conventionally processed beef. J. Food Sci. 47, 363–367.

Lee, D. T. F., and Rosenblatt, J. E. 1983. A comparison of four methods for detecting beta-lactamase in anaerobic bacteria. Diagn. Microbiol. Infect. Dis. 1, 173–175.
Lee, M. L., Smith, D. L., and Freeman, L. R. 1979. High-resolution gas chromatographic profiles of volatile organic compounds produced by microorganisms at refrigerated temperatures. Appl. Environ. Microbiol. 37, 85–90.
Lee, W. H., Harris, M. E., McClain, D., Smith, R. E., and Johnston, R. W. 1980. Two modified selenite media for the recovery of *Yersinia enterocolitica* from meats. Appl. Environ. Microbiol. 39, 205–209.
Leela, P. K., and Sankaran, R. 1981. Detection and enumeration of stressed coliforms. Nahrung 25, 435–440.
Leighton, I. 1978. Prospects of automation in microbiology. Med. Lab. Sci. 35, 213–214.
Leighton, P. M., and Little, J. A. 1983. Clinical comparison of the Enterotube II and API 20E systems for bacterial identification. Am. J. Clin. Pathol. 79, 367–369.
LeMinor, L. 1979. Tetrathionate reductase, β-glucuronidase, and ONPG-tests in the genus *Salmonella*. Zentralbl. Bakteriol. Parasitenkd. Infektionskr. Hyg. I Abt. A 243, 321–325.
Lennox, V. A., and Ackerman, V. P. 1984. Biochemical identification of bacteria by replicator methods on agar plates. Pathology. 16, 434–440.
Lepp, C. A., Nowlan, E. D., Mason, R. D. and Ramsey, W. S. 1979. Fluorophores as visualization aides in agar growth media. Experientia 35, 868–869.
Levin, J., and Bang, F. 1964. The role of endotoxin in the extracellular coagulation of *Limulus* blood. Bull. Johns Hopkins Hosp. 115, 265–274.
Levine, M. 1961. Facts and fancies of bacterial indices in standards for water and foods. Food Technol. 15, 29–34.
Licciardello, J. J. 1983. Botulism and heat-processed seafoods. Mar. Fish. Rev. 45, 1–7.
Lin, C. C., and Casida, Jr., L. E. 1984. GELRITE as a gelling agent in media for the growth of thermophilic microorganisms. Appl. Environ. Microbiol. 47, 427–429.
Ling, T. G., Ramstorp, M., and Mattiasson, B. 1982. Immunological quantitation of bacterial cells using a partition affinity ligand assay: A model study on the quantitation of streptococci. Anal. Biochem. 122, 26–32.
Littel, K. J., and Hartman, P. A. 1983. Fluorogenic selective and differential medium for isolation of fecal streptococci. Appl. Environ. Microbiol. 45, 622–627.
Loane, P., Tommereys, J., Stokoe, J., Newton, E., and Forbes, L. 1977. Coliform enumeration in butter. Aust. J. Dairy Technol. 32, 72–74.
Logan, N. A., and Berkeley, R. C. W. 1984. Identification of *Bacillus* strains using the API system. J. Gen. Microbiol. 130, 1871–1882.
Logan, N. A., Capel, B. J., Mellings, J., and Berkeley, R. C. W. 1979. Distinction between emetic and other strains of *Bacillus cereus* using the API system and numerical methods. FEMS Microbiol. Lett. 5, 373–375.
Lorenz, R. C., and Tuovinen, O. H. 1980. Inhibition by various membrane filters of *Escherichia coli* colony development. Microbios Lett. 12, 23–30.
Lovett, J., Francis, D. W., and Hunt, J. M. 1983. Isolation of *Campylobacter jejuni* from raw milk. Appl. Environ. Microbiol. 46, 459–462.
Lundholm, I. M. 1982. Comparison of methods for quantitative determinations of airborne bacteria and evaluation of total viable counts. Appl. Environ. Microbiol. 44, 179–183.
Luster, C. III. 1978. A rapid and sensitive monitoring technique for aseptically processed bulk tomato paste. J. Food Sci. 43, 1046–1048.
Maccani, J. E. 1979. Aerobically incubated medium for decarboxylase testing of *Enterobacteriaceae* by replica plating method. J. Clin. Microbiol. 10, 940–942.
MacDonell, M. T. 1983. Rapid estimation of MPN with a hand-held calculator. Can. J. Microbiol. 29, 621–623.
MacDonell, M. T., Singleton, F. L., Roszak, D. B., Hood, M. A., Tison, D. L., and Seidler,

R. J. 1983. Rapid GC mol % screening of primary culture lysates using horizontal slab gel electrophoresis. J. Microbiol. Methods 1, 81–88.

Mackey, B. M., and Derrick, C. M. 1982. A comparison of solid and liquid media for measuring the sensitivity of heat-injured S. typhimurium to selenite and tetrathionate media, and the time needed to recovery resistance. J. Appl. Bacteriol. 53, 233–242.

Mackey, B. M., Derrick, C. M., and Thomas, J. A. 1980. The recovery of sublethally injured Escherichia coli from frozen meat. J. Appl. Bacteriol. 48, 315–324.

Madden, J. M., McCardell, B. A., and Read, R. B., Jr. 1982. Vibrio cholerae in shellfish from U.S. coastal waters. Food Technol. 36 (3) 93–96.

Maerdh, P. A., Larsson, L., and Odham, G. 1981. Head-space gas chromatography as a tool in the identification of anaerobic bacteria and diagnosis of anaerobic infections. Scand. J. Infect. Dis. 26, 14–18.

Mafart, P., Bourgeois, C., Duteurtre, B., and Moll, M. 1978. Use of [^{14}C] lysine to detect microbial contamination in liquid foods. Appl. Environ. Microbiol. 35, 1211–1212.

Magalhaes, M., and Andrade, M. 1983. Efficacy of the Y medium for recovery of Shigella flexneri from stools. Rev. Microbiol. 14, 36–47.

Magee, J. T., Hindmarch, J. M., and Meechan, D. F. 1983. Identification of staphylococci by pyrolysis gas-liquid chromatography. J. Med. Microbiol. 16, 483–495.

Malcolm, S. 1984. A note on the use of the non-central t-distribution in setting numerical microbiological specifications for foods. J. Appl. Bacteriol. 57, 175–177.

Ma-Lin, C. F. A., and Beuchat, L. R. 1980. Recovery of chill-stressed Vibrio paraheamolyticus from oysters with enrichment broths supplemented with magnesium and iron salts. Appl. Environ. Microbiol. 39, 179–185.

Mallmann, W. L., Dawson, L. E., Sultzer, B., and Wright, H. 1958. Studies on microbiological methods for predicting shelf-life of dressed poultry. Food Technol. 12, 122–126.

Mandler, F., and Sfondrini, D. 1977. Evaluation of survival of bacteria on dry swabs and transport systems. Ann. Sclavo. Riv. Microbiol. Immunol. 19, 537–545.

Maness, D. D., Schneider, L. W., Sullivan, G., Gerald, J. Y., and Scholler, J. 1976. Fluorescence behavior of sterigmatocystin. J. Agric. Food Chem. 24, 961–963.

Mangels, J. I., Cox, M. E., and Lindberg, L. H. 1984. Methanol fixation. An alternative to heat fixation of smears before staining. Diagn. Microbiol. Infect. Dis. 2, 129–137.

Marcelis, J. H., Versteeg, H., Mansvelt Beck, H. J., and Vinke, D. 1980. Semielectronic turbidimeter for automated monitoring of bacterial growth in test tubes. Appl. Environ. Microbiol. 39, 281–284.

Marler, L., Allen, S., and Siders, J. 1984. Rapid enzymatic characterization of clinically encountered anaerobic bacteria with the API ZYM system. Eur. J. Clin. Microbiol. 3, 294–300.

Marold, L. M., Freedman, R., Chamberlain, R. E., and Miyashiro, J. J. 1981. New selective agent for isolation of Pseudomonas aeruginosa. Appl. Environ. Microbiol. 41, 977–980.

Marriott, N. G., Garcia, R. A. and Lee, D. R. 1978. Comparison of bacterial swab samples given different storage treatments. J. Food Prot. 41, 897–898.

Marshall, R. T., Hartman, P. A., Cannon, R. Y., Lambeth, L., Richardson, G. H., Spurgeon, K. R., Weddle, D. B., Wingfield, M., and White, C. H. 1978. Group comparative study of VRB-2 agar in the recovery of coliforms from raw milk, ice cream and cottage cheese. J. Food Prot. 41, 544–545.

Martin, W. T., Patton, C. M., Morris, G. K., Potter, M. E., and Puhr, N. D. 1983. Selective enrichment broth medium for isolation of Campylobacter jejuni. J. Clin. Microbiol. 17, 853–855.

4. Standard and Experimental Methods of Identification and Enumeration

Martinez, O. V., and Malinin. 1979. Effect of osmotic stabilizers on radiometric detection of cell wall-damaged bacteria. J. Clin. Microbiol. 10, 657–661.
Martins, S. B., and Shelby, M. J. 1980. Evaluation of a rapid method for the quantitative estimation of coliforms in meat by impedimetric procedures. Appl. Environ. Microbiol. 39, 518–524.
Martins, S. B., Hodapp, S., Dufour, S. W., and Kraeger, S. J. 1982. Evaluation of a rapid impedimetric method for determining the keeping quality of milk. J. Food Prot. 45, 1221–1226.
Marymont, J. H., III, Marymount, J. H., Jr., and Gavan, T. L. 1978. Performance of *Enterobacteriaceae* identification systems: an analysis of college of American Pathologists Survey Data. Am. J. Clin. Pathol. 70, 539–547.
Matches, J. R., and Abeyta, C. 1983. Indicator organisms in fish and shellfish. Food Technol. 37, 114–119.
Matsen, J. M., and Bale, M. J. 1981. Time-motion and cost comparison study of Micro-ID, API 20E, and conventional biochemical testing in identification of *Enterobacteriaceae*. J. Clin. Microbiol. 14, 665–670.
Matsumoto, M., Jinbo, K., Murakami, H., and Haruta, M. 1976. Detection method for injured coliform organisms in milk and dairy products. J. Food Hyg. Soc. Jap. 17, 85–88.
Matsunaga, T., Karube, I., and Suzuki, S. 1979. Electrode system for the determination of microbial populations. Appl. Environ. Microbiol. 37, 117–121.
Matsuyama, T. 1984. Staining of living bacteria with rhodamine 123. FEMS Microbiol. Lett. 21, 153–157.
Mattingly, J. A., and Gehle, W. D. 1984. An improved enzyme immunoassay for the detection of *Salmonella*. J. Food Sci. 49, 807–809.
Mavrommati, C., Xirouchaki, E., Vassiliadis, P., Trichopoulos, D., and Serie, C. 1981. Isolation of salmonellae from chicken carcasses with the use of enrichment in Rappaport-Vassiliadis medium. Med. Vet. Ec. Alfort, Paris. 157, 659–665.
Mavrommati, C., Kalapothaki, V., Trichopoulos, D., Vassiliadis, P., and Serie, C. 1984. Recovery of *Salmonella* from refrigerated preenrichment and refrigerated enrichment media. Int. J. Food Microbiol. 1, 5–11.
Mayhew, J. W., and Gorbach, S. L. 1975. Rapid gas chromatographic technique for presumptive detection of *Clostridium botulinum* in contaminated food. Appl. Microbiol. 29, 297–299.
Mayou, J. 1976. MPN—Most probable number. *In* "Compendium of Methods for the Microbiological Examination of Foods." M. L. Speck (Editor). Am. Public Health Assoc., Washington, D.C.
Mazza, G. 1983. Rapid assay for detection of microorganisms producing DNA-damaging metabolites. Appl. Environ. Microbiol. 45, 1949–1952.
McCarthy, L. R., Mayo, J. B., and Bell, G. 1978. Comparison of a commercial identification kit and conventional biochemical tests used for the identification of enteric gram-negative rods. Am. J. Clin. Pathol. 69, 161–164.
McCoy, J. H. 1961. The safety and cleanliness of waters and foods. J. Appl. Bacteriol. 24, 365–367.
McGibbon, L., Quail, E., and Fricker, C. R. 1984. Isolation of salmonellae using two forms of Rappaport-Vassiliadis medium and brilliant green agar. Int. J. Food Microbiol. 1, 171–177.
McMeekin, T. A., Gibbs, P. A., and Patterson, J. T. 1978. Detection of volatile sulfide-producing bacteria isolated from poultry-processing plants. Appl. Environ. Microbiol. 35, 1216–1218.

Mead, G. C., and Adams, B. W. 1977. A selective medium for the rapid isolation of *pseudomonas* associated with poultry meat spoilage. Br. Poult. Sci. 18, 661–670.

Mead, G. C., DeLeon, L. P., and Adams, B. W. 1981. A study of rapid and simplified confirmatory tests for *Clostridium perfringens.* J. Appl. Bacteriol. 51, 355–361.

Meadows, P. S., Anderson, J. G., and Patel, K. 1980. Synergistic inhibition of *Escherichia coli* growth and gas production in selective media. FEMS Microbiol. Lett. 8, 215–219.

Mehlman, I. J., Aulisio, C. C. G., and Sanders, A. C. 1978. Problems in the recovery and identification of *Yersinia* from food. J. Assoc. Off. Anal. Chem. 61, 761–771.

Mehlman, I. J., and Romero, A. 1982a. Enteropathogenic *Escherichia coli:* Methods for recovery from foods. Food Technol. 36 (3), 73–79.

Mehlman, I. J., and Romero, A. 1982b. Improved growth medium for *Campylobacter* species. Appl. Environ. Microbiol. 43, 615–618.

Metcalf, T. G. 1978. Indicators of viruses in shellfish. In "Indicators of Viruses in Water and Food." G. Berg (Editor). John Wiley and Sons Ltd., Baffins Lane, Chichester, W. Sussex, U.K.

Metcalf, T. G., Eckerson, D., and Moulton, E. 1980a. A method for recovery of viruses from oysters and hard and soft shell clams. J. Food Prot. 43, 89–90.

Metcalf, T. G., Moulton, E., and Eckerson, D. 1980b. Improved method and test strategy for recovery of enteric viruses from shellfish. Appl. Environ. Microbiol. 39, 141–152.

Michels, M. J. M., and Kagei, R. F. 1983. Egg yolk trypticase soy agar for the enumeration of heat-damaged spores of *Clostridium sporogenes.* J. Appl. Bacteriol. 55, 203–208.

Miller, L. F., Gress, H. S., and Jangaard, N. O. 1977. An ATP bioluminescence method for the quantification of viable yeast for fermentor pitching. J. Am. Brew. Chem. 36, 59–62.

Miller, M. L., and Koburger, J. A. 1985. *Plesiomonas shigelloides:* an opportunistic food and waterborne pathogen. J. Food Protection 48, 449–457.

Miller, M. W. 1979. Yeasts in food spoilage: an update. Food Technol. 33, 76–80.

Minor, L., Coynault, C., and Geuso, N. 1977. Positivity of ONPG test versus the presence of β-galactosidase in *Enterobacteriaceae* and other gram-negative bacilli. Ann. Microbiol. 128B (1), 35–43.

Mintzer-Morgenstern, L., and Katzenelson, E. 1982. A simple method for isolation of coagulase-positive staphylococci in a single step. J. Food Prot. 45, 218–222.

Miskimin, D. K., Berkowitz, K. A., Solberg, M., Riha, W. E., Jr., Franke, W. C., Buchanan, R. L., and O'Leary, V. 1976. Relationships between indicator organisms and specific pathogens in potentially hazardous foods. J. Food Sci. 41, 1001–1006.

Mislivec, P. B., and Bruce, V. R. 1977. Direct plating versus dilution plating in qualitatively determining the mold flora of dried beans and soybeans. J. Assoc. Off. Anal. Chem. 60, 741–743.

Moats, W. A. 1978. Comparison of four agar plating media with and without added novobiocin for isolation of salmonellae from beef and deboned poultry meat. Appl. Environ. Microbiol. 36, 747–751.

Moermans, R. J., and Bossuyt, R. 1978. A note on inverse transformation: an application on cell count numbers. Milchwissenschaft 33, 497–499.

Monford, J., and Thatcher, F. S. 1961. Comparison of four methods of isolating salmonellae from foods, and elaboration of a preferred procedure. J. Food Sci. 26, 510–517.

Monk, P. R. 1978. Microbial calorimetry as an analytical method. Process Biochem. 13 (12), 4–5.

Monk, P. R. 1979. Thermograms of *Streptococcus thermophilus bulgaricus* in single and mixed culture in milk medium. J. Dairy Res. 46, 485–496.

Monk, P. R., and Costello, P. J. 1983. Measurement of yeast growth in grape juice with a fibre optic nephelometer. J. Gen. Appl. Microbiol. 29, 467–475.
Monk, P. R., and Wadso, I. 1975. The use of microcalorimetry for bacterial classification. J. Appl. Bacteriol. 38, 71–74.
Moore, H. B. 1981. Rapid methods in microbiology: IV. Presumptive and rapid methods in anaerobic bacteriology. Am. J. Med. Technol. 47, 705–712.
Mor, J. R., 1976. Flow microcalorimetry in microbiology. In "Rapid Methods and Automation in Microbiology." M. M. Johnston and S. W. B. Newsom (Editors). Learned Information Ltd., Oxford, N.Y.
Morimoto, T., Itoh, H., and Chibata, I. 1980. An automated recording incubator for tube culture of microorganisms. Enzyme Microb. Technol. 2, 194–200.
Morring, K. L., Sorenson, W. G., and Attfield, M. D. 1983. Sampling for airborne fungi: A statistical comparison of media. Am. Ind. Hyg. Assoc. J. 44, 662–664.
Morris, G. K., and Feely, J. C. 1976. *Yersinia enterocolitica*: a review of its role in food hygiene. Bull. WHO 54, 79–85.
Morris, M. J., Young, V. M., and Moody, M. R. 1978. Evaluation of a multitest system for identification of saccharolytic pseudomonads. Am. J. Clin. Pathol. 69, 41–47.
Morris, S. C., and Nicholls, P. J. 1978. An evaluation of optical density to estimate fungal spore concentrations in water suspensions. Phytopathology 68, 1240–1242.
Morrison, S. M., Bartley, T. D., Quan, T. J., and Collins, M. T. 1982. Membrane filter technique for the isolation of *Yersinia enterocoliticia*. Appl. Environ. Microbiol. 43, 829–834.
Moss, C. W., Dees, S. B., and Guerrant, G. O. 1980. Gas-liquid chromatography of bacterial fatty acids with a fused-silica capillary column. J. Clin. Microbiol. 12, 127–130.
Moss, C. W., Shinoda, T., and Samuels, J. W. 1982. Determination of cellular fatty acid compositions of various yeasts by gas-liquid chromatography. J. Clin. Microbiol. 16, 1073–1079.
Mossel, D. A. A. 1979. The microbial associations of foods of animal origin. Arch. Lebensmittelhyg. 30, 82–84.
Mossel, D. A. A., and Eelderink, I. 1979. A simple test for the assessment of the suitability of pork plasma for incorporation into a Baird-Parker base medium for the enumeration of *Staphylococcus aureus* in foods and other specimens. Lab. Pract. 28, 623.
Mossel, D. A. A., and Harrewijn, G. A. 1977. Choice of indicator organisms for the assessment of the sanitary quality of foods and feeds. In "Global Impacts of Applied Microbiology. IVth International Conference." J. S., Furato (Editor). Sociedade Brasileira de Microbiologia, Revista de Microbiologia, Sao Paulo, Brazil.
Mossel, D. A. A., and Shennan, J. L. 1976. Microorganisms in dried foods: their significance, limitation and enumeration. J. Food Technol. 11, 205–220.
Mossel, D. A. A., Visser, M., and Cornelissen, A. M. R. 1963. The examination of foods for *Enterobacteriaceae* using a test of the type generally adopted for the detection of salmonellae. J. Appl. Bacteriol. 26, 444–452.
Mossel, D. A. A., Harrewyn, G. A., and Nesselrooy-van Zadelhoff, C. F. M. 1974. Standardization of the selective inhibitory effect of surface active compounds used in media for the detection of *Enterobacteriaceae* in foods and water. Health Lab. Sci. 11, 260–267.
Mossel, D. A. A., Bijker, P. G. H., and Eelderink, I. 1978a. Lancefield's group D streptococci in foods and water: their significance, enumeration, and control. Arch. Lebensmittelhyg. 29, 121–127.
Mossel, D. A. A., Eelderink, I., Koopmans, M., and Van Rossem, F. 1978b. Optimalisation

of a MacConkey-type medium for the enumeration of *Enterobacteriaceae*. Lab Pract. 27, 1049–1050.

Mossel, D. A. A., Richard, N., Gayral, J. P., and Brissuel, C. 1983. A comparative study of tests to be used in the preliminary taxonomic grouping of bacteria isolated from foods, drinking water, and medicinal preparations. Sci. Aliments. 3, 91–115.

Mugg, P. 1983. A rapid hippurate hydrolysis test for the presumptive identification of group B streptococci. Pathology 15, 251–252.

Mulindwa, D. K., and Pietzsch, O. 1979. Studies on the influence of competitive *Enterobacteriaceae* flora on *Salmonella* isolation. Zentralbl. Bakteriol. Parasitenkd. Infektionskr. Hyg., Abt. A 243, 336–348.

Mullan, M. A., and Walker, A. L. 1979. An agar medium and simple streaking technique for the differentiation of lactic streptococci. Dairy Ind. Int. 44(6), 13, 17.

Muller, H. E. 1979. Comparative tests of various liquid media for the preenrichment of salmonellae from milk powder. Zentralbl. Bakteriol. Parasitenkd. Infektionskr, Hyg. I Abt. B 168, 367–376.

Mundt, J. O. 1976. Streptococci in dried and frozen foods. J. Milk Food Technol. 39, 413–416.

Munford, R. S., and Hall, C. L. 1979. Radioimmunoassay for gram-negative bacterial lipopolysaccharide O antigens. Influence of antigen solubility. Infect. Immun. 26, 42–48.

Munson, T. E., Schrade, J. P., Bisciello, N. B., Jr., Fantasia, L. D., Hartung, W. H., and O'Connor, J. J. 1976. Evaluation of an automated fluorescent antibody procedure for detection of Salmonella in foods and feeds. Appl. Environ. Microbiol. 31, 514–521.

Murdock, D. I., and Hatcher, W. S., Jr. 1976. Plate loop method for determining total viable count of orange juice. J. Milk Food Technol. 39, 470–473.

Mukerjee, S. 1979. Principles and practice of typing *Vibrio cholerae*. In "Methods in Microbiology," Vol. 12. T. Bergan and J. R. Norris (Editors). Academic Press Inc., Ltd., Oval Road, London NW1 7DX, U.K.

Naguib, K., Park, D. L., Naguib, M. M., and Pohland, A. E. 1983. (Editors). "Proceedings of the International Symposium on Mycotoxins." National Research Center, Cairo, Egypt, 605 pp.

Naik, H. S., and Duncan, C. L. 1978. Detection of *Clostridium perfringens* enterotoxins in human fecal samples and anti-enterotoxin in sera. J. Clin. Microbiol. 7, 337–341.

Nair, G., Chowdhury, S., Das, P., Pal, S., and Pal, S. C. 1984. Improved preservation medium for *Campylobacter jejuni*. J. Clin. Microbiol. 19, 298–299.

Nambudripad, V. K. N. 1978. Development of a rapid coliform test for pasteurized milk. Indian J. Dairy Sci. 31, 282–284.

National Academy of Sciences, 1985. "An Evaluation of the Role of Microbiological Criteria for Foods and Food Ingredients." National Academy Press.

Nelson, F. E. 1943. Factors which influence the growth of heat-treated bacteria. I. A comparison of four agar media. J. Bacteriol. 45, 395–403.

Nelson, F. E. 1944. Factors which influence the growth of heat-treated bacteria. II. Further studies on media. J. Bacteriol. 48, 473–477.

Nelson, W. H. (Editor). 1985. Instrumental Methods for Rapid Microbiological Analysis. VCH Publishers, Deerfield Beach, Florida.

Newsom, S. W. B. 1978. A review of automation and rapid methods in microbiology. Med. Lab. Sci. 35, 215–222.

Newton, K. G. 1979. Value of coliform tests for assessing meat quality. J. Appl. Bacteriol. 47, 303–307.

Newton, S. B. 1978. Spiral system cuts cost of microbial counts by 50 percent. Food Prod. Dev. 12(11), 18, 22.

Ng, L. K., and Stiles, M. E. 1978. *Enterobacteriaceae* in ground meats. Can. J. Microbiol. 24, 1574–1582.

Ng, L. K., Stiles, M. E., and Taylor, D. E. 1985a. Comparison of basal media for culturing *Campylobacter jejuni* and *C. coli*. J. Clin. Microbiol. 21, 226–230.

Ng, L. K., Taylor, D. E., and Stiles, M. E. 1985b. Estimation of *Campylobacter* spp. in broth culture by bioluminescence assay of ATP. Appl. Environ. Microbiol. 49, 730–731.

Nishibuchi, M., Roberts, N. C., Bradford, Jr., H. B., and Seidler, R. J. 1983. Broth medium for enrichment of *Vibrio fluvialis* from the environment. Appl. Environ. Microbiol. 46, 425–429.

Niskanen, A., and Aalto, M. 1978. Comparison of selective media for coagulase-positive enterotoxigenic *Staphylococcus aureus*. Appl. Environ. Microbiol. 35, 1233–1236.

Niskanen, A., and Lindroth, S. 1976. Preparation of labeled Staphylococcal enterotoxin A with high specific activity. Appl. Environ. Microbiol. 32, 735–740.

Niskanen, A., Kitutamo, T., Raisanen, S., and Raevuori, M. 1978. Determination of fatty acid compositions of *Bacillus cereus* and related bacteria: a rapid gas chromatographic method using a glass capillary column. Appl. Environ. Microbiol. 35, 453–455.

Niven, C. F., Jr., Jeffrey, M. B., and Corlett, D. A., Jr. 1981. Differential plating medium for quantitative detection of histamine-producing bacteria. Appl. Environ. Microbiol. 41, 321–322.

Norberg, P. 1981. Enteropathogenic bacteria in frozen chicken. Appl. Environ. Microbiol. 42, 32–34.

Nortje, G. L., Swanepoel, E., Naude, R. T., Holzapfel, W. H., and Stevn, P. L. 1982. Evaluation of three carcass surface microbial sampling techniques. J. Food Prot. 45, 1016–1017.

Notermans, S., Hindle, V., and Kampelmacher, E. H. 1976. Comparison of cotton swabs versus alginate swab sampling method in the bacteriological examination of broiler chickens. J. Hyg. 77, 205–210.

Notermans, S., Van Leusden, F. M., and Van Schothorst, M. 1977. Suitability of different bacterial groups for determining fecal contamination during post-scalding stages in the processing of broiler chickens. J. Appl. Bacteriol. 43, 383–389.

Notermans, S., Dufrenne, J., and Kozaki, S. 1979. Enzyme-linked immunosorbent assay for detection of *Clostridium botulinum* Type-E toxin. Appl. Environ. Microbiol. 37, 1173–1175.

Nowotny, A. (Editor). 1966. Molecular Biology of Gram-Negative Bacterial Lipopolysaccharides. Annals of the New York Academy of Sciences. 133, 277–786.

Oberhofer, T. R. 1979. Comparison of the API 20E and Oxi/Ferm systems in identification of nonfermentative and oxidase-positive fermentative bacteria. J. Clin. Microbiol. 9, 220–226.

Oberhofer, T. R. 1983. Use of the API 20E Oxi-Ferm, and Minitek systems to identify nonfermentative and oxidase-positive fermentative bacteria: Seven years of experience. Diagn. Microbiol. Infect. Dis. 1, 241–256.

Oberhofer, T. R., Rowen, J. W., Cunningham, G. F., and Higbee, J. W. 1977. Evaluation of the Oxi/Ferm tube system with selected gram-negative bacteria. J. Clin. Microbiol. 6, 559–566.

Oblinger, J. L. 1983. Microbiology of hot-boned beef. Food Technol. 37, 86–94.

O'Donnell, A. G., Norris, J. R., Berkeley, R. C. W., Claus, D., Kaneko, T., Logan, N. A., and Nozaki, R. 1980. Characterization of *Bacillus subtilis*, *Bacillus pumilus*, *Bacillus licheniformis*, and *Bacillus amyloliquefaciens* by pyrolysis gas-liquid chromatography, deoxyribonucleic acid-deoxyribonucleic acid hybridization, biochemical tests, and API systems. Int. J. Syst. Bacteriol. 30, 448–459.

Oeding, P. 1979. Genus *Staphylococcus*. *In* "Methods in Microbiology." Vol. 12. T. Bergan and J. R. Norris (Editors). Academic Press Inc., Ltd., Oval Road, London NW1 7DX, U.K.

Olgaard, K. 1977. Determination of relative bacterial levels on carcasses and meats—a new quick method. J. Appl. Bacteriol. 42, 321–329.

Oliver, D. O., and Salo, R. J. 1978. Indicators of viruses in foods preserved by heat. *In* "Indicators of Viruses in Water and Food." G. Berg (Editor). John Wiley and Sons Ltd., Baffino Lane, Chichester, W. Sussex, U.K.

Olsen, R. L., and Richardson, G. H. 1980. A flooded plate loop count procedure. J. Food Prot. 43, 534–535.

Olson, N. A. 1980. The effects of milling on mold counts in tomato products. Food Technol. 34, 50–56.

Olsvik, O., and Berdal, B. P. 1982. Demonstration of *E. coli* heat-labile enterotoxin using bacterial cell sonicates. Acta Pathol. Microbiol. Immunol. Scand. Ser. B, 90, 319–321.

Olsvik, O., Granum, P. E., and Berdal, B. P. 1982. Detection of *Clostridium perfringens* type A enterotoxin by ELISA. Acta Pathol. Microbiol. Immunol. Scand. Ser. B, 90B, 445–447.

Oosterom, J., Vereijken, M. J. G. M., and Engels, G. B. 1981. *Campylobacter* isolation. Vet. Q. 3, 104.

Ordal, Z. J., Iandola, J. J., Ray, B., and Sinskey, A. G. 1976. Detection and enumeration of injured microorganisms. *In* "Compendium of Methods for the Microbiological Examination of Foods." M. L. Speck (Editor). Am. Public Health Assoc., Washington, D.C.

Ordonez, J. A. 1979. Random number sampling method for estimation of lactic acid bacteria. J. Appl. Bacteriol. 46, 351–353.

Orskov F., and Orskov, I. 1979. Serotyping of *Enterobacteriaceae*, with special emphasis on K antigen determination. *In* "Methods in Microbiology." Vol. II. T. Bergman and J. R. Norris (Editors). Academic Press Inc., Ltd., Oval Road, London, NW1 7DX, U.K.

Osborne, B. G. 1982. Mycotoxins and the cereals industry—A review. J. Food Technol. 17, 1–9.

O'Toole, D. K. 1983a. Weighing technique for determining bacterial dry mass based on rate of moisture uptake. Appl. Environ. Microbiol. 46, 506–508.

O'Toole, D. K. 1983b. Methods for the direct and indirect assessment of the bacterial content of milk. J. Appl. Bacteriol. 55, 187–201.

O'Toole, D. K. 1983c. A toluidine blue-membrane filter method for the quantitative staining of bacteria. Stain Technol. 58, 291–298.

O'Toole, D. K. 1983d. The toluidine blue-membrane filter method: Absorption spectra of toluidine blue stained bacterial cells and the relationship between absorbance and dry mass of bacteria. Stain Technol. 58, 357–364.

Ottaviani, F., and Aliani, A. 1977. Biochemical typing of salmonellae by a miniaturized system (API-50E). Ann. Sclavo. Riv. Microbiol. Immun. 19, 626–632.

Otte, I., Tolle, A., and Suhren, G. 1979. Microbial analysis of milk and milk products. 1. Cultivation of microflora and isolation of colonies to be identified. Milchwissenschaft 34, 85–88.

Ozsan, K., and Mercangoz, F. 1980. New media for the isolation of *Vibrio cholerae*. Zentralbl. Bakteriol. Mikrobiol. Hyg., I Abt. A 247, 71–73.

Pacova, Z., and Kocur, M. 1978. Phosphatase activity of aerobic and facultative anaerobic bacteria. Zentralbl. Bakteriol. Parasitenkd. Infektionskr. Hyg., I Abt. A 24, 481–487.

Pagel, J. E., and Hardy, J. M. 1980. Comparison of selective media for the enumeration and identification of fecal streptococci from natural sources. Can. J. Microbiol. 26, 1320–1327.

Pagel, J. E., Qureshi, A. A., Young, D. M., and Vlassoff, L. T. 1982. Comparison of four membrane filter methods and fecal coliform enumeration. Appl. Environ. Microbiol. 43, 787–793.

Pal, T., Pasca, A. S., Emoedy, L., Voeroes, S., and Selley, E. 1985. Modified enzyme-linked immunosorbent assay for detecting enteroinvasive *Escherichia coli* and virulent *Shigella* strains. J. Clin. Microbiol. 21, 415–418.

Palmer, W. J. 1981. (Editor). Rapid and Automated Methods in Microbiology and Immunology: a Bibliography 1976–1980. Information Retrieval Ltd., 1, Falconberg Court, London, W1V 5FG, U.K.

Palumbo, S. A. 1984. Heat injury and repair in *Campylobacter jejuni*. Appl. Environ. Microbiol. 48, 477–480.

Palumbo, S. A. 1986. *Campylobacter jejuni* in foods: Its occurrence, isolation from foods, and injury. J. Food Protection 49, 161–166.

Paradis, D. C., and Stiles, M. E. 1978. A study of microbial quality of vacuum-packaged, sliced bologna. J. Food Prot. 41, 811–815.

Paramasinan, C. N., Subramanian, S., Shanmugasundaram, N., and Sharma, K. B. 1979. Biological properties of *Salmonella typhi* used for further categorizing major phage types. Indian J. Med. Res. 69, 913–918.

Parisi, J. T. 1985. Coagulase-negative staphylococci and the epidemiological typing of *Staphylococcus epidermidis*. Microbiol Rev. 49, 126–139.

Park, C. E., El Derea, H. B., and Rayman, M. K. 1978. Evaluation of staphylococcal thermonuclease (TNase) assay as a means of screening foods for growth of staphylococci and possible enterotoxin production. Can. J. Microbiol. 24, 1135–1139.

Park, C. E., El Derea, H. B., and Rayman, M. K. 1979. Effect of non-fat dry milk on recovery of staphyloccoccal thermonuclease from foods. Can. J. Microbiol. 25, 44–46.

Park, C. E., Landgraf, M., and Stankiewicz, Z. K. 1981. A new sensitive and rapid procedure for the isolation of *Yersinia enterocolitica* from food, particularly from low calory food such as vegetables. In "Psychrotrophic Microorganisms in Spoilage and Pathogenicity." T. A. Roberts, G. Hoggs, J. H. B. Christian, and N. Skovgaard (eds.), Academic Press, P. 425–429.

Park, C. E., Stankiewicz, Z. K., Lovett, J., Hunt, J., and Francis, D. W. 1983. Effect of temperature, duration of incubation, and pH of enrichment culture on the recovery of *C. jejuni* from eviscerated market chickens. Can. J. Microbiol. 29, 803–806.

Patterson, J. T. 1972. Microbiological sampling of poultry carcasses. J. Appl. Bacteriol. 35, 569–575.

Patterson, R., and Damoglou, A. P. 1983. A note on the enumeration of low numbers of *Staphylococcus aureus* using a modified MPN method. Lab. Pract. 32, 73–74.

Paul, V. K., Balaya, S., Ghai, O. P., Dimah, C. G., Shriniwas, R. A., and Mohapatra, L. N. 1981. Comparison of response of Vero cells and rabbit ideal loops to the heat-labile enterotoxin of *Escherichia coli*. Indian J. Med. Res. 74, 799–804.

Pavlovskis, O. R., Iglewski, B. H., and Pollack, M. 1978. Mechanism of action of *Pseudomonas aeruginosa* exotoxin A in experimental mouse infections: Adenosine diphosphate ribosylation of Elongation Factor 2. Infection and Immunity 19, 29–33.

Peabody, F. R. 1963. Microbial indexes of food quality: The coliform group. In "Microbiological Quality of Foods." L. W. Slanetz, C. O. Chichester, A. R. Gaufin and Z. J. Ordal (Editors), Academic Press, New York, N.Y.

Peberdy, M. F., and Fryer, T. F. 1976. Improved selective media for the enumeration of propionibacteria from cheese. New Zeal. J. Dairy Sci. Technol. 11, 10–15.

Peeler, J. T., Gilchrist, J. E., Donnelly, C. B., and Campbell, J. E. 1977. A collaborative

study of the spiral plate method for examining milk samples. J. Food Prot. 40, 462–464.
Peeler, J. T., Leslie, J. E., Danielson, J. W., and Mosser, J. W. 1982. Replicate counting errors by analysts and bacterial colony counters. J. Food Prot. 45, 239–240.
Pennell, D. R., Rott-Petri, J. A., and Kurzynski, T. A. 1984. Evaluation of three commercial agglutination tests for the identification of *S. aureus*. J. Clin. Microbiol. 20, 614–617.
Perlman, D. 1978. (Editor). "Advances in Applied Microbiology." Academic Press, New York, N.Y.
Perry, B. F., Beezer, A. E., and Miles, R. J. 1980. Bioassay of phenol disinfectants by flow microcalorimetry. Microbios 29, 81–87.
Perry, B. F., Beezer, A. E., and Miles, R. J. 1983. Characterization of commercial yeast strains by flow microcalorimetry. J. Appl. Bacteriol. 54, 183–189.
Perutkova, S. 1981. *Salmonella* in meat products and in humans. In "Psychrotrophic Microorganisms in Spoilage and Pathogenicity." T. A. Roberts, G. Hobbs, J. H. B. Christian, and N. Skovgaard (eds.), Academic Press, p. 249–280.
Pestka, J. J., Gaur, P. K., and Chu, F. S. 1980. Quantitation of aflatoxin B_1 and Aflatoxin B_2 antibody by an enzyme-linked immunosorbent microassay. Appl. Environ. Microbiol. 40, 1027–1031.
Peterkin, P. I., and Sharpe, A. N. 1980. Membrane filtration of dairy products for microbiological analysis. 1980. Appl. Environ. Microbiol. 39, 1138–1143.
Peterkin, P. I., and Sharpe, A. N. 1981. Filtering out food debris before microbiological analysis. Appl. Environ. Microbiol. 42, 63–65.
Peterkin, P. I., and Sharpe, A. N. 1984. Rapid enumeration of *Staphylococcus aureus* in foods by direct demonstration of enterotoxigenic colonies on membrane filters by enzyme immunoassay. Appl. Environ. Microbiol. 47, 1047–1053.
Peterson, E. H., and Hsu, E. J. 1978. Rapid detection of selected gram-negative bacteria by aminopeptidase profiles. J. Food Sci. 43, 1853–1856.
Pettipher, G. L. 1982. Use of membrane filtration for assessing the hygienic quality of milk and milk products. J. Soc. Dairy Technol. 35, 59–63.
Pettipher, G. L., and Rodrigues, U. M. 1982. Rapid enumeration of microorganisms in foods by the direct epifluorescent filter technique. Appl. Environ. Microbiol. 44, 809–813.
Pettipher, G. L., Mansell, R., McKinnon, C. H., and Cousins, C. M. 1980. Rapid membrane filtration-epifluorescent microscopy technique for direct enumeration of bacteria in raw milk. Appl. Environ. Microbiol., 29, 423–429.
Pettipher, G. L., Fulford, R. J., and Mabbitt, L. A. 1983. Collaborative trial of the direct epifluorescent filter technique (DEFT), a rapid method for counting bacteria in milk. J. Appl. Bacteriol., 54, 177–182.
Pezzlo, M. 1981. Rapid methods in microbiology: III. Rapid methods for the identification of gram-negative organisms. Am. J. Med. Technol. 47, 705–712.
Phillips, S. B., and Amsterdam, D. 1977. API computer profiles: correlation of API 20E with API 10S. J. Clin. Microbiol. 6, 645–646.
Pietzsch, O., and Mulindwa, H. K. D. 1978. Special aspects concerning the pre-enrichment of salmonellae. Arch. Lebensmittelhyg. 29, 145–146.
Pitt, J. I., Hocking, A. D., and Glenn, D. R. 1983. An improved medium for the detection of *Aspergillus flavus* and *A. parasiticus*. J. Appl. Bacteriol. 54, 109–124.
Poelma, P. L., Romero, A., and Andrews, W. H. 1978. Comparative accuracy of five biochemical systems for identifying *Salmonella* and related food-borne bacteria: collaborative study. J. Assoc. Off. Anal. Chem. 6, 1043–1049.

Poelma, P. L., Andrews, W. H., and Wilson, C. R. 1981. Comparison of methods for the isolation of *Salmonella* species from lactic casein. J. Food Sci. 46, 804–809.
Polvino, D. A., and Bernard, D. T. 1982. Media comparison for the enumeration and recovery of *Clostridium sporogenes* P.A. 3679 spores. J. Food Sci. 47, 579.
Pons, W. A., Jr. 1976. Resolution of aflatoxins B_1, B_2, and G_2 by high pressure liquid chromatography. J. AOAC 59, 101–105.
Poole, R. K. 1982. Rapid estimates of sizes of microorganisms with the Coulter Nano-sizer. Microbios Lett. 19, 109–117.
Powell, J. C., Moore, A. R., and Gow, J. A. 1979. Comparison of EC broth and medium A-1 for the recovery of *Escherichia coli* from frozen shucked snow crab. Appl Environ. Microbiol. 37, 836–840.
Presswood, W. G., and Strong, D. K. 1978. Modification of M-FC medium by eliminating rosolic acid. Appl. Environ. Microbiol. 36, 90–94.
Previte, J. J. 1972. Radiometric detection of some food-borne bacteria. Appl. Microbiol. 24, 535–539.
Previte, J. J., Rudenauer, P., and Rowley, D. 1977. Development of a specific radiometric coliform assay. Abstr. Ann. Meet. Am. Soc. Microbiol., p. 261.
Pronovost, A. D., Baumgarten, A., and Andiman, W. A. 1982. Chemiluminescent immunoenzymatic assay for rapid diagnosis of viral infections. J. Clin. Microbiol. 16, 345–349.
Prpic, J. K., Robins-Browne, R. M., and Davey, R. B. 1983. Differentiation between virulent and avirulent *Yersinia enterocolitica* isolates by using Congo red agar. J. Clin. Microbiol. 18, 486–490.
Puckey, D. J., Norris, J. R., and Cutteridge, C. S. 1980. Discrimination of microorganisms using direct probe mass spectrometry. J. Gen. Microbiol. 118, 535–538.
Pugsley, A. P., and Evison, L. M. 1975. A fluorescent antibody technique for the enumeration of faecal streptococci in water. J. Appl. Bacteriol. 38, 63–65.
Pusch, D. J., Busta, F. F., Moats, W. A., and Schulze, A. E. 1976. Direct microscopic count. *In* "Compendium of Methods for the Microbiological Examination of Foods." M. L. Speck (Editor). Am. Public Health Assoc., Washington, D.C.
Qadri, S. M. H., and Nichols, C. W. 1978. Evaluation of a commercial multitest system for identification of yeasts. Am. J. Med. Technol. 44, 368–372.
Qadri, S. M. H., deSilva, M. J., Qadri, S. G. M., and Villarreal, A. 1979. Presumptive identification of enterococci from other D streptococci by a rapid sodium chloride tolerance test. Med. Microbiol. Immunol. 167, 197–203.
Qadri, S. M. H. deSilva, M. I., and Zubairi, S. 1980. Rapid test for determination of esculin hydrolysis. J. Clin. Microbiol. 12, 472–474.
Rale, V. B., and Vakil, J. R. 1984. A note on an improved molybdate agar for the selective isolation of yeasts from tropical fruits. J. Appl. Bacteriol. 56, 409–413.
Ramia, S., Neter, E., and Brenner, D. J. 1982. Production of enterobacterial common antigen as an aid to classification of newly identified species of the families Enterobacteriaceae and Vibrionaceae. Int. J. Syst. Bacteriol. 32, 395–398.
Ramsey, W. S., Nowlan, E. D., Simpson, L. B., Messing, R. A., and Takequchi, M. M. 1980. Applications of fluorophore-containing microbial growth media. Appl. Environ. Microbiol. 29, 373–375.
Rapp, M., and Ihle, P. 1979. A new method for the detection of coliform bacteria with a reduced incubation time. Milchwissenschaft 34, 471–474.
Rappold, H., and Bolderdijk, R. F. 1979. Modified lysine iron agar for isolation of *Salmonella* from food. Appl. Environ. Microbiol. 38, 162–163.

Rappold, H., Bolderdijk, R. F., and DeSmedt, J. M. 1984. Rapid cultural method to detect *Salmonella* in foods. J. Food Protection 47, 46–48.

Ray, B. 1986. Impact of bacterial injury and repair in food microbiology: Its past, present, and future. J. Food Protection 49, 651–655.

Ray, B., and Johnson, C. 1984. Survival and growth of freeze-stressed *Campylobacter jejuni* cells in selective media. J. Food Saf. 6, 183–185.

Ray, B., and Speck, M. L. 1973. Freeze-injury in bacteria. C.R.C. Critical Rev., Clinical Laboratory Sciences, pp. 161–213.

Ray, B., and Speck, M. L. 1978. Plating procedure for the enumeration of coliforms from dairy products. Appl. Environ. Microbiol. 35, 820–822.

Ray, B., Hawkins, S. M., and Hackney, C. R. 1978. Method for the detection of injured *Vibrio Parahaemolyticus* in seafoods. Appl. Environ. Microbiol. 35, 1121–1127.

Rayman, M. K., and Aris, B. 1981. The Anderson-Baird-Parker direct plating method versus the most probable number procedure for enumeration *Escherichia coli* in meats. Can. J. Microbiol. 27, 147–149.

Rayman, M. K., Devoyed, J. J., Purvis, U., Kusch, D., Lanier, J., Gilbert, R. J., Till, D. G., and Jarvis, G. A. 1978. An international comparative study of four media for the enumeration of *Staphylococcus aureus* in foods. Can. J. Microbiol. 24, 274–281.

Rayman, M. K., Jarvis, G. A., Davidson, C. M., Long, S. Allen, J. M., Tong, T., Dodsworth, P., McLaughlin, S., Greenberg, S., Shaw, B. G., Beckers, H. J., Quist, S., Noltingham, P. M., and Stewart, B. J. 1979. An international comparative study of the MPN procedure and the Anderson-Baird-Parker direct plating method for the enumeration of *Escherichia coli* biotype 1 in raw meats. Can. J. Microbiol. 25, 1321–1327.

Reber, C. L., and Marshall, R. T. 1982. Comparison of VRB and VRB-2 agars for recovery of stressed coliforms from stored acidified half-and-half. J. Food Prot. 45, 584–585.

Reinbold, G. W. 1983. Indicator organisms in dairy products. Food Technol. 37, 111–113.

Reiner, E. 1967. Studies on differentiation of microorganisms by pyrolysis-gas-liquid chromatography. J. Gas Chromatogr. 5, 65–70.

Reiner, E., and Bayer, F. L. 1978. Botulism: a pyrolysis-gas-liquid chromatographic study. J. Chromatogr. Sci. 16, 623–629.

Reisner, G. S. 1978. Detection of bacterial nitrite production from nitrate by a nitrate-starch-iodide agar medium. Appl. Environ. Microbiol. 36, 384–385.

Restaino, L., Jeter, W. S., and Hill, W. M. 1980. Thermal injury of *Yersinia enterocolitica*. Appl. Environ. Microbiol. 40, 939–949.

Restaino, L., Komatsu, K. K., and Syracuse, M. J. 1982. A note on novobiocin in XLD and HE agars: The optimum levels required in two commercial sources of media to improve isolation of salmonellas. J. Appl. Bacteriol. 53, 285–288.

Reusse, U. 1978. Methods for the detection of salmonellae and *Enterobacteriaceae*. Arch. Lebensmittelhyg. 29, 138–145.

Reusse, U., Meyer, A., and Tillack, J. 1976. Methods for isolation of salmonellae from frozen poultry. Arch. Lebensmittelhyg. 27, 98–100.

Reuter, G. 1978. Selective cultivation of "enterococci" from food of animal origin. Arch. Lebensmittelhyg. 29, 128–131.

Rey, C. R., Halaby, G. A., Lovgren, E. V., and Wright, T. A. 1982. Evaluation of a membrane filter test kit for monitoring bacterial counts in cannery cooling waters. J. Food Prot. 45, 1087–1090.

Rey, C. R., Halaby, G. A., Reed, T. J., and Lovgren, E. V. 1985. Simple method of sample preparation for bacterial counts in quality control of frozen vegetables. J. Food Protection 48, 210–214.

Richard, C. 1977. Tetrathionate-reductase (TTR) in gram-negative bacteria: diagnostic and epidemiological value. Bull. Inst. Pasteur. 75, 369–382.

4. Standard and Experimental Methods of Identification and Enumeration

Richard, C. 1978. Research techniques of enzymes used in the diagnosis of gram-negative bacteria. Ann. Biol. Clin. 35, 407–424.

Richard, J. 1980. Effect of several shaking methods on plate counts of raw-milk bacteria. Lait 60, 211–225.

Richard, J., Andersen, H. M., and Gratadoux, J. J. 1983. Rapid method for selecting appropriate solid media for the enumeration of aerobic mircoorganisms. J. Appl. Bacteriol. 54, 329–334.

Richards, G. P. 1978. Comparative study of methods for the enumeration of total and fecal coliforms in the eastern oyster, *Crassostrea virginica*. Appl. Environ. Microbiol. 36, 975–978.

Richards, G. P. 1985. Outbreaks of shellfish-associated enteric virus illness in the United States: Requisite for development of viral guidelines. J. Food Protection 48, 815–823.

Richter, E. R., and Banwart, G. J. 1982. Evaluation of an X-ray microprobe technique as a possible aid to detect salmonellae. Can. J. Microbiol. 28, 650–653.

Richter, E. R., Burns, M. C., Banwart, G. J., and Rheins, M. S. 1982. Gas-liquid chromatographic differentiation between *Salmonella gallinarum* and *Salmonella pullorum*. J. Food Prot. 45, 919–922.

Richter, R. 1979. Psychrotrophic bacteria and shelf-life. Am. Dairy Rev. 41 (7) 50J.

Rigby, C. E. 1982. Most probable number cultures for assessing *Salmonella* contamination of eviscerated broiler carcasses. Can. J. Comp. Med. 46, 279–282.

Rigby, C. E. 1984. ELISA for detection of *Salmonella* lipopolysaccharide in poultry specimens. Appl. Environ. Microbiol. 47, 1327–1330.

Rigby, C. E., and Pettit, J. R. 1980. Delayed secondary enrichment for the isolation of salmonellae from broiler chickens and their environment. Appl. Environ. Microbiol. 40, 783–786.

Rizzo, A. F. 1980. Rapid gas-chromatographic method for identification of metabolic products of anaerobic bacteria. J. Clin. Microbiol. 11, 418–421.

Roberts, D. 1982. Bacteria of public health significance. *In* "Meat Microbiology." M. H. Brown (Editor). Applied Science Publishers LTD, New York, N.Y.

Roberts, G. D., Horstmeier, C. D., Land, G. A., and Foxworth, J. H. 1978. Rapid urea broth test for yeasts. J. Clin. Microbiol. 7, 584–588.

Roberts, T. A. 1979. Cured meats. Arch. Lebensmittelhyg. 30, 92–94.

Roberts, T. A., Macfie, H. J. H., and Hudson, W. R. 1980. The effect of incubation temperature and site of sampling on assessment of the numbers of bacteria on red meat carcasses at commercial abattoirs. J. Hyg. 85, 371–380.

Robin, T., and Janda, J. M. 1984. Enhanced recovery of *Pseudomonas aeruginosa* from diverse clinical specimens on a new selective agar. Diagn. Microbiol. Infect. Dis. 2, 207–211.

Robison, B. J. 1984. Evaluation of a fluorogenic assay for detection of *E. coli* in foods. Appl. Environ. Microbiol. 48, 285–288.

Robison, B. J., Pretzman, C. I., and Mattingly, J. A. 1983. Enzyme immunoassay in which a myeloma protein is used for detection of salmonellae. Appl. Environ. Microbiol. 45, 1816–1821.

Rodricks, J. V. 1976. The occurrence and control of mycotoxins and mycotoxicoses. Food and Nutrition 2, 9–14.

Rodricks, J. V., and Lovett, J. 1976. Toxigenic fungi. *In* "Compendium of Methods for the Microbiological Examination of Foods." M. L. Speck (Editor). Am. Public Health Assoc., Washington, D.C.

Rogol, M., Sechter, I., and Gerichter, B. C. 1981. Improved method for cholera diagnosis. J. Clin. Microbiol. 13, 696–697.

Roig, J. M., Dorronsoro, I., and Diaz, R. 1983. Rapid identification of *Serratia marcescens* by coagglutination. J. Clin. Microbiol. 18, 741–742.
Rollof, J., Hedstroem, S. A., and Nilsson-Ehle, P. 1984. A simple turbidimetric method for specific measurement of *S. aureus* lipase activity. Acta Pathol. Microbiol. Immunol. Scand. Sect. B, 92B, 155–158.
Ronnberg, B., Carlsson, J., and Wadstrom, T. 1984. Development of an ELISA for detection of *E. coli* heat-stable enterotoxin. FEMS Microbiol. Lett. 23, 275–279.
Rosen, R. T., Rosen, J. D., and Di Prossimo, V. P. 1984. Confirmation of aflatoxins B_1 and B_2 in peanuts by gas chromatography/mass spectrometry/selected ion monitoring. J. Agric. Food Chem. 32, 276–278.
Rosenthal, S. L., Freundlich, L. F., and Washington, W. 1978. Laboratory evaluation of a multitest system for identification of gram-negative organisms. Am. J. Clin. Pathol. 70, 914–917.
Ross, G. D., Lunklater, P. M., and Hall, J. R. 1980. Measurement of the growth of lactic streptococci in skim milk by direct estimation of cell dry weight. Aust. J. Dairy Technol. 35, 89–92.
Rothenberg, P. J., Stern, N. J., and Westhoff, D. C. 1984. Selected enrichment broths for recovery of *Campylobacter jejuni* from foods. Appl. Environ. Microbiol. 48, 78–80.
Rowley, D. B., Previte, J. J., and Srinivasa, H. P. 1978a. A radiometric method for rapid screening of cooked foods for microbial acceptability. J. Food Sci. 43, 1720–1722.
Rowley, D. B., Sullivan, R., and Josephson, E. S. 1978b. Indicators of viruses in foods preserved by ionizing radiation. *In* "Indicators of Viruses in Water and Food." G. Berg (Editor). John Wiley and Sons Ltd., Baffins Lane, Chichester, W. Sussex, U.K.
Rowley, D. B., Vandemark, P., Johnson, D., and Shattuck, E. 1979. Resuscitation of stressed fecal coliforms and their subsequent detection by radiometric and impedance techniques. J. Food Prot. 42, 335–341.
Rubin, S. J., and Woodard, M. 1983. Enhanced isolation of *Campylobacter jejuni* by cold enrichment in Campy-thio broth. J. Clin. Microbiol. 18, 1008–1010.
Ruoff, K. L., Ferraro, M. J., Jerz, M. E., and Kissling, J. 1982. Automated identification of gram-positive bacteria. J. Clin. Microbiol. 16, 1091–1095.
Ruoff, K. L., and Kunz, L. J. 1983. Use of the Rapid STREP system for identification of viridans streptococcal species. J. Clin. Microbiol. 18, 1138–1140.
Russek, E., and Colwell, R. R. 1983. Computation of most probable numbers. Appl. Environ. Microbiol. 45, 1646–1650.
Russell, R. R. B. 1976. Free endotoxin—a review. Microbios 2, 125–135.
Rutherford, I., Moody, V., Gavan, T. L., Ayers, L. W., and Taylor, D. L. 1977. Comparative study of three methods of identification of *Enterobacteriaceae*. J. Clin. Microbiol. 5, 458–464.
Rycroft, J. A. 1980. Membrane filtration media for the enumeration of coliform organisms and *Escherichia coli* in water: Comparison of Tergitol 7 and lauryl sulfate with Teepol 610. J. Hyg. 85, 181–191.
Sacks, L. E., and Menefee, E. 1972. Thermal detection of spoilage in canned foods. J. Food Sci. 37, 518–523.
Sacks, L. E., and Thomas, R. S. 1979. High yields of coatless spores of *Clostridium perfringens* strain 8-6 in a defined medium. Can. J. Microbiol. 25, 642–645.
Sacks, L. E., and Thompson, P. A. 1978. Clear, defined medium for the sporulation of *Clostridium perfringens*. Appl. Environ. Microbiol. 35, 405–410.
Sadovski, A. Y. 1977. Acid sensitivity of freeze-injured salmonellae in relation to their isolation from frozen vegetables by pre-enrichment procedures. J. Food Technol. 12, 85–91.

Sakaguchi, G. 1982. *Clostridium botulinum* toxins. Pharmacol. Ther. 19, 165–194.
Sakazaki, R., Karashimada, T., Yuda, K., Sakai, S., Asakawa, Y., Yamazakix, M., Nakanishi, H., Kobayashi, K., Nishio, T., Okazaki, N., Doke, T., Shimada, T., and Tamura, K. 1979. Enumeration of, and hygienic standard of food safety for *Vibrio parahaemolyticus*. Arch. Libensmittelhyg. 30, 103–106.
Salveson, A., and Bergan, T. 1981. Enterobacteria differentiated by gas-liquid chromatography of metabolites. Zentralbl. Bakteriol. Microbiol. Hyg., I Abt. A 250, 104–112.
Salzman, G. C., and Gregg, C. T. 1984. Current and experimental methods of rapid microbial identification. Bio/Technology. 2, 243–248.
Salzman, G. C., Griffith, J. K., and Gregg, C. T. 1982. Rapid identification of microorganisms by circular-intensity differential light scattering. Appl. Environ. Microbiol. 44, 1081–1085.
Sanders, A. C., Bryan, F. L., and Olson, Jr., J. C. 1976. Foodborne illness—suggested approaches for the analysis of foods and specimens obtained in outbreaks. *In* "Compendium of Methods for the Microbiological Examination of Foods." M. L. Speck (Editor). Am. Public Health Assoc., Washington, D.C.
Sankaran, R., and Leela, R. K. 1982. Comparison of coagulase and thermonuclease activity with enterotoxin production by staphylococci from foods. Nahrung. 26, 549–553.
Sanville, C. 1985. Luminescence as an analytical tool. Clinical Prod. Rev. 4, 28–32.
Satta, G., Grazi, G., Varaldo, P. E., and Fontana, R. 1979. Detection of bacterial phosphatase activity by means of an original and simple test. J. Clin. Pathol. 32, 391–395.
Saunders, G. C., and Bartlett, M. L. 1977. Double-antibody solid-phase enzyme immunoassay for detection of staphylococcal enterotoxin A. Appl. Environ. Microbiol. 34, 518–522.
Sayem El Daher, N., and Simard, R. E. 1985. Putrefactive amine changes in relation to microbial counts of ground beef during storage. J. Food Protection 48, 54–58.
Sayem El Daher, N., and Simard, R. E., Fillion, J., and Roberge, A. G. 1984. Extraction and determination of biogenic amines in ground beef and their relation to microbial quality. Lebensm.-Wiss. Technol. 17, 20–23.
Schade, J. E., and King, Jr., A. D. 1984. Analysis of the major *Alternaria* toxins. J. Food Protection 47, 978–995.
Schafer, M. L., Peeler, J. T., Bradshaw, J. G., Hamilton, C. H., and Carver, R. B. 1982. A rapid gas chromatographic method for the identification of sporeformers and nonsporeformers in swollen cans of low-acid foods. J. Food Sci. 47, 2033–2037.
Schau, H. P. 1979. *Clostridium perfringens* as a foodstuff poisoner—recommendations for the bacteriological examination of foodstuffs for *Clostridium perfringens*. Z. Gesamte Hyg. Grenzget., 25, 521–525.
Scheuber, P. H., Mossmann, H., Beck, G., and Hammer, D. K. 1983. Direct skin test in highly sensitized guinea pigs for rapid and sensitive determination of staphylococcal enterotoxin B. Appl. Environ. Microbiol. 46, 1351–1356.
Schiemann, D. A. 1979a. Effect of dyes on the quantitative recovery of *Yersinia enterocolitica*. Appl. Environ. Microbiol. 38, 205–211.
Schiemann, D. A. 1979b. Synthesis of a selective agar medium for *Yersinia enterocolitica*. Can. J. Microbiol. 25, 1298–1304.
Schiemann, D. A. 1980. *Yersinia enterocolitica*: observations on some growth characteristics and response to selective agents. Can. J. Microbiol. 26, 1232–1240.
Schiemann, D. A., and Toma, S. 1978. Isolation of *Yersinia enterocolitica* from raw milk. Appl. Environ. Microbiol. 35, 54–58.

Schmidt, R., and Dose, K. 1984. HPLC: A tool for the analysis of T-2 toxin and HT-2 toxin in cereals. J. Anal. Toxicol. 8, 43–45.

Schweighardt, H., and Leibetseder, J. 1981. Analysis of mycotoxins by high-pressure liquid chromatography (HPLC). Wien. Tierarztl. Monatsachr. 68, 302–305.

Scott, E., Bloomfield, S. F., and Barlow, C. G. 1984. A comparison of contact plate and calcium alginate swab techniques for quantitative assessment of bacteriologic contamination of environmental surfaces. J. Appl. Bacteriol. 56, 317–320.

Scott, P. M., Lawrence, J. W., and Van Walbeek, W. 1970. Detection of mycotoxins by thin-layer chromatography: Application to screening of fungal extracts. Appl. Microbiol. 20, 839–842.

Seenappa, M., and Kempton, A. G. 1981. A simple key for the identification of *Bacillus* species common in foods. J. Food Sci. Technol. 18, 131–133.

Seng, L. Y., and Jegathesan, M. 1978. Occurrence of coliform bacteria other than *Escherichia coli* in foods. Southeast Asian J. Trop. Med. Public Health 9, 529–533.

Senior, B. W., 1981. A rapid and simple method for distinguishing colonies of *Proteus* from those of *Salmonella* and *Shigella*. J. Med. Microbiol. 14, 151–152.

Seydel, J. K., and Wempe, E. 1980. A simple, fast and inexpensive kinetic method to differentiate between total and viable bacteria using Coulter Counter technique. Arzneim.-Forsch. 30, 298–301.

Shanholtzer, C. J., and Peterson, L. R. 1982. Clinical laboratory evaluation of the thermonuclease test. Am. J. Clin. Pathol. 77, 587–591.

Sharf, J. M. 1966. Recommended methods for the microbiological examination of foods. Am. Public Health Assoc., Inc., New York, N.Y.

Sharpe, A. N., and Jackson, A. K. 1972. Stomaching: a new concept in bacteriological sample preparation. Appl. Microbiol. 24, 175–178.

Sharpe, A. N., Woodrow, M. N., and Jackson, A. K. 1970. Adenosine triphosphate levels in foods contaminated by bacteria. J. Appl. Bacteriol. 33, 758–767.

Sharpe, A. N., Diotte, M. P., Dudas, I., and Michaud, G. L. 1978. Automated food microbiology: potential for the hydrophobic grid-membrane filter. Appl. Environ. Microbiol. 36, 76–80.

Sharpe, A. N., Peterkin, P. I., and Dudas, I. 1979a. Membrane filtration of food suspensions. Appl. Environ. Microbiol. 37, 21–35.

Sharpe, A. N., Peterkin, P. I., and Malik, N. 1979b. Improved detection of coliforms and *Escherichia coli* in foods by a membrane filter method. Appl. Environ. Microbiol. 38, 431–435.

Sharpe, A. N., Diotte, M. P., Dudas, I., Malcolm, S., and Peterkin, P. I. 1983a. Colony counting on hydrophobic grid-membrane filters. Can. J. Microbiol. 29, 797–802.

Sharpe, A. N., Rayman, M. K., Burgener, D. M., Conley, D., Loit, A., Milling, M., Peterkin, P. I., Purvis, U., and Malcolm, S. 1983b. Collaborative study of the MPN, Anderson-Baird-Parker direct plating, and hydrophobic grid-membrane filter methods for the enumeration of *Escherichia coli* biotype 1 in foods. Can. J. Microbiol. 29, 1247–1252.

Shelly, D. C., Warner, I. M., and Quarles, J. M. 1983. Characterization of bacteria by mixed-dye fluorometry. Clin. Chem. 29, 290–296.

Shihabi, Z. K., and Wasilauskas, B. L. 1979. Presumptive identification of the bacterium *Escherichia coli* by high performance liquid chromatography. J. Liq. Chromatogr. 2, 851–860.

Shinoda, S., Nakahara, N., Ninomiya, Y., Itoh, K., and Kane, H. 1982. Serological method for identification of *Vibrio parahaemolyticus* from marine samples. Appl. Environ. Microbiol. 45, 148–152.

Shungu, D., Valiant, M., Tutlane, V., Weinberg, E., Weissberger, B., Koupal, L.,

Gadebusch, H., and Stapley, E. 1983. Gelrite as an agar substitute in bacteriological media. Appl. Environ. Microbiol. 46, 840–845.

Shute, L. A., Gutteridge, C. S., Norris, J. R., and Berkeley, R. C. W. 1984. Curie-point pyrolysis mass spectrometry applied to characterization and identification of selected *Bacillus* species. J. Gen. Microbiol. 130, 343–355.

Siems, H. 1978. Experience in the isolation of salmonellae in foods of different risk categories. Zentralbl. Bakteriol. Parasitenkd. Infektionskr. Hyg., I Abt. B. 167, 120–128.

Sil, J., and Bhattacharyya, F. K. 1978. A rapid test for the identification of all serotypes of *Vibrio cholerae* (including "non-agglutinating" vibrios). J. Med. Microbiol. 12, 63–70.

Silliker, J. H. 1963. Total counts as indexes of food quality. In "Microbiological Quality of Foods." L. W. Slanetz., C. W. Chichester, A. R. Gaufin and Z. J. Ordal (Editors), Academic Press, New York, N.Y.

Silliker, J. H. 1980. Status of *Salmonella*—ten years later. J. Food Prot. 43, 307–313.

Silliker, J. H., and Gabis, D. A. 1973. Comparison of analytical schemes for detection of *Salmonella* in dried foods. Can. J. Microbiol. 19, 475–479.

Silliker, J. H., and Gabis, D. A. 1974. The influence of selective enrichment media and incubation temperatures on the detection of salmonellae in frozen meats. Can. J. Microbiol. 20, 813–816.

Silliker, J. H., and Gabis, D. A. 1976. Indicator tests as substitutes for direct testing of dried foods and feeds for *Salmonella* Can. J. Microbiol. 22, 971.

Silliker, J. H. Gabis, D. A., and May, A. 1979. Collaborative/comparative studies on determination of coliforms using the most probable number procedure. J. Food Prot. 42, 638–644.

Sinell, H. J. 1978. Experience with selective cultivation of staphylococci from foods. Arch. Lebensmittelhyg. 29, 136.

Singh, S. N., Rose, J. B., and Berga, C. P. 1983. Concentration of viruses from tap water and sewage with a charge-modified filter aid. J. Virol. Methods 6, 329–336.

Skalka, B., and Smola, J. 1983. Selective diagnostic medium for pathogenic *Listeria* spp. J. Clin. Microbiol. 18, 1432–1433.

Skillern, J. K., and Overman, T. L. 1983. Oxidase testing from Kligler's iron agar and triple sugar iron agar slants. Curr. Microbiol. 8, 269–271.

Slabyj, B. M., Martin, R. E., and Ramsdell, G. E. 1981. Reproducibility of microbiological counts on frozen cod: a collaborative study. J. Food Sci. 46, 716–719.

Slack, M. P. E., and Wheldon, D. B. 1978. A simple and safe volumetric alternative to the method of Miles, Misra and Irwin for counting viable bacteria. J. Med. Microbiol. 11, 541–545.

Slanetz, L. W., Chichester, C. O. Gaufin, A. R., and Ordal, Z. J. 1963. (Editors). "Microbiological Quality of Foods." Academic Press, New York, N.Y.

Smeltzer, T. I., and Duncalfe, F. 1979. Secondary selective enrichment of salmonellae from naturally contaminated specimens by using a selective motility system. Appl. Environ. Microbiol. 37, 725–728.

Smith, J. L., and Palumbo, S. A. 1978. Injury to *Staphylococcus aureus* during sausage fermentation. Appl. Environ. Microbiol. 36, 857–860.

Smith, J. L., and Rockliff, S. 1982. Rapid single-tube confirmatory test for *Escherichia coli*. J. Hyg. 89, 149–154.

Smith, J. L., Bencivengo, M. M., and Buchanan, R. L. 1984. Enterotoxin biosynthesis by progeny of repaired heat-injured cells of *S. aureus*. J. Food Saf. 6, 203–209.

Smith, P. B., Gavan, T. L., Isenberg, H. D., and Sonnenwirth, A. 1978. Multi-laboratory evaluation of an automated microbial detection/identification system. J. Clin. Microbiol. 8, 657–666.

Snygg, B. G., Andersson, J. E., Krall, C. A., Stollman, U. M., and Akesson, C. A. 1979.

Separation of botulism-positive and -negative fish samples by means of a pattern recognition method applied to headspace gas chromatograms. Appl. Environ. Microbiol. 38, 1081–1085.

Sobsey, M. D., Carrick, R. J., and Jensen, H. R. 1978. Improved methods for detecting enteric viruses in oysters. Appl. Environ. Microbiol. 36, 121–128.

Solberg, M., Miskimin, D. K., Martin, B. A., Page, G., Goldner, S., and Libfeld, M. 1976. What do microbiological indicator tests tell us about the safety of foods? Food Prod. Develop. 10 (9), 72–80.

Solomon, W. R. 1984. Sampling techniques for airborne fungi. In "Mould Allergy," Y. Al-Doory and J. F. Domson (eds.), Lea and Febiger, Philadelphia, PA., pp. 41–65.

Soltesz, L. V. Schalem, C., and Mardh, P. A. 1980. An effective, selective medium for *Yersinia enterocolitica* containing sodium oxalate. Acta. Pathol. Microbiol. Scand., Ser. B 88, 11–16.

Soncini, G., Ponti, W., Balsari, A., and Poli, G. 1979. Development and control of a technique for the detection of enteroviruses in edible molluscs. Boll. Ist. Sieroter. Milan 58, 216–219.

Southern, P. M. Jr. 1979. New methods in automation and rapid methods in microbiology. Food Technol. 33, 54–56.

Speck, M. L. 1976. (Editor). "Compendium of Methods for the Microbiological Examination of Foods." Am. Public Health Assoc., Washington, D.C.

Speck, R. V. 1981. Thermophilic organisms in food spoilage: sulfide spoilage anaerobes. J. Food Prot. 44, 149–153.

Spengler, M., Rodeheaver, G. T., Richter, L., Edgerton, M. T., and Edlich, R. F. 1978. The Gram stain—the most important diagnostic test of infection. J. Am. Coll. Emergency Physicians 7, 434–438.

Sperber, W. H. 1977. The identification of staphylococci in clinical and food microbiology laboratories. CRC Critical Reviews in Clin. Lab. Sci. 7, 121–184.

Splittstoesser, D. F. 1976. Enumeration of heat resistant mold. In "Compendium of Methods for the Microbiological Examination of Foods." M. L. Speck (Editor). Am. Public Health Assoc., Washington, D.C.

Splittstoesser, D. F. 1983. Indicator organisms on frozen, blanched vegetables. Food Technol. 37, 103–105.

Splittstoesser, D. F., Bowers, J., Kerschner, L., and Wilkison, M. 1980. Detection and incidence of *Geotrichum candidum* in frozen blanched vegetables. J. Food Sci. 45, 511–513.

Splittstoesser, D. F., Stewart, J. D., and Wilkison, M. 1982. Survival of fecal coliforms in frozen vegetable homogenates. J. Food Prot. 45, 1041–1043.

Stamer, J. R. 1979. The lactic acid bacteria: microbes of diversity. Food Technol. 33, 60–65.

Staneck, J. L., Vincelette, J., Lamothe, F., and Polk, E. A. 1983. Evaluation of the Sensititre system for identification of *Enterobacteriaceae*. J. Clin. Microbiol. 17, 647–654.

Stanndard, C. J., and Wood, J. M. 1983. The rapid estimation of microbial contamination of raw meat by measurement of ATP. J. Appl. Bacteriol. 55, 429–438.

Stanndard, C. J., Patel, P. D., Wood, J. M., and Gibbs, P. A. 1982. Rapid microbiological methodology. Food 4, 18–24.

Stavric, S., and Jeffrey, D. 1977. A modified bioassay for heat-stable *Escherichia coli* enterotoxin. Can. J. Microbiol. 23, 331–336.

Stec, E. 1977. A comparison of media for quantitative determination of fungi in food products. Rocz. Panstu. Zakl. Hig. 28, 9–14.

Stengren, S. R., and Starzyk, M. J. 1984. A modified medium for the recovery of *Staphylococcus* from water. Microbios 41, 191–203.

4. Standard and Experimental Methods of Identification and Enumeration

Stephen, J., and Pietrowski, R. A. 1981. "Bacterial Toxins." American Society for Microbiology (Publishers).
Stern, N. J. 1981a. *Campylobacter Fetus* ssp. *jejuni:* recovery methodology and isolation from lamb carcasses. J. Food Sci. 46, 660–661.
Stern, N. J. 1981b. Discrimination of *Bacillus cereus* from selected *Bacillus* spp. by pyrolysis gas-liquid chromatography. J. Food Sci. 46, 1427–1429.
Stern, N. J. 1982a. The inability of pyrolysis gas-liquid chromatography to differentiate selected foodborne bacteria. J. Food Prot. 45, 229–233.
Stern, N. J. 1982b. Methods for recovery of *Campylobacter jejuni* from foods. J. Food Prot. 45, 1332–1337.
Stern, N. J. 1982c. Selectivity and sensitivity of three media for recovery of inoculated *Campylobacter jejuni* from ground beef. J. Food Saf. 4, 169–175.
Stern, N. J. 1982d. *Yersinia enterocolitica:* Recovery from foods and virulence characterization. Food Technol. 36 (3) 84–88.
Stern, N. J., and Damare, J. M. 1982. Comparison of selected *Yersinia enterocolitica* indicator tests for potential virulence. J. Food Sci. 47, 582.
Stern, N. J., and Kotula, A. W. 1982. Survival of *Campylobacter jejuni* inoculated into ground beef. Appl. Environ. Microbiol. 44, 1150–1153.
Stern, N. J., Kotula, A. W., and Pierson, M. D. 1979. Differentiation of selected *Enterobacteriaceae* by pyrolysis gas-liquid chromatography. Appl. Environ. Microbiol. 38, 1098–1102.
Stern. N. J., Kotula, A. W., and Pierson, M. D. 1980. Virulence prediction of *Yersinia enterocolitica* by pyrolysis-gas liquid chromatography. Appl. Environ. Microbiol. 40, 646–651.
Stern, N. J., Rothenberg, P. J., and Stone, J. M. 1985. Enumeration and reduction of *Campylobacter jejuni* in poultry and red meats. J. Food Protection. 48, 606–610.
Stersky, A. K., Riedel, G. W., and Thacker, C. 1980. Control of foaming during microbiological analysis of foods and recovery of indicator organisms. J. Food Prot. 43, 40–43.
Stevens, M., Feltham, R. K. A., Schneider, F., Grasmick, C., Schaak, F., and Roos, P. 1984. A collaborative evaluation of a rapid automated bacterial identification system: The autobac IDX. Eur. J. Clin. Microbiol. 3, 419–423.
Stewart, B. J., Eyles, M. J. and Murrell, W. G. 1980. Rapid radiometric method for detection of *Salmonella* in foods. Appl. Environ. Microbiol. 40, 223–230.
Stiles, M. E., and Ng, L. K. 1980. Estimation of *Escherichia coli* in raw ground beef. Appl. Environ. Microbiol. 40, 346–351.
Stiles, M. E., Ramji, N. W., Ng, L. K., and Paradis, D. C. 1978. Incidence and relationship of group D streptococci with other indicator organisms in meats. Can. J. Microbiol. 24, 1502–1508.
Stinson, E. E. 1985. Mycotoxins—their biosynthesis in *Alternaria*. J. Food Protection. 48, 80–91.
St. John, W. D., Matches, J. R., and Wekell, M. M. 1982. Use of iron milk medium for enumeration of *Clostridium perfringens*. J. Assoc. Off. Anal. Chem. 65, 1129–1133.
Stoloff, L., Nesheim, S., Yin, L., Rodricks, J. V., Stack, M., and Campbell, A. D. 1971. A multimycotoxin detection method for aflatoxins, ochratoxins, zearalenone, sterigmatocystin, and patulin. J. AOAC 54, 91–97.
Stringer, M. J., Turnbull, P. C. B., and Gilbert, R. J. 1980. Application of serological typing to the investigation of outbreaks of *Clostridium perfringens* food poisoning, 1970–1978. J. Hyg. 84, 443–456.

Sudarsanam, T. S., and Nambudripad, V. K. N. 1978. Development of a rapid coliform test for pasteurized milk. Indian J. Dairy Sci. 31, 282–284.

Sudarsanam, T. S., Sinha, R. N., and Nambudripad, V. K. N. 1978. Relationship between methylene blue, resazurin and nitrate reduction times of fresh and stored pasteurized milk. Indian J. Dairy Sci. 31, 179–180.

Sugiyama, H., Mills, D. C., and Kuo, L. J. C. 1978. Number of *Clostridium botulinum* spores in honey. J. Food Prot. 41, 848–850.

Sukroongreung, S., Nilakul, C., and Sirinanunta, S. 1981. Free coagulase and thermostable nuclease reactions of staphylococci: a comparison. Am. J. Med. Technol. 47, 237–239.

Sullivan, J. D., Jr., Ellis, P. C., Lee, R. G., Combs, W. S., Jr., and Watson, S. W. 1983. Comparison of the *Limulus* amebocyte lysate test with plate counts and chemical analyses for assessment of the quality of lean fish. Appl. Environ. Microbiol. 45, 720–722.

Sullivan, J. D., Jr., Valois, F. W., and Watson, S. W. 1976. Endotoxins: The limulus amoebocyte lysate system. *In* "Mechanisms in Bacterial Toxinology." A. W. Bernheimer (Editor). John Wiley & Sons, New York, N.Y.

Sullivan, J. D., Jr., and Watson, S. W. 1974. Factors affecting the sensitivity of Limulus Lysate. Appl. Microbiol. 28, 1023–1026.

Sullivan, N. M., Mayhew, J., DiTullio, D., and Tally, F. P. 1978. Argon detector: alternative detection system for gas-liquid chromatographic analysis of short-chain organic acids. J. Clin. Microbiol. 8, 369–373.

Sullivan, R., and Peeler, J. T. 1982. Evaluation of a method for recovery of virus from oysters. J. Food Prot. 45, 636–637.

Sveum, W. H., and Hartman, P. A. 1977. Timed-release capsule method for the detection of salmonellae in foods and feeds. Appl. Environ. Microbiol. 33, 630–634.

Sveum, W. H., and Kraft, A. A. 1981. Recovery of salmonellae from foods using a combined enrichment technique. J. Food Sci. 46, 94–99.

Swaminathan, B., and Ayres, J. C. 1980. A direct immunoenzyme method for the detection of salmonellae in foods. J. Food Sci. 45, 352–355.

Swaminathan, B., Ayres, J. C., and Williams, J. E. 1978a. Control of nonspecific staining in the fluorescent antibody technique for the detection of salmonellae in foods. Appl. Environ. Microbiol. 35, 911–919.

Swaminathan, B., Denner, J. B., and Ayres, J. C. 1978b. Rapid detection of salmonellae in foods by membrane filter-disc immunoimmobilization technique. J. Food Sci. 43, 1444–1447.

Szakal, S. 1979. Rapid diagnostic tests and their use in veterinary microbiological food investigation (compilatory article). Magy. Allatorv. Lapja 34, 393–399.

Szita, G., Takacs, J., and Lendvai, I. 1979. Comparative study on surface count methods for viable coliforms in milk samples. Acta Vet. Acad. Sci. Hung. 27, 331–335.

Taguchi, M., Tsukiji, M., and Tsuchiya, T. 1979. Rapid identification of yeasts by serological methods. A combined serological and biological method. Sabouraudia 17, 185–191.

Takacs, J., and Domjan, H. K. 1979. Comparative studies on the detectability of *Staphylococcus aureus* by means of the media commonly used in food microbiology. Acta. Vet. Acad. Sci. Hung. 27, 323–329.

Takacs, J., Lendvai, I., and Abd-El-Bakey, A. 1979a. Comparative study for coliform-count on solid and liquid media. Arch. Lebensmittelhyg. 30, 41–43.

Takacs, J., Lendvai, I., and Abd-El-Bakey, A. 1979b. Correlation between coliform and *Enterobacteriaceae* count in raw and retailed milk. Arch. Lebensmittelhyg. 30, 44–46.

Takano, M., and Tsuchido, T. 1982. Availability of growth delay analysis for the evaluation of total injury of stressed bacterial populations. J. Fement. Technol. 60, 189–198.

Tamminga, S. K., Beumer, R. R., and Kampelmacher, E. H. 1980. Bacteriological examination of ice-cream in the Netherlands: comparative studies on methods. J. Appl. Bacteriol. 49, 239–253.

Tamplin, M., Rodrick, G. E., Blake, N. J., and Cuba, T. 1982. Isolation and characterization of *Vibrio vulnificus* from two Florida estuaries. Appl. Environ. Microbiol. 44, 1466–1470.

Tan, S. T., Maxcy, R. B., and Thompson, T. L. 1983. Paper replication method for isolation of radiation-sensitive mutants. Appl. Environ. Microbiol. 46, 233–236.

Tang, C. C., and Jackson, H. 1979. Minimal medium recovery of chilled *Salmonella heidelberg*. J. Appl. Bacteriol. 46, 143–146.

Tarrand, J. J., and Groeschel, D. H. M. 1982. Rapid modified oxidase test for oxidase-variable bacterial isolates. J. Clin. Microbiol. 16, 772–774.

Taylor, S. L., and Woychik, N. A. 1982. Simple medium for assessing quantitative production of histamine by Enterobactericeae. J. Food Prot. 45, 747–751.

Teixeira, L. M., Moss, C. W., and Fackham, R. R. 1983. Gas-liquid chromatography of the fatty acids of *Streptococcus faecalis* with a fused silica capillary column. FEMS Microbiol. Lett. 17, 257–260.

Tejedor, C., and Chordi, A. 1984. A computer program for teaching microbial identification. Am. J. Pharm. Educ. 48, 149–154.

Terplan, G., and Zaadhof, K. J. 1978. Methods for the detection of *S. aureus* in foods. Arch. Lebensmittelhyg. 29, 132–135.

Tham, W., and Danielsson, M. L. 1980. Reliability of VRB agar and BGLB broth for enumeration of 44C coliforms in food. Nord. Vet. Med. 32, 325–331.

Tharrington, G., Jr., Ashton, D. H., Hatfield, J. R., and Fry, F. H. 1978. Non-specific staining of a *Lactobacillus* by *Salmonella* fluorescent antibodies. J. Food Sci. 43, 548–552.

Thatcher, F. S., and Clark, D. S. 1973. (Editors). Microorganisms in Foods: Their Significance and Methods of Enumeration. University of Toronto Press, Toronto, Canada.

Thatcher, F. S., and Clark, D. S. 1974. (Editors). Microorganisms in Foods: Sampling for Microbiological Analysis: Principles and Specific Applications. University of Toronto Press, Toronto, Canada.

Thomas, C. J., and McMeekin, T. A. 1980. A note on scanning microscopic assessment of stomacher action on chicken skin. J. Appl. Bacteriol. 49, 339–344.

Thomas, S. B., and Thomas, B. F. 1978. The bacterial content of milking machines and pipeline milking plants. Part 6. Psychrotrophic bacteria. Dairy Ind. Int. 43 (10), 5–10.

Thomason, B. M. 1976. Fluorescent antibody detection of salmonellae. In "Compendium of Methods for the Microbiological Examination of Foods." M. L. Speck (Editor). Am. Public Health Assoc., Washington, D.C.

Thomason, B. M., and Dodd, D. J. 1978. Enrichment procedures for isolating salmonellae from raw meat and poultry. Appl. Environ. Microbiol. 36, 627–628.

Thomason, B. M., Hebert, G. A., and Cherry, W. B. 1975. Evaluation of a semiautomated system for direct fluorescent antibody detection of salmonellae. Appl. Microbiol. 30, 557–564.

Thompson, J. S., and Borczyk, A. A. 1984. Use of a single-tube medium for screening of pathogenic members of the family Enterobacteriaceae. J. Clin. Microbiol. 20, 136–137.

Thompson, P. J. 1981. Thermophilic organisms involved in food spoilage: aciduric flat-sour sporeforming aerobes. J. Food Prot. 44, 154–156.

Thompson, M. R., Brandwein, H. J., LaBine-Racke, M., and Giannella, R. A. 1984. Simple and reliable ELISA with monoclonal antibodies for detection of *E. coli* heat-stable enterotoxins. J. Clin. Microbiol. 20, 59–64.

Thompson, W. L., and Wannemacher, Jr., R. W. 1984. Detection and quantitation of T-2 mycotoxin with a simplified protein synthesis inhibitor assay. Appl. Environ. Microbiol. 48, 1176–1180.

Thomson, J. E., and Bailey, J. S. 1978. Innovations in sampling and culturing methods for bacteriological examination of broiler carcasses. J. Food Sci. 43, 1301–1302.

Thomson, J. E., Cox, N. A., Bailey, J. S., Holladay, J. H., and Richardson, R. L. 1976. Bacteriological sampling of poultry carcasses by a template-swab method. Poult. Sci. 55, 459–462.

Tierney, J. T., Fassolitis, A., Van Donsel, D., Rao, V. C., Sullivan, R., and Larkin, E. P. 1980. Glass wool-hydroextraction method for recovery of human enteroviruses from shellfish. J. Food Prot. 43, 102–104.

Tilbury, R. H. 1981. Xerotolerant (osmophilic) yeasts. *In* "Biology and Activities of Yeasts." F. A. Skinner, S. M. Passmore and R. R. Davenport (Editors). Academic Press for the Society of Applied Bacteriology.

Tilton, R. C. 1982. (Editor). Rapid Methods and Automation in Microbiology. Am. Soc. Microbiol., Washington, D.C.

Tomasulo, M., and Braunstein, H. 1979. Late deoxyribonuclease activity of *Salmonella enteritidis*. Am. J. Clin. Pathol. 72, 82–84.

Tomlins, R. I., Pierson, M. D., and Ordal, Z. J. 1971. Effect of thermal injury on the TCA cycle enzymes of *Staphylococcus aureus* MF31 and *Salmonella typhimurium* 7136. Can. J. Microbiol. 17, 759–765.

Tompkin, R. B. 1983. Indicator organisms in meat and poultry products. Food Technol. 37, 107–110.

Tongpim, S., Beumer, R. R., Tamminga, S. K., and Kampelmacher, E. H. 1984. Comparison of modified Rappaport's medium (RV) and Muller-Kauffmann medium (MK-ISO) for the detection of *Salmonella* in Meat Products. Int. J. Food Microbiol. 1, 33–42.

Torangos, G. A., Gerba, C. P., and Hanssen, H. 1984. Simple field method for concentration of viruses from large volumes of water. Appl. Environ. Microbiol. 48, 431–432.

Tortora, J. C. O. 1984. Alternative medium for *Clostridium perfringens* sporulation. Appl. Environ. Microbiol. 47, 1172–1174.

Trepeta, R. W., and Edberg, S. C. 1984. Methylumbelliferyl-beta-D-gluronide-base medium for rapid isolation and identification of *Escherichia coli*. J. Clin. Microbiol. 19, 172–174.

Trevors, J. T. 1983. A note on viability measurements in *Saccharomyces* spp. Biotechnol. Lett. 5, 363–364.

Trevors, J. T., Merick, R. L., Russell, I., and Stewart, G. G. 1983. A comparison of methods for assessing yeast viability. Biotechnol. Lett. 5, 131–134.

Truscott, R. B. 1983. A comparison of two enrichment and two plating media for the isolation of *Salmonella* sp. from broilers. Can. J. Comp. Med. 47, 373–374.

Tsan, Y.-C. 1978. Nucleic acid hydrolysis by *Vibrio cholerae*. Chin. J. Microbiol. 11, 114–115.

Tsuji, K., and Steindler, K. A. 1983. Use of Magnesium to increase sensitivity of *Limulus* amebocyte lysate for detection of endotoxin. Appl. Environ. Microbiol. 45, 1342–1350.

Tsuji, K., Martin, P. A., and Bussey, D. M. 1984. Automation of chromogenic substrate *Limulus* amebocyte lysate assay method for endotoxin by robotic system. Appl. Environ. Microbiol. 48, 550–555.

Turnbull, P. C. B. 1980. Principles of food poisoning—and its control. S. Afr. Food Rev. 7, 113–124.

4. Standard and Experimental Methods of Identification and Enumeration 189

Turnbull, P. C. B., and Rose, P. 1982. *Campylobacter jejuni* and *Salmonella* in raw red meats. A public health laboratory service survey. J. Hyg. 88, 29–38.
Ueno, Y., and Kubota, K. 1976. DNA-attacking ability of carcinogenic mycotoxins in recombination-deficient mutant cells of *Bacillus subtillis*. Cancer Res. 36, 445–451.
Ulitzer, S. 1979. A sensitive bioassay for lipase using bacterial bioluminescence. Biochim. Biophys. Acta 572, 211–217.
Ulitzur, S., and Heller, M. 1978. A new, fast, and very sensitive bioluminescence assay for phospholipases A and C. Anal. Biochem. 91, 421–431.
Ulitzur, S., Yagen, B., and Rottem, S. 1979. Determination of lipopolysaccharide by a bioluminescence technique. Appl. Environ. Microbiol. 37, 782–784.
Urban, T., and Jarstrand, C. 1979. Nitroblue tetrazolium (NBT) reduction by bacteria. Some properties of the reaction and its possible use. Acta Pathol. Microbiol. Scand., Ser. B 87, 227–233.
Van den Broek, M. J. M., Mossel, D. A. A., and Eggenkamp, A. E. 1979a. Occurrence of *Vibrio Parahaemolyticus* in Dutch mussels. Appl. Environ. Microbiol. 37, 438–442.
Van den Broek, M. J. M., Mossel, D. A. A., Eggenkamp, A. E., Elderink, I., and Verouden, M. E. 1979b. The microbial association of bivalve shellfish intended for consumption in the raw state with particular reference to safety. Arch. Lebensmittelhyg. 30, 98–102.
Van Dilla, M. A., Langlois, R. G., Pinkel, D., Yajko, D., and Hadley, W. K. 1983. Bacterial characterization by flow cytometry. Science (Wash.), 220, 620–622.
Van Dilla, M. A., Dean, P. N., Laerum, O. D., and Melamed, M. R. 1985. (Eds.), "Flow Cytometry." Instrumentation and Data Analysis. Academic Press, Inc., Orlando, Fla.
Van Doorne, H., Baird, R. M., Hendriksz, D. T., van der Kreek, D. M., and Pauwels, H. P. 1981. Liquid modification of Baird-Parker's medium for the selective enrichment of *Staphylococcus aureus*. Antonie van Leeuwenhoek 47, 267–278.
Van Doorne, H., Pauwels, H. P., and Mossel, D. A. A. 1982. Selective isolation and enumeration of low numbers of *Staphylococcus aureus* by a procedure that relies on elevated-temperature culturing. Appl. Environ. Microbiol. 44, 1459–1462.
Van Dyke, K. 1985. (Editor). "Bioluminescence and Chemiluminescence: Instruments and Applications." Volumes 1 and 2. CRC Press, Inc., Boca Raton, Fla.
Van Netten, P., van der Zee, H., and Mossel, D. A. A. 1984. A note on catalase enhanced recovery of acid injured cells of gram negative bacteria and its consequence for the assessment of the lethality of L-lactic acid decontamination of raw meat surfaces. J. Appl. Bacteriol. 57, 169–173.
Van Pee, W., and Stragier, J. 1979. Evaluation of some cold enrichment and isolation media for the recovery of *Yersinia enterocolitica*. Antonie Van Leeuwenhoek 45, 465–477.
Van Schothorst, M., Gilbert, R. J., Harvey, R. W. S., Pietzsch, O., and Kampelmacher, E. H. 1978. Comparative studies on the isolation of *Salmonella* from minced meat. Zentralbl. Bakteriol. Parasitenkd. Infektionskr. Hyg., I ABT. B 167, 138–145.
Van Schothorst, M., and Renaud, A. M. 1983. Dynamics of *Salmonella* isolation with modified Rappaport's medium (R10). J. Appl. Bacteriol. 54, 209–215.
Van Soestbergen, A. A., and Lee, C. H. 1969. Pour plates or streak plates? Appl. Microbiol. 18, 1092–1093.
Vanstaen, H. 1980. Applicability of bioluminescence for rapid detection of viable microorganisms. Lab. Pract. 29, 1281–1283.
Van Vuuren, H. J. J., Toerien, D. F., and Lategan, P. M. 1978. Identification of some

Enterobacteriaceae by gas chromatography of volatile metabolites formed in a standardized growth medium. S. Afr. J. Sci. 74, 387–388.
Vanyi, A., Bata, A., and Lasztity, R. 1982. Quantitative determination of some *Fusarium* toxins by gas chromatography. Acta Vet. (Hung.) 30, 65–69.
Vassilev, P. 1978. Studies on growth properties on solid nutrient media when proving sanitary-indicative microorganisms by replica plating. Meat Ind. Bull. 11 (3), 13–16.
Vassiliadis, P. 1983. The Rappaport-Vassiliadis (RV) enrichment medium for the isolation of salmonellas: An overview. J. Appl. Bacteriol. 54, 77–83.
Vassiliadis, P., Trichopoulos, D., Pateraki, E., and Papaiconomou, N. 1977. Isolation of *Salmonella* from minced meat by the use of a new procedure of enrichment. Zentralbl. Bakteriol. Parasitenkd. Infektionskr. Hyg., I Abt. B. 166, 81–86.
Vassiliadis, P., Trichopoulos, D., Papadakis, J., and Kalapothaki, V. 1979a. Brilliant green deoxycholate agar as an improved selective medium for the isolation of *Salmonella*. Ann. Soc. Belge Med. Trop. 59, 117–120.
Vassiliadis, P., Trichopoulos, D., Papoutsakis, G., and Pallandiou, E. 1979b. A note on the comparison of two modifications of Rappaport's medium with selenite broth in the isolation of salmonellas. J. Appl. Bacteriol. 46, 567–569.
Vassiliadis, P., Trichopoulos, D., Kalapothaki, V., and Serie, C. 1981. Isolation of *Salmonella* with the use of 100 ml. of the R-10 Modification of Rappaport's enrichment medium. J. Hyg. 87, 35–41.
Vassiliadis, P., Kalapothaki, V., Mavrommati, C., and Trichopoulos, D. 1984. A comparison of the original Rappaport medium and the Rappaport-Vassiliadis medium in the isolation of salmonellae from meat products. J. Hyg. 93, 51–58.
Vaughn, J. M., Landry, E. F., Vicale, T. J., and Dahl, M. C. 1979. Modified procedure for the recovery of naturally accumulated poliovirus from oysters. Appl. Environ. Microbiol. 38, 594–598.
Velan, B., and Halmann, M. 1978. Chemiluminescence immunoassay; a new sensitive method for determination of antigens. Immunochemistry 15, 331–333.
Velin, D., and Emody, L. 1982. The stability of enterotoxin production in *Yersinia enterocolitica* and the methanol solubility of heat-stable enterotoxin. Acta Microbiol. Acad. Sci. Hung. 29, 227–233.
Verrips, C. T., and Kwast, R. H. 1982. Recovery of heat-injured *Citrobacter freundii* cells. J. Appl. Bacteriol. 52, 15–20.
Vesely, D., Vesela, D., and Jelinek, R. 1983. Use of chick embryo in screening for toxin-producing fungi. Mycopathologia 88, 135–140.
Vidon, D. J. M., and Delmas, C. L. 1981. Incidence of *Yersinia enterocolitica* in raw milk in eastern France. Appl. Environ. Microbiol. 41, 355–359.
Von Graevenitz, A., and Bucher, C. 1983. Accuracy of the KOH and vancomycin tests in determining the gram reaction of non-enterobacterial rods. J. Clin. Microbiol. 18, 983–985.
Von Rheinbaben, K. E., and Hadlock, R. M. 1981. Rapid distinction between micrococci and staphylococci with furazolidone agars. Antonie Van Leeuwenhoek 47, 41–51.
Wait, D. A., and Sobsey, M. D. 1983. Method for recovery of enteric viruses from estuarine sediments with chaotropic agents. Appl. Environ. Microbiol. 46, 379–385.
Waite, W. M. 1981. Coliform test for cheese. Lancet 2, 1174–1175.
Waitkins, S. A., Ball, C. L., and Fraser, C. A. M. 1980. Use of the API-ZYM system in rapid identification of non-haemolytic streptococci. J. Clin. Pathol. 33, 53–57.
Wakisaka, Y., and Koizumi, K. 1982. A selective isolation procedure for *Pseudomonas* bacteria. J. Antibiot. 35, 622–623.

Wakisaka, Y., Koizumi, K., and Nishimoto, Y. 1982. A preferential isolation procedure for asporogenous gram-positive bacteria. J. Antibiot. 35, 441–449.
Waldman, S. A., O'Hanley, P., Falkow, S., Schoolnik, G., and Murad, F. 1984. A simple, sensitive, and specific assay for the heat-stable enterotoxin of *E. coli* J. Infect. Dis. 149, 83–89.
Waller, J. R., Hodel, S. L., and Nuti, R. N. 1985. Improvement of two toluidine blue O-mediated techniques for DNase detection. J. Clin. Microbiol. 21, 195–199.
Wang, H., Wang, D. I. C., and Cooney, C. L. 1978. The application of dynamic calorimetry for monitoring growth of *Saccharomyces cerevisiae*. Eur. J. Appl. Microbiol. Biotechnol. 5, 207–214.
Wang, W. L., Blaser, M., and Cravens, J. 1978. Isolation of *Campylobacter*. Br. Med. J. 2, 57.
Ward, B. K., Chenoweth, C. M., and Irving, L. G. 1982. Recovery of viruses from vegetable surfaces. Appl. Environ. Microbiol. 44, 1389–1394.
Wasserfallen, K. A., and Rinderknecht, F. 1978. Characterisation of bacteria by Curie-point-pyrolysis, gas chromatography and computerisation of the pyrograms. Chromatographia 11, 128–136.
Waters, J. R. 1972. Sensitivity of the $^{14}CO_2$ radiometric method for bacterial detection. Appl. Microbiol. 23, 198–199.
Watson, D. C., and Walker, A. P. 1978. A modification of Brilliant Green Agar for improved isolation of *Salmonella*. J. Appl. Bacteriol. 45, 195–204.
Watson, D. H., 1984. An accessment of food contamination by toxic products of *Alternaria*. J. Food Protection 47, 485–488.
Watson, D. H., and Lindsay, D. G. 1982. A critical review of biological methods for the detection of fungal toxins in foods and foodstuffs. J. Sci. Food Agric. 33, 59–67.
Weagant, S. D. 1983. Medium for presumptive identification of *Yersinia enterocolitica*. Appl. Environ. Microbiol. 45, 472–473.
Weihe, J. D., Seibt, S. L., and Hatcher, Jr., W. S. 1984. Estimation of microbial population in frozen concentrated juice using automated impedance measurements. J. Food Sci. 49, 243–246.
Wentsel, R. S., O'Neill, P. E., and Kitchens, J. R. 1982. Evaluation of coliphage detection as a rapid indicator of water quality. Appl. Environ. Microbiol. 43, 430–434.
Wesley, R. D., Swaminathan, B., and Stadelman, W. J. 1983. Isolation and enumeration of *Campylobacter jejuni* from poultry products by a selective enrichment method. Appl. Environ. Microbiol. 46, 1097–1102.
Weymann, L. H., Stager, C. E., Qadri, S. G. M., Villarreal, A., and Qadri, S. M. H. 1979. Evaluation of a modified dye pour-plate auxanographic method for the rapid identification of clinically significant yeasts. Comparison with two commercial systems. Med. Microbiol. Immunol. 167, 11–20.
White, C. A., and Hall, L. P. 1984. The effect of temperature abuse on *Staphylococcus aureus* and salmonellae in raw beef and chicken substrates during frozen storage. Food Microbiol. 1, 29–38.
Whittaker, B. L., and Chipley, J. R. 1979. Conditions for induction of bacteriophage from lysogenic *Bacillus megaterium* with aflatoxin B_1. Appl. Environ. Microbiol. 37, 554–558.
Wickstrom, E., and Mischke, C. F. 1980. Deoxynucleoside composition of DNA determined by reverse-phase LC. Amer. Laboratory 12, 35–38.
Wiegel, J., and Quandt, L. 1982. Determination of the gram type using the reaction between polymyxin B and lipopolysaccharides of the outer cell wall of whole bacteria. J. Gen. Microbiol. 128, 2371–2378.

Wilkins, J. R. 1978. Use of platinum electrodes for the electrochemical detection of bacteria. Appl. Environ. Microbiol. 36, 683–687.
Williams, M. L. B. 1971. The limitations of the DuPont Luminescent Biometer in microbiological analysis of food. Can. Inst. Food Technol. J. 4, 187.
Williams, R. R., Wehr, H. M., Stroup, J. R., Park, M., and Poindexter, B. E. 1980. Effects of freezing and laboratory procedures on the recovery of bacteria from ground beef. J. Food Sci. 45, 757–759.
Wilson, C. R., Andrews, W. H., and Poelma, P. L. 1980. Recovery of *Salmonella* from milk chocolate using a chemically defined medium and five nondefined broths. J. Food Sci. 45, 310–313, 316.
Winblad, S. 1979. *Yersinia enterocolitica* (synonyms: *Pasteurella* X, *Bacterium enterocoliticum* for serotype 0-8). In "Methods in Microbiology." Vol. 12. T. Bergan and J. R. Norris (Editors). Academic Press Inc. Ltd., Oval Road, London NW1 7DX, U.K.
Wojton, B., and Kossakowska, A. 1977. The use of different bacteriological tests for the evaluation of sanitary quality of carcasses. Bull. Vet. Inst. Pulawy 21, 69–74.
Wood, W. A. 1981. Physical methods. In "Manual of Methods for General Bacteriology." P. Gerhardt, R. G. E. Murray, R. N. Costilow, E. W. Nester, W. A. Wood, N. R. Krieg, and G. B. Phillips (Editors). Am. Soc. Microbiol., Washington, D.C.
Woodward, J. D. 1978. Medium for *Lactobacillus* and *Pediococcus*. J. Inst. Brew. 84, 293.
Woolfrey, B. F., Lally, R. T., and Ederer, M. N. 1984. An evaluation of three rapid coagglutination tests: Sero-STAT, Accu-Staph, and Staphyloslide, for differentiating *Staphylococcus aureus* from other species of staphylococci. Am. J. Clin. Pathol. 81, 345–348.
World Health Organization. 1981. Rapid laboratory techniques for the diagnosis of viral infections. World Health Organization Technical Report Series 66, 1–60.
Wren, M. W. D. 1979. An improved cooked meat medium for the growth of anaerobic bacteria. Med. Lab. Sci. 36, 197–199.
Wright, R. C. 1984. A new selective and differential agar medium for *E. coli* and coliform organisms. J. Appl. Bacteriol. 56, 381–388.
Wyatt, C. J., and Timm, E. M. 1982. Occurrence and survival of *Campylobacter jejuni* in milk and turkey. J. Food Prot. 45, 1218–1220.
Wyatt, L. E., Nickelson, R., II, and Vanderzant, C. 1980. Effects of sampling procedures on *Salmonella* recovery from fresh water catfish. J. Food Prot. 43, 44–45.
Wyatt, P. J. 1970. Cell wall thickness, size distribution, refractive index ratio and dry weight content of living bacteria. (*Staphylococcus aureus*). Nature 226 (5242), 277–279.
Yates, I. E., and Porter, J. K. 1982. Bacterial bioluminescence as a bioassay for mycotoxins. Appl. Environ. Microbiol. 44, 1072–1075.
Yde, M., and Ghysels, G. 1984. Performance of several enrichment media in the isolation of salmonellae from liquid egg products. J. Food Protection 47, 217–219.
Zaremba, M. 1978. PLP test as a rapid method for preliminary identification of *Yersinia enterocolitica* and *Yersinia pseudotuberculosis*. Med. Dosw. Mikrobiol. 30, 243–246.
Zavanella, M., Tagliabue, S., and Lodetti, E. 1980. A rapid microdilution method for total count of bacteria in foods. Ind. Aliment. 19, 226–229.
Zavate, O., Cotor, F., and Ivan, A. 1979. The role of foodstuffs in the transmission of some viral diseases. Hgiena 28, 105–110.
Zink, D. L., Lachica, R. V., and Dubel, J. R. 1982. *Yersinia enterocolitica* and *Yersinia enterocolitica*-like species: their pathogenicity and significance in foods. J. Food Saf. 4, 223–241.
Zipkes, M. R., Gilchrist, J. E., and Peeler, J. T. 1981. Comparison of yeast and mold counts by spiral, pour, and streak plate methods. J. Assoc. Off. Anal. Chem. 64, 1465–1469.

CHAPTER 5

Contamination of Poultry during Processing

J. S. Bailey
J. E. Thomson
N. A. Cox

U.S. Department of Agriculture
Agricultural Research Service
Richard B. Russell Agricultural Research Center
Athens, Georgia 30613

I. INTRODUCTION

Fresh poultry meat has a diverse natural microflora. Fortunately, most of these organisms are not pathogenic to humans. The major area of concern in the processing of poultry is to remove as much of this contamination as possible and to prevent cross-contamination of any potential pathogens which might be present. May (1974) stated, "There exists in all plants numerous points of potential cross-contamination, examples of which are listed in [Table I]. In fact, anything that touches a single bird might cause contamination and anything that touches more than one bird might have cross-contamination." In this chapter, the bacteria associated with live poultry entering the processing plant will be discussed, and those processing operations which most influence the final microbiological quality of fresh processed poultry either by removal of the organisms or by contributing to cross-contamination will be closely examined.

II. BACTERIA ASSOCIATED WITH LIVE POULTRY

Under current broiler production practices, healthy chickens ready for processing harbor a tremendous number and variety of bacteria. Exter-

TABLE I

Some Points of Potential Cross-Contamination in Poultry Processing Plants[a]

Receiving and hanging	Chilling
bird to bird in coops	chill water
air in holding sheds	ice
coops	bird to bird
hands of hangers	air
dust and air in hanging area	elevators
shackles and rail dust	belts and chutes
Killing	giblet to giblet, neck to neck
bird to bird	Grading
air	employees hands
killing machine or knife	belts
shackles and rail dust	shackles and rail dust
Scalding and defeathering	bird to bird
scald water	air
picking fingers	Ice packing
condensate	employees hands
air	packing bins
bird to bird	bird to bird
pinners hands	air
hock cutter	ice
belt for rehang	packing material
shackles and rail dust	giblet or neck to carcass (or vice versa)
rehang operators hands	Cut-up
Evisceration	employees hands
employees hands, inspectors hands	saws or power knives
knives and other cutting instruments	bird to bird
machine contact surfaces (oil sac, lung machines, head cutters, etc.)	part to part
air	air
shackles and rail dust	belts, bins, pans, etc.
bird to bird	shackles and rail dust
noncutting instruments (lung guns, lung rakes, head pullers, etc.)	
belts and chutes	
giblet flumes and water	
hang back racks	

[a] From May (1974).

nally, these microorganisms are present on the surfaces of feet, feathers and skin in association with adhering soil, dust, and fecal material internally they are present, for the most part, in the intestinal tract an respiratory system. In addition to these "natural" contaminants healthy birds, a host of other organisms, including viruses, mycoplas mas, and pathogenic fungi, are sporadically introduced into the process ing plant environment by undetected diseased or carrier birds (Mercuri et al., 1974). The principal genera of bacteria associated with the intestinal tract of healthy birds are *Lactobacillus* (Shapiro and Sarles, 1949; Fuller, 1973), *Corynebacterium* (Barnes, 1960; Schefferle, 1966), *Escherichia*—predominantly *E. coli*, *Streptococcus*—predominantly *S. facecalis*, and *Clostridium*—predominantly *C. perfringens* (Shapiro and Sarles, 1949). Bacteria found in the respiratory tract of healthy birds, principally in the nasal cavity, trachea, and lungs, include *Streptococcus, Staphylococcus, Corynebacterium, Lactobacillus, Escherichia,* and *Bacillus* (Smibert et al., 1958). Air sacs were found to be relatively free of microorganisms in the study by Smibert et al. The psychrophilic bacteria isolated from live chickens at processing plants belong primarily to the genera *Flavobacterium, Acinetobacter, Achromobacter,*[1] and *Corynebacterium* (Barnes, 1960: Barnes and Melton, 1971: and Clark and Lentz, 1969). A comprehensive review of the bacteria found on the surface of poultry carcasses at various steps of dressing and processing was conducted by Brune and Cunningham (1971).

III. SCALDING

Water in immersion systems might be expected to be contaminated. However, total aerobic counts of scald water, in most instances, rarely exceed 50,000 per milliliter (Walker and Ayres, 1956; Mead and Impey, 1970; Mulder and Veerkamp, 1974). After an initial increase, bacterial counts of scald water remain relatively constant throughout the processing day (Mulder and Veerkamp, 1974). Total bacterial counts of skin of broilers immediately after scalding are also low (usually less than 10,000/cm^2) and do not differ markedly from the counts of the skin of the live bird (Walker and Ayres, 1956: Clark and Lentz, 1969; Surkiewicz et al., 1969). *Clostridium perfringens* can be isolated rather routinely in relatively low numbers from scald water samples and from birds immediately after scalding (Mead and Impey, 1970; Lillard, 1971). *Staphylococcus* can be isolated rather frequently from scald water or scalded birds, but usually in low numbers (Walker and Ayres, 1956; Surkiewicz et al., 1969). Salmonellae are rarely isolated from scald water samples (Browne, 1949: Morris

---

[1] *Achromobacter* is no longer recognized as a genus in "Bergey's Manual of Determinative Bacteriology" (Buchanan and Gibbons, 1974).

and Ayres, 1960; Surkiewicz et al., 1969). Psychrotrophic bacteria are present on the feet, feathers, and skin of live birds—primarily *Achromobacter (Acinetobacter), Corynebacterium,* and *Flavobacterium*—but their numbers decrease markedly after scalding. Pseudomonads, the principal spoilers of eviscerated poultry, are not usually found on the live bird or in the intestinal tract, so that their isolation from scald water is rare (Barnes, 1960; Clark and Lentz, 1969; Lahallec, 1974). Scald water in immersion systems and its contaminants can enter the trachea and can invade air sacs and may contaminate internal organs and edible tissues (Thomson and Kotula, 1959: Lillard, 1971). The degree of such contamination is less when slaughter is by the Kosher cut (trachea severed), if birds are electrically stunned, and when bleeding time prior to scalding is 2 min or more (Thomson and Kotula, 1959: Tarver and May, 1963a,b).

Although the external surfaces of slaughtered birds to be scalded are heavily contaminated, the continuous overflow of contaminated scald water and the simultaneous introduction of fresh water, plus the destruction of some bacteria by heat, prevent the excessive accumulation of bacteria in commercial tank scalders. There is opportunity for bacteria to be transferred from one carcass to another via the scald water, i.e., cross-contamination, but it is highly unlikely to result in significant differences in either the nature or degree of contamination of the external surfaces among birds. Contamination of the skin, in particular, apparently does not increase during immersion scalding. The great reduction in numbers of most psychrophilic bacteria during scalding and the absence of pseudomonads indicate that scald water contamination plays a relatively minor role in spoilage of broiler carcasses.

The possible inspiration of contaminated scald water by the birds, with subsequent contamination of air sacs, lungs, and possible other internal tissues by pathogenic bacteria is the major objection to immersion scalding. The incidence of viable salmonellae in scald water is low, but other potentially harmful contaminants, particularly the spores of *C. perfringens,* can be expected to be routinely present, regardless of the scald water temperature, and can thus contaminate air sacs and other internal organs (Lillard, 1971).

Although not all reports agree, the preponderance of evidence indicates that shelf life of broilers is reduced by scalding at temperatures above 58°C. This can be attributed to the changes that occur in the outer epidermal layer (cuticle) of the bird's skin. Scalding at about 58–60°C (subscald) and above, followed by mechanical picking, results in removal of this cuticle, whereas scalding at 52–53°C (semiscald) does not. The cuticle-free and slightly denatured skin of hard-scalded broilers apparently serves as a more suitable substrate for spoilage microorganisms—particularly *Pseudomonas* species whose primary sources are water and ice (Barnes, 1960). The work of Clark (1968) in a significant basic study that supports the practical findings of Ziegler and Stadelman (1955), Essary *et al.* (1958), and Berner *et al.* (1969) demonstrated that growth of *Pseudomonas* on skin was greater for broilers scalded at 59°C than for those scalded at 54°C. *Achromobacter*, on the other hand, grew better on 54°C- than on 50°C-scalded skin. When birds are scalded at 58–60°C the skin surface must be kept moist, or discoloration occurs. With dry-picked or when 52–54°C scald temperatures are used, the skin can dry without discoloration.

IV. BACTERIOLOGY OF OTHER SCALDING METHODS

Patrick *et al.* (1972) reported that steam–hot water spray–scalded birds (53–55°C, 131 sec) showed lower total coliform counts than immersion-scalded birds (55°C, 120 sec) immediately after scalding and defeathering, but no differences were found between the two methods after chilling. In later work, Patrick *et al.* (1973) were unable to show a statistically significant difference in the percentage of isolations of *Salmonella* between the two systems, although there was a trend toward fewer isolations from the spray-scalded birds.

An experimental batch-type scalder using subatmospheric steam produced a lower degree of air sac and lung bacterial contamination of broilers when compared to immersion water scalders, but the difference was absent after the birds had been conventionally slush ice–chilled (Klose *et al.*, 1971; Kaufman *et al.*, 1972: Lillard, 1973).

The question as to whether high-temperature (ca 60°C) or low-temperature (ca 54°C) scalding should be used during processing of broilers, regardless of whether immersion or other systems are employed, is of concern to many processors. With regard to the claim that high-temperature scalding is more hygenic, there is certainly theoretical and experimental evidence that temperatures of about 60°C are more effective in reducing numbers of bacteria in scald water than lower temperatures. Grossklaus and Lessing (1964) reported that destruction of *Salmonella typhimurium* (artificially inoculated) required 2 hr in 52°C scald water but only 30 minutes in 60°C scald water. That this difference in bactericidal activity at this point in the processing line does result in a significantly lower incidence of pathogen-contaminated carcasses at the end of the line has not been clearly established. The bulk of the evidence suggests that other processes, e.g., defeathering, evisceration, and chilling, are of greater importance than scalding in cross-contamination of carcasses.

V. PICKING

Little work has been done to evaluate the effect of picking and pinning (feather removal) on the microbiological quality of poultry meat. May (1961) found counts to be relatively low after the bird had been picked and washed; the count then increased as the carcass progressed through various stations to the final wash. Surkiewicz et al. (1969) found total aerobic plate counts to be between $1200/cm^2$ and $35,000/cm^2$ after picking, but in six of eight plants surveyed, counts were higher than when birds exited the scald tank. Feather removal resulted in cross-contamination when both an external and internal marker strain of *E. coli* was used to artificially contaminate broiler carcasses (Mulder et al., 1978). More cross-contamination occurred during scalding and plucking when lower scalding temperatures (52–54°C) were used than when a higher scalding temperature (60°C) was used (Mulder et al., 1978).

VI. EVISCERATION

Any location where carcasses must be handled or are allowed to rub against each other has the potential for cross-contamination. May (1961) studied, at 13 different plants, the contamination at intermediate steps during evisceration (Table II). He found counts to be relatively low until after the birds had been transferred to the evisceration line: the counts then increased as the carcass progressed through various stations to the final wash. Schuler and Badenhop (1972) showed that the transfer belt and equipment which touched the carcasses had increased bacterial buildup through the day and could have very high counts which greatly contributed to cross-contamination. Spray washing (Table III) and washing with the combined scrubbing action of flexible water fingers have been shown to reduce the microbial load on the skin of broilers (Gunderson et al., 1954; Keel and Parmelee, 1968). A chlorinated spray wash of

TABLE II

Mean[a] Total Counts of Bacteria per Square Centimeter at Various Work Stations[b]

Sample	Number of organisms		
	Lowest	Usual range	Highest
After pick and wash	162	273–4317	4840
Feet and preen removal	272	455–3803	6480
Transfer to evisceration line	320	525–14780	16600
Opening body cavity	1089	1089–18039	23580
Draw viscera	1089	1160–15950	28419
Government inspection	840	1420–16913	22793
Giblet removal	1790	1950–15550	16793
Lung removal	1222	1390–14625	20670
Crop removal	1000	1379–8280	27010
Final wash	54	360–4920	6990
House inspector	72	340–5373	6800

[a] Mean of 10 bird samples per station.
[b] From May (1961).

TABLE III

Total Viable Bacterial Counts per Square Centimeter of Surface of Freshly Killed Warm Eviscerated Poultry[a]

Sample	Number of organisms		
	Lowest	Usual range	Highest
Before singer	910	910–4640	12900
After singer	80	1700–1750	7700
After Bloom-o-matic	360	430–770	860
After pinning	490	600–680	4500
After neck cutting	900	1090–5740	6290
After opening cut	850	1150–2140	2320
After evisceration	2430	3680–13130	15940
After head, neck, kidney, and ovary removal	1540	14650–16130	26960
After inside wash	1660	1670–3670	4090
Before final wash	5590	6360–7100	21500
After final wash	250	500–790	1200

[a] From Gunderson et al. (1954).

40 ppm greatly reduced the number of viable bacteria on the eviscerator, and a wash of 70 ppm chlorine almost totally eliminated any buildup of bacteria (Thomson et al., 1984). The microbiological quality of inspection-passed and fecal contamination–condemned broiler carcasses was indistinguishable when condemned carcasses were spray-washed for 5 sec with about 200 ml of water (Blankenship et al., 1975).

Increased demand of fast-food service establishments for boned poultry meat and the potential to save time, labor, and expense of all conventional processing points beyond defeathering have increased the popularity of hot-boned or unconventionally harvested poultry meat. Barnes and Impey (1975) found that uneviscerated broiler carcasses stored at 10 and 4°C had significantly longer shelf life than conventionally eviscerated broilers stored at the same temperatures. Lillard et al. (1984a) found that poultry meat which was spray-washed immediately after defeathering and before hot-boning had similar bacterial quality to meat from fully processed chilled carcasses. Incidence of coagulase-positive staphylococci and C. perfringens was not significantly different between conventionally eviscerated broilers and uneviscerated, spray-washed broilers (Lillard et al., 1984b), and the incidence of salmonellae was significantly reduced by reducing the number of stages of processing.

VII. CHILLING

In evaluating the effect of continuous-immersion chillers, in which carcasses and water flow in the same direction, with 0.5 gal overflow per carcass, Brewer et al. (1961), Kotula et al. (1962), Farrell and Barnes (1964), Surkiewicz et al. (1969), Keel and Parmelee (1968), Knoop et al. (1971), Veerkamp et al. (1972), Vacinek (1972), Brant (1973), and May (1974) found that total microbial counts were reduced. Vacinek (1972) found that even with a buildup of 4000–6000 microorganisms per milliliter in chill water during the day, there was a 60–98% reduction in microbial counts on the carcass skin during continuous chilling. A reduction in numbers of a tracer organism (E. coli strain) was found during continuous immersion chilling (van Schothorst et al., 1972). Thomson et al. (1965, 1966) found no significant difference in carcass microbial counts at up to five locations in commercial immersion chillers. In contrast, Clark and Lentz (1969), Lillard (1971), and Peric et al. (1971b) found that continuous-immersion chilling could be related to an increase in carcass total microbial counts. Grossklaus and Lessing (1964), Berner and Scholtyssek (1968), and Grossklaus (1972) claimed that immersion chilling is unhygienic. Important factors contributing to microbial counts of immersion chilled poultry are (1) bacterial contamination on carcasses before chilling, (2) the amount of water overflowed and replaced per carcass, and (3) the ratio of birds to water in the chiller (Peric et al., 1971b). These factors could contribute to differences found by various investigators.

Continuous-immersion chilling systems thus can be operated in a manner that reduces total microbial count, but criticism has been leveled at these systems on the basis that food-poisoning bacteria, some types of which may be present in significant numbers on relatively few carcasses before chilling, may be distributed to many other carcasses by the fluid chilling media during chilling (so-called cross-contamination) (Bulling and Pietzsch, 1966; Morris and Wells, 1970).

Surkiewicz et al. (1969) found no reduction in incidence of salmonellae contamination after immersion chilling, and Morris and Wells (1970) found higher incidence of salmonellae after immersion chilling than before. A tracer microorganism inoculated on some carcasses before chilling could be found on others after chilling, verifying cross-contamination (van Schothorst et al., 1972). The "seeding" of carcasses with spoilage types of bacteria may also occur during immersion chilling. Knoop et al. (1971) presented evidence that initial psychrophile counts are higher on wet-chilled than on dry-chilled carcasses, and that shelf life of dry-chilled birds was longer.

Campbell et al. (1983) surveyed 9 federally inspected poultry processing plants and found 5.5% of carcasses entering and 11.6% of the carcasses exiting the chiller to be contaminated with salmonellae. In a 10-year follow-up survey of 15 poultry processing plants 36.9% of carcasses exiting the chiller were found to be contaminated with salmonellae (Green et al., 1982).

Mead and Impey (1970) found C. perfringens in chill water in both chicken and turkey processing plants, and Lillard (1971) found increased incidence of C. perfringens on broilers after immersion chilling, compared to immediately before chilling. Busta et al. (1973) found C. perfringens in 53%, S. aureus in 22%, Salmonella in 17.6%, and coliforms in 100% in chiller water samples from 3 turkey processing plants. The incidence of these organisms on turkey skin differed very little before and after chilling; C. perfringens 87% before, 83% after; S. aureus 71% before, 67% after; salmonellae 28% before, 24% after; and coliforms 100% both before and after (Busta et al., 1973).

Chlorination of spray-water or chiller water has been investigated as a means to reduce microbial counts or to prevent spreading of contamination from bird to bird. In-plant chlorination significantly improves sanitation in processing plants (Goresline et al., 1951; Gunderson et al., 1954; Drewniak et al., 1954; Barnes 1972; Ranken, 1973). Smith (1970) reported that Blackshear and Sanders found that 40–60 ppm chlorine from gas chlorinators added to final spray washer at pH 6.5–7.5 reduced microbial counts about 0.7 log compared to plain water. Sanders and Blackshear (1971) found that hypochlorite solutions at 40–60 ppm chlorine in the spray washer also effectively reduced carcass microbial counts. Kotula et al. (1967) determined that postchill carcass spray washing with water chlorinated at up to 50 ppm did not extend shelf life. A 5-min postchill dip in 10 or 20 ppm chlorine water was ineffective in extending carcass shelf life, but chilling for 2 hr in 10 or 20 ppm chlorinated water significantly increased shelf life (Ziegler and Stadelman, 1955). Lillard (1980) found that 20 ppm Cl_2 and 3 ppm ClO_2 significantly reduced the bacterial counts of carcasses compared to those chilled in untreated water. Thiessen et al. (1984) found that increasing the concentration of ClO_2 from 0. to 1.39 mg/liter resulted in reducing the bacteria count to the point where salmonellae could not be isolated from the chill water or the chilled broiler carcasses.

Dawson et al. (1956) found that chlorine at 140 ppm in chill water for 2 hr increased shelf life by 5 days, and 20–40 ppm, by 2–4 days beyond controls. McVicker et al. (1958) found that 20 ppm chlorine in chill water could extend shelf life. Mallman et al. (1959) determined that chlorine concentrations between 50 and 100 ppm extended shelf life several days. With an initial concentration of available chlorine of 200 ppm in chiller water, shelf life at 1°C (34°F) was extended by about 20% (Ranken et al., 1965). Wabeck et al. (1968a) found no reduction in total counts on chicken drumsticks treated with 20 ppm chlorine. Patterson (1968b) found an increase of 2 days in carcass shelf life with 200 ppm chlorine, and of 4 days with 400 ppm.

Mead and Thomas (1973) determined that 45–50 ppm chlorine maintained in a continuous-immersion chiller with addition of 5 liters of fresh water per carcass destroyed most bacteria present in the chill water. Chlorine at 25–30 ppm and overflow of 8 liters per carcass were equally effective. Brant (1973) reported that total counts were reduced about 90% during chilling in some plants chlorinating chill water at 18–25 ppm. In other plants, reduction in total counts in breast skin area during chilling in unchlorinated water was about 60%, and in water chlorinated at 50 ppm, was 85%.

The effect of chlorination on salmonellae and other food-poisoning microorganisms on poultry carcasses has been studied, with variable results. Dixon and Pooley (1961) found that 200 ppm chlorine treatment for 10 min was effective when fewer than 1000 *Salmonella* organisms had been inoculated per carcass. Nilsson and Regner (1963) determined that 20 ppm free chlorine usually eliminated up to 10^3 cells of artificially inoculated *S. typhimurium* on chicken carcasses, but an inoculum of 10^6 cells did not. Barnes (1965) found that even 200 ppm hardly reduced fecal organisms on poultry carcasses but did prevent cross-contamination and multiplication of spoilage organisms. Thomson *et al.* (1967) showed that 100 and 200 ppm chlorine significantly reduced artificially inoculated salmonellae on chicken skin. Kotula *et al.* (1967) found significantly fewer *Salmonella*-contaminated carcasses among those washed after chilling with chlorine in concentrations up to 50 ppm than without chlorine in water spray. Total aerobic counts and incidence of fecal streptococci and *S. aureus* were not significantly reduced by chlorination of continuous immersion chill water at up to 20 ppm chlorine, but coli-aerogenes numbers were reduced at 20 ppm (Patterson, 1968a). Wabeck *et al.* (1968b) showed that 20 and 40 ppm chlorine destroyed salmonellae when organic matter was not present but only slightly reduced salmonellae on chicken meat. Thomson *et al.* (1979) found that 0.25 gal of fresh water input per carcass was not significantly different from the normal 0.50 gal per carcass on either cross-contamination or elimination of *Salmonella* from inoculated carcasses, and that fewer carcasses showed cross-contamination after chilling with 50 ppm chlorine than 0 ppm, but cross-contamination was not eliminated.

Dye and Mead (1972) found that chlorine treatment during processing probably would not eliminate clostridial spores from poultry carcasses, but even low levels of free available chlorine (about 0.5 ppm) are more effective in destroying spores in chill water than are high levels of combined available chlorine (such as chloramine).

The addition of other chemicals to the chill water to destroy bacteria has been investigated. Perry *et al.*, (1964) and Mountney and O'Malley (1964, 1965), Juven *et al.* (1974), Cox *et al.* (1974) and Thomson *et al.* (1976) found that acids added to chill water improved carcass shelf life. Elliot *et al.* (1964), Steinhauer and Banwart (1964), Taylor *et al.* (1965), Panda (1970), and Thomson *et al.* (1976) inhibited microbial growth on poultry by addition of polyphosphates to the chill water. *Salmonella* (250 cells per carcass) were eliminated, and cross-contamination of broiler carcasses was prevented when gluteraldehyde (Thomson *et al.*, 1977) and poly(hexamethylenebiguanide) hydrochloride) (Islam and Islam, 1979; Thomson *et al.*, 1981) were added to prechill water.

Countercurrent immersion chilling, at which water flow is in the opposite direction from carcass flow, was shown to give an efficient chilling and good microbiological quality when water overflow was $2\frac{1}{2}$ liters per carcass (Mulder *et al.*, 1976).

The effectiveness of spray and dry chilling in reducing microbial counts in comparison to immersion chilling has been studied. Grossklaus and Lessing (1964) suggested spray or cold-air cooling to remedy hygienic problems of immersion chilling. Grossklaus and Levetzow (1967) found that water-spray chilling reduced bacteria counts and extended shelf life of chicken carcasses. Peric et al. (1971a) developed a model prototype spray chilling system with a capacity of 1080 birds per hour. A reduction of about 90% in total and Enterobacteriaceae counts could be effected. When *Serratia marcescens*-inoculated birds were passed through the spray chiller, cross-contamination was limited to birds adjacent to highly contaminated carcasses when strong air circulation was present. Veerkamp et al. (1972), in comparing two types of commercial continuous-immersion chillers and a prototype spray cooling apparatus set at either high or low water-usage rate, concluded that the cleaning effect of continuous-immersion chilling is equal to that of spray chilling, provided sufficient water replacement is made in the immersion chiller. Leistner et al. (1972) described two chilling systems, the first a water spray system, and the second a combination water spray and air blast system. Reductions of 90% or greater in total and Enterobacteriaceae counts were found in both systems.

Dry chilling with air has been used commercially (Anonymous, 1969, 1970a,b, 1973; Stephens, 1970; Timmons, 1969). Knoop et al. (1970, 1971) found that initial psychrophile counts of wet-chilled birds were higher than dry-chilled. However, Berner et al. (1969) found significantly lower bacteria counts, both immediately after chilling and during storage up to 32 days at $-1°C$ (30°F), on air-chilled than on water immersion–chilled carcasses; a difference of more than log 2 of the microbial count was evident at 16 days between dry- and wet-chilled carcasses. Chilling with a standard prechill followed by crust freezing in a cryogenic CO_2 tunnel extended the shelf life by 3 days when carcasses were held at 37°C (Thomson et al., 1983).

Immersion chilling as commonly practiced by the poultry industry has often come under scrutiny because of the potential for cross-contamination. The findings of most researchers indicate that there is a potential for cross-contamination, but that with properly used equipment and adequate water replacement, the washing effect of commercial immersion chilling of broilers will reduce total bacterial counts.

VIII. BACTERIA ASSOCIATED WITH GIBLETS

The microbiological profile of giblets which had been through standard processing, water flumed, and batch ice chilled are shown in Table IV. Initial total aerobic counts per gram ranged from log 3.5/g to log

5. Contamination of Poultry during Processing

TABLE IV

Microbiological Profile of Broiler Chicken Giblets[a]

Ingredients	Replicate	Total aerobic count/g[b]	Coliform count/g[b]	E. coli count/g[b]	Salmonella found	S. aureus count/g[b]	KF streptococcal count/g[b]
Heart	1	4.5	2.7	2.4	Yes	1.7	2.3
	2	3.5	>3.4	2.4	Yes	1.0	1.5
Liver	1	3.8	>3.4	>3.4	Yes	1.6	2.5
	2	5.0	>3.4	.5	Yes	1.8	2.3
Gizzard	1	5.0	>3.4	1.3	Yes	None	1.6
	2	4.4	>3.4	2.0	Yes	1.0	1.0

[a] From Cox et al. (1983).
[b] Counts expressed as $\log_{10}$ of number per gram.

5.0/g. Coliform counts were greater than log 3.4 [three-tube most-probable number (MPN)]. The MPN of *E. coli* per gram was similar in each replication in all ingredients except liver samples. *Salmonella* was recovered from all giblets in each replication. *Staphylococcus aureus* counts ranged from not detected to log 2/g. Streptococcal counts ranged from log 1.0/g to log 2.5/g.

Initial total aerobic counts were similar to the log 4.2/g reported by Mead and Adams (1980). Mead and Adams (1980) and May et al. (1962) had reported counts of livers to be log 4.7/g and log 4.2/g, respectively. Gizzard counts reported by Cox et al. (1983) were similar to those reported by May et al. (1962), Charoenpong and Chen (1979), and Mead and Adams (1980). Charoenpong and Chen (1979) reported log 1.7 coliforms per gram on chicken gizzards, while Mead and Adams (1980) found coliforms on giblets to be approximately log 4.0. Salzer et al. (1965) found *Salmonella* in only 5% of turkey giblets, but they used a less sensitive swab technique. Charoenpong and Chen (1979) found log 2.3 *S. aureus* per gram of gizzard tissue.

IX. SUMMARY

Fresh poultry enters the processing plant carrying a diverse natural microflora, most of which are not pathogenic to humans. It is the goal of the poultry processor to try to reduce the total number of microorganisms in order to assure an adequate shelf life during distribution and to try to prevent the cross-contamination of any pathogenic bacteria which might be present.

All of the processing plant operations can affect the microbiological quality of the fully processed carcass. Cross-contamination may result from bird-to-bird contact, handling of the carcasses, or contact with processing equipment, tools, or the water bath of the scald and chill tanks. Good manufacturing practices (minimum handling of carcasses: good sanitation of processing equipment and tools; air, water, and ice free of contamination) and proper fresh water input and overflow in scald and chill tanks will lead to minimal cross-contamination, and the resulting carcass will have an adequate shelf life to satisfy both the processor and the consumer.

REFERENCES

Anonymous. 1969. Hillcrest's dry chill is "dry" indeed. Broiler Industry 32(10): 57.
Anonymous. 1970a. Chilling equipment must go, Hillcrest's Stephens insists. Broiler Industry 33(3): 53.
Anonymous. 1970b. New chill system promises dry pack, good yield. Poultry Meat 21(10): 10.
Anonymous. 1973. Air-chilled broilers fetch a premium. Poultry International 12(2): 12.
Barnes, E. M. 1960. The sources of different psychrophilic spoilage organisms on chilled eviscerated poultry. Progress in Refrigeration, Sci. Technol. 3: 97.
Barnes, E. M. 1965. The effect of chlorinating chill tanks on the bacteriological condition of processed chickens. Institute Internationale du Froid. Bulletin Commission 4-Karlsruhe 1-7.
Barnes, E. M. 1972. Food poisoning and spoilage bacteria in poultry processing. Vet. Rec. 90: 720.
Barnes, E. M., and Impey, C. S. 1975. The shelf-life of uneviscerated and eviscerated chicken carcasses stored at 10°C and 4°C. Br. Poultry Sci. 16: 319.
Barnes, E. M., and Melton, W. 1971. Extracellular enzymic activity of poultry spoilage bacteria. J. Appl. Bact. 34: 599.
Berner, H., and Scholtyssek, S. 1968. Moglichkelten zur Verbesserung der Schlacthygiene beim Geflugel. [Chilling possibilities for the improvement of slaughter hygiene of poultry.] Die Fleischwirtschaft 48: 422.
Berner, H., Kleeberger, A., and Busse, M. 1969. Untersuchungen uber ein neues Kuhlverfahren fur Schlachtgefluge. III. Hygienische uberprufung des Verfahrens hygiene for the new method). Die Fleischwirtschaft 49: 1617.
Blankenship, L. C., Cox, N. A., Craven, S. E., Mercuri, A. J., and Wilson, R. L. 1975. Comparison of the microbiological quality of inspection-passed and fecal contamination-condemned broiler carcasses. J. Food Sci. 40: 1236.
Brant, A. W. 1973. Controlled sanitary immersion chilling of poultry in the U.S.A. Poultry meat symposium, World's Poultry Science Association (European Federation), Roskilde, Denmark.
Brewer, R. N., Edgar, S. A., Mora, E. C., and Pruett, J. 1961. A study of the sanitation of chilling operations in poultry processing plants in Alabama. Poultry Sci. 40: 1383.

Browne, A. S. 1949. The public health significance of *Salmonella* on poultry and poultry products. Ph.D. Thesis, Univ. of Calif. p. 17.
Brune, H. E., and Cunningham, F. E. 1971. A review of microbiological aspects of poultry processing. World's Poultry Sci. 27: 223.
Buchanan, R. E., and Gibbons, N. E. 1974. Bergey's Manual of Determinative Bacteriology, 8th ed. The Williams and Wilkins Co., Baltimore, MD.
Bulling, E., and Pietzsch, O. 1966. Salmonella-Bakterien in Schlactgeflugel (*Salmonella* bacteria in poultry). Berl. Munch. Tierarztl. Wschr. 79: 132.
Busta, F. F., Zottola, E. A., Arnold, E. A., and Hagborg, M. M. 1973. Research Report. Incidence and control of unwanted microorganisms in turkey products. I. Influence of handling and freezing on viability of bacteria in and on products. Dept. of Food Science and Nutrition. Monograph. Univ. of Minnesota, St. Paul, MN.
Campbell, D. F., Johnston, R. W., Campbell, G. S., McClain, D., and Macaluso, J. F. 1983. The microbiology of raw, eviscerated chickens: A ten year comparison. Poultry Sci. 62: 437.
Charoenpong, C., and Chen, T. C. 1979. Microbiological quality of refrigerated chicken gizzard from different sources as related to their shelf-lives. Poultry Sci. 58: 824.
Clark, D. S. 1968. Growth of psychrotolerant pseudomonads and *Achromobacter* on chicken skin. Poultry Sci. 47: 1575.
Clark, D. S., and Lentz, C. P. 1969. Microbiological studies in poultry processing plants in Canada. Can. Inst. Food Technol. Journal 2: 33.
Cox, N. A., Mercuri, A. J., Juven, B. J., Thomson, J. E., and Chew, V. 1974. Evaluation of succinic acid and heat to improve the microbiological quality of poultry meat. J. Food Sci. 39: 985.
Cox, N. A., Bailey, J. S., Lyon, C. E., Thomson, J. E., and Hudspeth, J. P. 1983. Microbiological profile of chicken patty products containing broiler giblets. Poultry Sci. 62: 960.
Dawson, L. E., Mallman, W. L., Frang, M., and Walters, S. 1956. The influence of chlorine treatments on bacterial population and taste panel evaluation of chicken fryers. Poultry Sci. 35: 1140.
Dixon, J. M. S., and Pooley, F. E. 1961. The effect of chlorination on chicken carcasses infected with salmonellae. J. Hyg. Camb. 59: 343.
Drewniak, E. E., Howe, M. A., Goresline, H. E., and Baush, E. R. 1954. Studies of sanitizing methods for use in poultry processing. U.S. Department of Agriculture Circular 930.
Dye, M., and Mead, G. C. 1972. The effect of chlorine on the viability of clostridial spore. J. Food Technol. 7: 173.
Elliot, R. P., Straka, R. P., and Garibaldi, J. A. 1964. Polyphosphate inhibition of growth of pseudomonads from poultry meat. Applied Microbiol. 12: 517.
Essary, E. O., Moore, W. E. C., and Kramer, C. Y. 1958. Influence of scald temperatures, chill times, and holding temperatures on the bacterial flora and shelf life of freshly chilled, tray-packed poultry. Food Technol. 11: 684.
Farrell, A. J., and Barnes, E. M. 1964. The bacteriology of chilling procedures used in poultry processing plants. British Poultry Sci. 5: 89.
Fuller, R. 1973. Ecological studies on the *Lactobacillus* flora associated with the crop epithelium of the fowl. J. Appl. Bact. 36: 131.
Goresline, H. E., Howe, M. A., Baush, E. R., and Gunderson, M. F. 1951. In-plant chlorination does a 3-way job. U.S. Egg Poultry Mag. 57(4): 12, 13, 29.
Green, S. S., Moran, A. B., Johnston, R. W., Uhler, P., and Chris, J. 1982. The incidence of *Salmonella* species and serotypes in young whole chicken carcasses in 1979 as compared with 1967. Poultry Sci. 61: 288.

Grossklaus, D. 1972. Hygiene-probleme beim Schlachtgeflugel Zucht, Haltung, Schlactung, Verpackung (Poultry hygiene problems. Breeding, maintenance, slaughter, packing). Die Fleischwirtschaft 52: 1011.

Grossklaus, D., and Lessing, G. 1964. Hygiene-Probleme in Geflugelschlachtereien (Problems of hygiene in poultry slaughterhouses). Die Fleischwirtschaft 44: 1253.

Grossklaus, D., and Levetzow, R. 1967. Die Kuhlunf des Schlachtgeflugels—ein hygien isches und lebensmittelrechtliches Problem (The cooling of slaughtered poultry—a problem of hygiene and food legislation). Berl. Munch. Tierarzt. Wschr. 78: 187.

Gunderson, M. R., McFadden H. W., and Kyle, T. S. 1954. Bacteriology of Commercial Poultry Processing. Burgess Publishing Co., Minneapolis, MN. 98 pp.

Islam, M. N., and Islam, N. B. 1979. Extension of poultry shelf-life by poly-(hexamethylenebiguanide hydrochloride). J. Food Prot. 42: 416.

Juven, B. J., Cox, N. A., Mercuri, A. J., and Thomson, J. E. 1974. A hot acid treatment for eliminating *Salmonella* from chicken meat. J. Milk Food Technol. 37: 237.

Kaufman, V. F., Klose, A. A., Bayne, H. G., Poole, M. F., and Lineweaver, H. 1972. Plant processing of sub-atmospheric steam scalded poultry. Poultry Sci. 51: 1188.

Keel, J. E., and Parmelee, C. F. 1968. Improving the bacteriological quality of chicken fryers. J. Milk and Food Technol. 31: 377.

Klose, A. A., Kaufman, V. F., and Poole, M. F. 1971. Scalding poultry by steam at subatmospheric pressures. Poultry Sci. 50: 302.

Knoop, G. N., Hale, K. K., Jr., and Stadelman, W. J. 1970. Microbiological characteristics of wet- and dry-chilled poultry. Poultry Sci. 49: 1405.

Knoop, G. N., Parmelee, C. E., and Stadelman, W. J. 1971. Microbiological characteristics of wet- and dry-chilled poultry. Poultry Sci. 50: 530.

Kotula, A. W., Banwart, G. J., and Kinner, J. A. 1967. Effect of postchill washing on bacterial counts of broiler chickens. Poultry Sci. 46: 1210.

Kotula, A. W., Thomson, J. E., and Kinner, J. A. 1962. Bacterial counts associated with the chilling of fryer chickens. Poultry Sci. 41: 818.

Lahellec, C. 1974. The shelf-life of fresh poultry: origin of psychrotophic microorganisms, quantitative evolution of flora on carcasses stored under refrigeration. World Poultry Sci. J. 30. 53.

Leistner, L., Rossmanith, E., and Woltersdorf, W. 1972. Rationalisierung des Spruh-Kuhlverfahrens fur Schlachthahnchen (Rationalizing the spray method of chilling poultry). Die Fleischwirtschaft 52: 362.

Lillard, H. S. 1971. Occurrence of *Clostridium perfringens* in broiler processing and further processing operations. J. Food Sci. 38: 1008.

Lillard, H. S. 1973. Contamination of blood system and edible parts of poultry with *Clostridium perfringens* during water scalding. J. Food Sci. 38: 151.

Lillard, H. S. 1980. Effect on broiler carcasses and water of treating chiller water with chlorine or chlorine dioxide. Poultry Sci. 59: 1761.

Lillard, H. S., Hamm, D., and Thomson, J. E. 1984a. Effect of reduced processing on recovery of foodborne pathogens from hot-boned broiler meat and skin. J. Food Prot. 47: 209.

Lillard, H. S., Hamm, D., and Thomson, J. E. 1984b. Effect of spray-washing uneviscerated carcasses on microbiological quality of hot-boned poultry meat harvested after reduced processing. Poultry Sci. 63: 661.

Mallman, W. L., Dawson, L. E., and Jones, E. 1959. The influence of relatively high chlorine concentrations in chill tanks in shelf-life of fryers. Poultry Sci. 38: 1225.

May, K. N. 1961. Skin contamination of broilers during commercial evisceration. Poultry Sci. 40: 531.

May, K. N. 1974. Changes in microbial numbers during final washing and chilling of commercially slaughtered broilers. Poultry Sci. 53: 1282.
May, K. N., Irby, J. D., and Carmon, J. L. 1962. Shelf-life and bacterial counts of excised poultry tissue. Food Technol. 16: 66.
McVicker, R. J., Dawson, L. E., Mallman, W. L., Walters, S., and Jones, E. 1958. Effect of certain bacterial inhibitors on shelf-life of fresh fryers. Food Technol. 12: 147.
Mead, G. C., and Adams, B. W. 1980. A note on the shelf-life and spoilage of commercially-processed chicken giblets. Br. Poultry Sci. 21: 411.
Mead, G. C., and Impey, C. S. 1970. The distribution of clostridia in poultry processing plants. British Poultry Sci. 11: 407.
Mead, G. C., and Thomas, N. L. 1973. Factors affecting the use of chlorine in the spin-chilling of eviscerated poultry. British Poultry Sci. 14: 99.
Mercuri, A. J., Cox, N. A., and Thomson, J. E. 1974. Microbiological aspects of poultry scalding. Proceedings, XV World Poultry Congress, New Orleans, Louisiana, 543–545.
Morris, T. G., and Ayres, J. C. 1960. Incidence of salmonellae on commercially processed poultry. Poultry Sci. 39: 1131.
Morris, G. K., and Wells, J. G. 1970. *Salmonella* contamination in a poultry-processing plant. Applied Microbiol. 19: 795.
Mountney, G. J., and O'Malley, J. E. 1964. Acids as poultry meat preservatives. Poultry Sci. 43: 1345.
Mountney, G. J., and O'Malley, J. E. 1965. Acids as poultry meat preservatives. Poultry Sci. 44: 582.
Mulder, R. W. A. W., and Veerkamp, C. H. 1974. Improvements in poultry slaughterhouse hygiene as a result of cleaning before cooling. World Poultry Sci. J. 30: 49.
Mulder, R. W. A. W., Dooresteijn, L. W. J., Hofmans, G. J. P., and Veerkamp, C. H. 1976. Experiments with continuous immersion chilling of broiler carcasses according to the code of practice. J. Food Sci. 41: 438.
Mulder, R. W. A. W., Dooresteijn, L. W. J., and Van Der Broek, J., 1978. Cross-contamination during the scalding and plucking of broilers. British Poultry Sci. 19: 61.
Nilsson, T., and Regner, B. 1963. The effect of chlorine in the chilling water on salmonellae in dressed chicken. Acta Vet. Scand. 4: 307.
Panda, P. C. 1970. Bacteriological condition and shelf-life of dressed poultry chilled in slush ice tanks containing polyphosphates. Indian Food Packer 24(6): 30.
Patrick, T. E., Goodwin, T. L., Collins, J. A., Wyche, R. C., and Love, B. E. 1972. Steam versus hot-water scalding in reducing bacterial loads on the skin of commercially processed poultry. Appl. Microbiol. 23: 796.
Patrick, T. E., Collins, J. A., and Goodwin, T. L. 1973. Isolation of *Salmonella* from carcasses of steam and water scalded poultry. J. Milk and Food Technol. 36: 34.
Patterson, J. T. 1968a. Bacterial flora of chicken carcasses treated with high concentrations of chlorine. J. Appl. Bacteriol. 31: 544.
Patterson, J. T. 1968b. Chlorination of water used for poultry processing. British Poultry Sci. 9: 129.
Peric, M., Rossmanith, E., and Leistner, L. 1971a. Verbesserung der mikrobiologischen Qualitat von Schluchthahnchen durch die Spruh-Kuhlung (Improving the microbiological quality of chickens by spray chilling). Die Fleischwirtschaft 51: 574.
Peric, M., Rossmanith, E., and Leistner, L. 1971b. Undersuchungen uber die Beeinflussung des Oberflachenkeimgehaltes von Schlacthahnchen durch die Spinchiller-Kuhlung (Investigations into the influence of spin chiller cooling on the surface bacterial content of poultry). Die Fleischwirtschaft 51: 216.

Perry, G. A., Lawrence, R. L., and Melnick, D. 1964. Extension of poultry shelf life by processing with sorbic acid. Food Technol. 18: 891.

Ranken, M. D. 1973. Chlorination and the hygienic operation of spin chillers. 4th European Poultry Conference. Proceedings. London 353–358.

Ranken, M. D., Clewlow, G., Shrimpton, D. H., and Stevens, B. J. H. 1965. Chlorination in poultry processing. British Poultry Sci. 6: 331.

Salzer, R. H., Kraft, A. A., and Ayres, J. C. 1965. Effect of processing on bacteria associated with turkey giblets. Poultry Sci. 44: 952.

Sanders, D. H., and Blackshear, C. D. 1971. Effect of chlorination in the final washer on bacterial counts of broiler chicken carcasses. Poultry Sci. 50: 215.

Schefferle, H. E. 1966. Coryneform bacteria in poultry deep litter. J. Appl. Bact. 29: 147.

Schuler, G. A., and Badenhop, A. F. 1972. Microbiology survey of equipment in selected poultry processing plants. Poultry Sci. 51: 830.

Shapiro, S. K., and Sarles, W. B. 1949. Microorganisms in the intestinal tract of normal chickens. J. Bact. 58: 542.

Smibert, R. M., DeVolt, H. M., and Faber, J. E. 1958. Studies on "air-sac" infection in poultry. Poultry Sci. 38: 1393.

Smith, C. H. 1970. Spray-in-line process clobbers bacteria. Broiler Industry 33(2): 38.

Steinhauer, J. E., and Banwart, G. J. 1964. The effect of food grade polyphosphates on the microbial population of chicken meat. Poultry Sci. 43: 618.

Stephens, A. 1970. Packages Poultry better way. Food Engineering 42: 123.

Surkiewicz, B. F., Johnston, R. W., Moran, A. B., and Krumm, G. W. 1969. A bacteriological survey of chicken eviscerating plants. Food Technol. 23: 1066.

Tarver, F. R., and May, K. N. 1963a. Effect of bleed time prior to scald and refrigerated storage upon bacterial counts in the auxillary diverticula of the interclavicular air sac of chickens. Food Technol. 17: 198.

Tarver, F. R., and May, K. N. 1963b. Effect of slaughter technique, immersion scald, and refrigerated storage on bacterial counts of poultry air sacs. Poultry Sci. 42: 1141.

Taylor, M. H., Smith, L. T., and Mitchell, J. D. 1965. The effect of packaging materials and a tripolyphosphate (Kena) on the shelf life of turkey steaks. Poultry Sci. 44: 297.

Thiessen, G. P., Usborne, W. R., and Orr, H. L. 1984. The efficacy of chlorine dioxide in controlling *Salmonella* contamination and its effect on product quality of chicken broiler carcasses. Poultry Sci. 63: 647.

Thomson, J. E., and Kotula, A. W. 1959. Contamination of the air sac areas of chicken carcasses and its relationship to scalding and method of killing. Poultry Sci. 38: 1433.

Thomson, J. E., Mercuri, A. J., Kinner, J. A., and Sanders, D. H. 1965. Effect of time and temperature of continuous chilling of fryer chickens on bacterial counts. Poultry Sci. 44: 1421.

Thomson, J. E., Mercuri, A. J., Kinner, J. A., and Sanders, D. H. 1966. Effect of time and temperature of commercial continuous chilling of fryer chickens on carcass temperatures, weight and bacterial counts. Poultry Sci. 45: 363.

Thomson, J. E., Banwart, G. J., Sanders, D. H., and Mercuri, A. J. 1967. Effect of chlorine, antibiotics, B-propiolactone, acids and washing on *Salmonella typhimurium* on eviscerated fryer chickens. Poultry Sci. 46: 146.

Thomson, J. E., Cox, N. A., and Bailey, J. S. 1976. Chlorine, acid and heat treatment to eliminate *Salmonella* on broiler carcasses. Poultry Sci. 55: 1513.

Thomson, J. E., Cox, N. A., and Bailey, J. S. 1977. Control of *Salmonella* and extension of shelf-life of broiler carcasses with a glutaraldehyde product. J. Food Sci. 42: 1353.

Thomson, J. E., Bailey, J. S., Cox, N. A., Posey, D. A., and Carson, M. O. 1979. *Salmonella*

on broiler carcasses as affected by fresh water input rate and chlorination of chiller water. J. Food Prot. 42: 954.
Thomson, J. E., Cox, N. A., Bailey, J. S., and Islam, M. N. 1981. Minimizing *Salmonella* contamination on broiler carcasses with poly(hexamethylenebiguanide hydrochloride). J. Food Prot. 44: 440.
Thomson, J. E., Bailey, J. S., and Cox, N. A. 1983. Weight change and spoilage of broiler carcasses—effect of chilling and storage methods. Poultry Sci. 63: 510.
Thomson, J. E., Bailey, J. S., Cox, N. A., and Shackelford, A. D. 1984. Chlorine wash to reduce bacteria on poultry processing equipment. Proceedings, 44th Annual Meeting, Institute of Food Technologists, Anaheim, CA.
Timmons, D. 1969. Hillcrest scores "dry chill" breakthrough. Poultry Meat 20(11): 10, 11, 14, 15.
Vacinek, A. A. 1972. Transient cooling, organoleptic quality and moisture exchange in crust frozen air blast chilled poultry carcasses. M.S. Thesis, Univ. of Georgia, Athens, GA.
van Schothorst, M., Notermans, S., and Kampelmacher, E. H. 1972. Einige hygienische Aspekte der Geflügelschlachtung (Hygiene in poultry slaughter). Die Fleischwirtschaft 52: 749.
Veerkamp, C. H., Mulder, R. W. A. W., and Gerrits, A. R. 1972. Kuhlung und Reinigung von Schlachtgeflugel (Chilling and cleaning poultry). Die Fleischwirtschaft 52: 612.
Wabeck, C. J., Parmelee, C. E., and Stadelman, W. J. 1968a. Carbon dioxide preservation of fresh poultry. Poultry Sci. 47: 468.
Wabeck, C. J., Schwall, D. F., Evancho, G. M., Heck, J. G., and Rogers, A. B. 1968b. Salmonella and total count reduction in poultry treated with sodium hypochlorite solutions. Poultry Sci. 47: 1090.
Walker, W. H., and Ayres, J. C. 1956. Incidence and kinds of organisms associated with commercially dressed poultry. Appl. Microbiol. 4: 345.
Zeigler, F., and Stadelman, W. J. 1955. The effect of different scald water temperatures on the shelf life of fresh, non-frozen fryers. Poultry Sci. 34: 237.

CHAPTER 6

U.S. Department of Agriculture Standards for Processed Poultry and Poultry Products

George J. Mountney

U.S. Department of Agriculture
Cooperative State Research Service
Washington, D.C. 20250

The practical application of the discipline of food microbiology in poultry processing and marketing is to ensure clean, wholesome products of uniform quality. Since it is not possible to run bacterial determinations on each poultry carcass, U.S. Department of Agriculture (USDA) inspection and grading offer the most practical substitute for individual sets of counts.

The discipline of food microbiology is concerned with three general areas. They are prevention of food spoilage measured by visual appearance, taste, odor and texture; food safety measured by the presence or absence of human pathogens in or on the food; and alteration of the functional properties of traditional human foods by the use of microorganisms. Cheeses, wines, and sauerkraut are examples of the latter process. At present, there is little application of microorganisms to alter poultry meats, with the possible exception of fermented sausages.

Under commercial production, processing, handling, and marketing conditions, it is not possible or desirable to run microbiological counts to determine the presence or absence of pathogens on all birds handled. For this reason, poultry lots are sampled on a statistical basis at critical points and with the expectation that the results from the samples will represent the particular lot of poultry under scrutiny.

Over the years, a considerable body of knowledge and experience has been accumulated, which when practiced offers the most practical methods for control of microorganisms on poultry. Some of the more obvious methods include selecting and processing only normal, healthy birds and processing them in a clean, sanitary environment. This body of knowledge has been evaluated, standardized, tested and incorporated into federal, state, and municipal laws to ensure that consumers receive safe, clean poultry of identified levels of quality.

At one time most poultry was marketed live and slaughtered in thousands of establishments, such as the back alley of the local butcher shop or at best in a local slaughter plant. This process gradually gave way to marketing poultry bled out and with the feathers removed (New York dressed) and then to eviscerated poultry as we know it today. Centralization, modernization, mechanization, and concentration of processing plants occurred at the same time changes in the forms of poultry offered for sale were taking place. As a result of these changes and concentration of the industry, 1–day's output of a typical poultry processing plant will be eaten by about 150,000 consumers. If such poultry becomes contaminated or unsafe for human consumption, massive foodborne outbreaks will occur scattered over a distribution area of several states. For these reasons, national uniform laws and policies had to be established with provisions for monitoring and regulating the sales of poultry meats.

To meet these needs, Congress, in 1957, passed legislation to "provide for the compulsory inspection by the United States Department of Agriculture of poultry and poultry products to prevent the movement in interstate or foreign commerce or in a major comsuming area of poultry products which are unwholesome, adulterated, or otherwise unfit for human food." This law was followed in 1968 by an amendment which extended federal wholesomeness to interstate operations, which for all practical purposes brought the dependence of state and local governments for inspection of poultry and poultry products to an end. The following is an excerpt from the 1968 "Poultry Products Inspection Act" (Anonymous, 1968).

451. Congressional statement of findings.

Poultry and poultry products are an important source of the Nation's total supply of food. They are consumed throughout the Nation and the major portion thereof moves in interstate or foreign commerce. It is essential in the public interest that the health and welfare of consumers be protected by assuring that poultry products distributed to them are wholesome, not adulterated, and properly marked, labeled, and packaged. Unwholesome, adulterated, or misbranded poultry products impair the effective regulation of poultry products in interstate or foreign commerce, are injurious to the public welfare, destroy markets for wholesome, not adulterated, and properly labeled and packaged poultry products, and result in sundry losses to poultry producers and processors of poultry and poultry products, as well as injury to consumers. It is hereby found that all articles and poultry which are regulated under the chapter are either in interstate or foreign commerce or substantially affect such commerce, and that regulation by the Secre-

6. U.S.D.A. Standards for Processed Poultry

tary of Agriculture and cooperation by the States and other jurisdictions as contemplated by this chapter are appropriate to prevent and eliminate burdens upon such commerce, to effectively regulate such commerce, and to protect the health and welfare of consumers.

452. Congressional declaration of policy.

It is hereby declared to be the policy of the Congress to provide for the inspection of poultry and poultry products and otherwise regulate the processing and distribution of such articles as hereinafter prescribed to prevent the movement or sale in interstate or foreign commerce, or of burdening of such commerce, by poultry products which are adulterated or misbranded. It is the intent of Congress that when poultry and poultry products are condemned because of disease, the reason for condemnation in such instances shall be supported by scientific fact, information or criteria, and such condemnation under this chapter shall be achieved through uniform inspection standards and uniform applications thereof.

The enforcement of these acts was delegated to the Secretary of the USDA, who in turn, delegated them to the Food Safety and Inspection Service (FSIS) for day-to-day administration. The official procedural guidelines and instructions to assist FSIS employees in enforcing the laws and regulations relating to federal poultry inspection are published in MPI-7 "Meat and Poultry Inspection Manual" (Anonymous, 1979), available from the Superintendant of Documents, Washington, D.C. Administrative guidelines include procedures for applications for inspection and assignment and authorities of FSIS employees as well as operational procedures which include requirements for facilities and equipment, sanitation requirements, antemortem and postmortem inspection procedures as well as slaughter, dressing, and chilling requirements. Other instructions include procedures for handling and reporting inedible and condemned products, labeling (Fig. 1), reinspection, standards of identity and composition, and export and import requirements. The following example describes the requirements for microbial control and monitoring.

Fig. 1. These logos can be displayed on the wrappers, boxes, and labels of poultry carcasses processed, inspected, graded, and meeting USDA Grade A carcass requirements. [*USDA photo.*]

(g) Microbiological Control and Monitoring

(1) Plant's responsibility. Establishments conforming with all other provisions of this section may be considered in compliance if they implement an approved microbiological control and monitoring program in lieu of a midshift cleanup. Plant management desiring to develop such a program shall:

1. Request the inspector in charge, in writing, a 30-day exemption from midshift cleanup to collect preliminary test data, implement the program and submit all related information. Such exemption may be granted, provided required provisions are in compliance.

2. Provide the inspector in charge with copies of all subsequent information.

3. Collect preliminary test data by swab sampling three or more sites on each operating line for 5 days. Such days shall be selected at random during a 2-week period. Site selection for each line should include at least one at beginning, one mid-way, and the third at the end of processing line. Each site should be swabbed 30 minutes after, and at each subsequent 2-hour period. A swab dilution technique should be used to sample each measure site in product contact zone. Data should be reported as microbes per square inch.

4. Develop a microbiological control and monitoring program. Record all information as follows:

a. Microbiological controls (preoperative sanitation instructions, sanitary operating specifications, environmental control).

b. Microbiological monitoring (sampling, laboratory procedure, actions). Written instructions should be directed to involve plant employees.

5. Submit proposed program—with plant's sanitation requirements and procedures—through the inspector in charge, to STS-SDS for approval.

(2) Changes. Variations to improve a microbiological control and monitoring program should be submitted to STS-SDS for approval.

(h) Midshift Cleanup Requirement

A program approval does not relieve the plan from other required sanitary practices. Plant failure to comply with all provisions of this subpart will require a midshift cleanup.

(i) Inspector's Responsibility

Approval of a program does not relieve the inspector from his inspection duties. He should be familiar with the program and periodically determine whether the plan (1) follows all procedures, (2) uses the monitoring program to identify potential weaknesses or deviations, (3) makes appropriate corrections if necessary.

Listed below are examples of procedures for inspecting and handling live birds and carcasses.

(b) Poultry

Poultry with signs of abnormalities or diseases—dirty, ruffled feathers: swollen sinuses and/or wattles: eye and/or nostril discharge; off-color diarrhea and pasty vent: swellings; lameness; ascites; cachexia; CNS disorder (wry neck), etc.—shall be handled as suspects.

Each suspect may be retained and slaughtered at the end of the day's operation, if practicable and adequate facilities are available, or all poultry in the lot may be slaughtered and handled as suspect.

In either case, line speed shall be reduced to allow adequate post-mortem inspection.

9.15 CONDEMNED

When ante-mortem inspection of abnormal animals (livestock or poultry) reveals a dying condition, a disease or condition requiring carcass condemnation on post-mortem inspection, or a disease or condition requiring further observation or treatment, such animals must be identified as "U.S. Condemned" and must be withheld from slaughter.

(1) Carcass-viscera inspection.
1. Observe and palpate tibia (drumstick).
2. Observe hock joints.
3. Open body cavity and observe inner surfaces and organs.
4. Observe and palpate liver, heart, and spleen. Crush spleen of mature poultry.
5. Observe other viscera and carcass exterior.
6. Instruct trimmer on disposition of abnormal or diseased carcasses (hang back, trim, remove viscera, condemn, etc.).

(2) Facilities and procedures. The following facilities and procedures are required:
1. Lighting—of enough intensity, uniform, and properly directed at work levels.
2. Hand-washing facilities—adequate and properly located.
3. Lines with two or more inspection stations—with dividers of marked shackles to prevent inspector's confusion.
4. Shackle suspensions—suitable for poultry carcass.
5. Conveyor belts or pans (when used)—synchronized with overhead conveyors and sanitized when saving viscera for edible purpose.
6. Line start and stop control—within inspector's reach.
7. Inspector's worksheet holder—conveniently located for inspector and helper.
8. Trained inspector's helper.
9. Carcass and viscera—adequately presented for inspection to allow prompt examination of entire carcass (inner and outer surfaces), and all organs. Visceral organs—heart, liver gizzard, etc.—must be presented close to the carcass, (not farther than 6 inches and preferably suspended by natural attachments below the carcass opening).
10. Foreman's cooperation—close cooperation between foreman and inspector is always necessary.

In summary, since it is not possible to determine microbiological counts and species for each bird or carcass, it is necessary to use other indicators of wholesomeness which are more easily observed. It should also be emphasized that inspection for wholesomeness is one to determine whether the product is for human consumption and that such inspection is compulsory.

Whereas inspection for wholesomeness is compulsory, grading of poultry and poultry products is voluntary, with the exception that if a plant agrees to voluntary inspection it is their responsibility to follow the regulations or lose the services of the USDA and the benefits derived from grading. The legal basis for USDA grading of poultry are listed in the Agricultural Marketing Act of 1946, as amended, (7 U.S.C. 1621 *et. seq.*). The following advantages of such a grading service are cited in this Act.

The voluntary programs provide for interested parties a national grading service based on official U.S. classes, standards, and grades. The costs involved in furnishing these grading programs are paid by the user of the service.

The grading programs and regulations established a basis for quality and price relationship and enable more orderly marketing. Consumers can purchase officially graded product with the confidence of receiving quality in accordance with the official identification.

The "Regulations Governing the Voluntary Grading of Poultry Products and Rabbit Products and U.S. Classes, Standards, and Grades" (Anonymous, 1982) are administered by the USDA, Agricultural Marketing Service (AMS), Poultry Division.

The following excerpts from these regulations define the breadth and scope of the services activities.

70.4 Grading service available.

The regulations in this part provide for the following kinds of service; and any or more of the different services applicable to official plants may be rendered in an official plant:

(a) Grading of ready-to-cook poultry and rabbits in an official plant or at other locations with adequate facilities.

(b) Grading of specific poultry food products in official plants.

Basis of Service

70.10 Grading service.

Any grading service in accordance with the regulations in this part shall be for class, quality, quantity, or condition or any combination thereof. Grading service with respect to determination of quality of products shall be on the basis of United States classes, standards, and grades as contained in Subpart B and C of this part. However, grading service may be rendered with respect to products which are bought and sold on the basis of institutional contract specifications or specifications of the applicant, and such service, when approved by the Administrator shall be rendered on the basis of such specifications.

70.12 Supervision.

All grading service shall be subject to supervision at all times by the responsible State supervisor, regional director, and national supervisor. Such service shall be rendered in accordance with instructions issued by the Administrator where the facilities and conditions are satisfactory for the conduct of the service and the requisite graders are available. Whenever the supervisor of a grader has evidence that such grader incorrectly graded a product, such supervisor shall take such action as is necessary to correct the grading and to cause any improper grademarks which appear on the product prior to shipment of the product from the place of initial grading.

70.13 Ready-to-cook poultry and rabbits and specified poultry food products; eligibility.

Only ready-to-cook poultry and rabbits and specified poultry food products which are inspected and passed by the poultry inspection service of the U.S. Department of Agriculture or by any other official inspection system acceptable to the Department may be graded.

Poultry carcasses are graded as A, B, or C quality depending upon conformation, flesh-

6. U.S.D.A. Standards for Processed Poultry

ing, fat covering, presence or absence of pinfeathers, exposed flesh, discolorations, miscellaneous defects such as broken bones and freezing defects.

The following quotes are descriptions of the several classes of ready-to-cook poultry:

UNITED STATES CLASSES OF READY-TO-COOK POULTRY

The kinds of poultry are as follows: chickens, turkeys, ducks, geese, guineas, and pigeons.

70.201 Chickens.
The following are the various classes of chickens:

(a) Rock Cornish game hen or Cornish game hen. A Rock Cornish game hen or Cornish game hen is a young immature chicken (usually 5 to 6 weeks of age), weighing not more than 2 pounds ready-to-cook weight, which was prepared from a Cornish chicken or the progeny of a Cornish chicken crossed with another breed of chicken.

(b) Rock Cornish fryer, roaster, or hen. A Rock Cornish fryer, roaster, or hen is the progeny of a cross between a purebred Cornish and a purebred Rock chicken, without regard to the weight of the carcass involved; however, the term "fryer," "roaster," or "hen," shall apply only if the carcasses are from birds with ages and characteristics that qualify them for such designation under paragraphs (c) and (d) of this section.

(c) Broiler or fryer. A broiler or fryer is a young chicken (usually under 13 weeks of age), of either sex, that is tender-meated with soft, pliable, smooth textured skin and flexible breastbone cartilage.

(d) Roaster or roasting chicken. A bird of this class is a young chicken (usually 3 to 5 months of age), of either sex, that is tender-meated with soft, pliable, smooth-textured skin and breastbone cartilage that may be somewhat less flexible than that of a broiler or fryer.

(e) Capon. A capon is a surgically unsexed male chicked (usually under 8 months of age) that is tender-meated with soft, pliable, smooth-textured skin.

(f) Hen, fowl, or baking or stewing chicken. A bird of this class is a more mature female chicken (usually more than 10 months of age), with meat less tender than that of a roaster or roasting chicken and nonflexible breastbone tip.

(g) Cock or rooster. A cock or rooster is a mature male chicken with coarse skin, toughened by darkened meat, and hardened breastbone tip.

70.202 Turkeys
The following are the various classes of turkeys:

(a) Fryer-roaster turkey. A fryer-roaster turkey is a young immature turkey (usually under 16 weeks of age), of either sex, that is tender-meated with soft, pliable, smooth-textured skin and flexible breastbone cartilage.

(b) Young turkey. A young turkey is a turkey (usually under 8 months of age) that is tender-meated with soft, pliable smooth-textured skin, and breastbone cartilage that is somewhat less flexible than in a fryer-roaster turkey. Sex designation is optional.

(c) Yearling turkey. A yearling turkey is a fully matured turkey (usually under 15 months of age), that is reasonably tender-meated and with reasonable smooth-textured skin. Sex designation is optional.

(d) Mature turkey or old turkey (hen or tom). A mature or old turkey is an old turkey of either sex (usually in excess of 15 months of age), with coarse skin and toughened flesh.

TABLE I

Summary of Specifications of Quality for Individual Carcasses of Ready-to-Cook Poultry and Parts Therefrom (Minimum Requirements and Maximum Defects Permitted)

FACTOR	A QUALITY			B QUALITY			C QUALITY
CONFORMATION Breastbone Back Legs and Wings	Normal Slight curve or dent Normal (except slight curve) Normal			Moderate deformities Moderately dented, curved or crooked Moderately crooked Moderately misshapen			Abnormal Seriously curved or crooked Seriously crooked Misshapen
FLESHING	Well fleshed, moderately long, deep and rounded breast			Moderately fleshed, considering kind, class and part			Poorly fleshed
FAT COVERING	Well covered—especially between heavy feather tracts on breast and considering kind, class and part			Sufficient fat on breast and legs to prevent distinct appearance of flesh through the skin			Lacking in fat covering over all parts of carcass
PINFEATHERS Nonprotruding pins and hair Protruding pins	Free Free			Few scattered Free			Scattering Free
EXPOSED FLESH[1] Carcass Weight Minimum / Maximum None / 1½ lbs. Over 1½ lbs. / 6 lbs. Over 6 lbs. / 16 lbs. Over 16 lbs. / None	Breast and Legs None None None None	Elsewhere ¼" 1½" 2" 3"	Part Slight trim on edge	Breast and Legs[2] ¼" 1½" 2" 3"	Elsewhere[2] 1½" 3" 4" 5"	Part Moderate amount of flesh normally covered	No Limit
DISCOLORATIONS[1] None / 1½ lbs. Over 1½ lbs. / 6 lbs. Over 6 lbs. / 16 lbs. Over 16 lbs. / None	½" 1" 1½" 2"	1" 2" 2½" 3"	¼" ¼" ½" ½"	1" 2" 2½" 3"	2" 3" 4" 5"	½" 1" 1½" 1½"	No Limit[4]
Disjointed bones Broken bones Missing parts	1 None Wing tips and tail[5]			2 disjointed and no broken or 1 disjointed and 1 non-protruding broken Wing tips, 2nd wing joint and tail Back area not wider than base of tail and extending half way between oase of tail and hip joints.			No limit No limit Wing tips, wings and tail Back area not wider than base of tail extending to area between hip joints.
FREEZING DEFECTS (When consumer packaged)	Slight darkening over the back and drumsticks. Few small ⅛" pockmarks for poultry weighing 6 lbs. or less and ¼" pockmarks for poultry weighing more than 6 lbs. Occasional small areas showing layer of clear or pinkish ice.			Moderate dried areas not in excess of ½" in diameter. May lack brightness. Moderate areas showing layer of clear, pinkish or reddish colored ice.			Numerous pockmarks and large dried areas.

[1]Total aggregate area of flesh exposed by all cuts and tears and missing skin, not exceeding the area of a circle of the diameters shown.
[2]A carcass meeting the requirements of A quality for fleshing may be trimmed to remove skin and flesh defects, provided that no more than one-third of the flesh is exposed on any part and the meat yield is not appreciably affected.
[3]Flesh bruises and discolorations such as blue back are not permitted on breast and legs of A quality birds. Not more than one-half of total aggregate area of discolorations may be due to flesh bruises or blue back (when permitted), and skin bruises in any combination.
[4]No limit on size and number of areas of discoloration and flesh bruises if such areas do not render any part of the carcass unfit for food.
[5]In ducks and geese, the parts of the wing beyond the second joint may be removed, if removed at the joint and both wings are so treated.

(e) For labeling purposes, the designation of sex within the class name is optional, and the two classes of young turkeys may be grouped and designated as "young turkeys."

70.203 Ducks.

The following are the various classes of ducks:

(a) Broiler duckling or fryer duckling. A broiler duckling or a fryer duckling is a young duck (usually under 8 weeks of age), of either sex, that is tender-meated and has a soft bill and a soft windpipe.

(b) Roaster duckling. A roaster duckling is a young duck (usually under 16 weeks of age), of either sex, that is tender-meated and has a bill that is not completely hardened and a windpipe that is easily dented.

(c) Mature duck or old duck. A mature duck or an old duck is a duck (usually over 6 months of age), of either sex, with toughened flesh, hardened bill, and hardened windpipe.

70.204 Geese.
The following are the various classes of geese:

(a) Young goose. A young goose may be of either sex, is tender-meated, and has a windpipe that is easily dented.

(b) Mature goose or old goose. A mature goose or old goose may be either sex and has toughened flesh and hardened windpipe.

70.205 Guineas.
The following are the various classes of guineas:

(a) Young guinea. A young guinea may be of either sex, is tender-meated and has a flexible breastbone cartilage.

(b) Mature guinea or old guinea. A mature guinea or an old guinea may be of either sex, has toughened flesh, and a hardened breastbone.

70.206 Pigeons.
The following are the various classes of pigeons:

(a) Squab. A squab is a young, immature pigeon of either sex, and is extra tender-meated.

(b) Pigeon. A pigeon is a mature pigeon of either sex, with coarse skin and toughened flesh.

Table I is a "Summary of Specifications for Standards of Quality for Individual Carcasses of Ready-to-Cook Poultry and Parts Therefrom."

In summary the main advantage of consumer grades for poultry is the fact that consumers cannot be expert in knowing all levels of all food commodities they may wish to purchase. For that reason they must rely on an impartial agency, the USDA, AMS, to represent their interests.

REFERENCES

Anonymous. 1968. The Poultry Products Inspection Act, as amended (Pub. L. 85–172, approved August 28, 1957, 71 Stat. 441. as amended by the Act of June 25, 1962, 76 Stat. 110 and the Wholesome Poultry Products Act of August 18, 1968, Pub. L. 90–492, 82 Stat. 791–808; 21 U.S.C. 451 *et seq.*). United States Congress, Washington, D.C.

Anonymous. 1979. Meat and Poultry Inspection Manual, MPI-7, USDA, Food Safety and Quality Service, Washington, D.C.

Anonymous. 1982. Regulations governing the voluntary grading of poultry products and rabbit products and U.S. Classes, Standards, and Grades (7 CFR part 70) USDA, Agricultural Marketing Service, Poultry Division, Washington, D.C.

CHAPTER 7

Packaging of Processed Poultry

Larry E. Dawson

Department of Food Science and Human Nutrition
Michigan State University
East Lansing, Michigan 48824

I. INTRODUCTION

Packaging is the use of containers and components, including label and decoration, to contain, protect, identify, and assist in the distribution and merchandising of products. Most processed poultry products today require some packaging during production and distribution. Modern packaging requires specialized materials, equipment, skills, and operational facilities to meet fundamental needs. In this discussion, major emphasis will be placed on the protection function as a means of minimizing microbial problems.

As reported in earlier chapters, poultry products are contaminated by different levels of different species of bacteria. Processing practices are designed to minimize numbers of organisms on or in the product at completion of processing, and the products must then be protected from recontamination or cross-contamination. This is especially important for cooked, ready-to-eat-products which may not be reheated sufficiently to destroy all pathogenic organisms which may be present.

Packaging poultry products should encompass an entire distribution system from processor to consumer. This is usually a three-phase system consisting of primary packaging, secondary packaging, and tertiary packaging.

The primary package is usually the plastic film immediately surrounding an individual bird, part, or combination of parts, frequently overwrapped in a styrofoam or cardboard tray. This is the area of packaging which has received most research attention and concern, but it certainly is influenced by secondary and tertiary packaging systems.

Secondary packaging consists of a rigid container such as a cardboard shipping box surrounding and protecting a single primary unit or a small number of primary units. This lends itself to improved efficiency of handling relatively small units of product.

The tertiary package refers to a bundled package or somehow fastened group of boxes of product designed for pallet or mechanical transportation during storage, truck transport, and wholesale or retail storage before sales.

At this point, one-half of the packaging system has been utilized. From the palletized product in some phase of distribution, the products move to a central warehouse where the bundled or palleted products are divided into one or more secondary units for delivery to retail outlets. In some systems, the secondary package is discarded at the retail level, sometimes at the time of product display, and sometimes products remain in, and are displayed in, opened secondary packages. In either situation, only products in primary packages are purchased for home consumption. The primary package thus becomes the final and perhaps the most important protection against microbial contamination or cross-contamination. These primary packages should provide protection during transport and home storage until removed for consumption preparation.

II. SPECIFIC PACKAGE FUNCTION

A. Protection from Physical Damage

Since poultry products are now consumed in areas some distance from where they are produced they must be protected from physical damage which could lead to microbial contamination. This is especially true for those poultry products which are packaged ready for retail sales. Any damaged units need to be repackaged before appropriate display allows microbial cross-contamination. Thus the primary and secondary packaging materials must be flexible, strong, puncture resistant, and lightweight.

B. Protection from Dehydration and Gas Exchange

Processed poultry products are distributed either frozen for long-term holding or unfrozen and refrigerated for rapid movement through channels of trade. Frozen poultry must be packaged and sealed in low moisture-vapor transmission materials. Sealability is an essential attribute of the film. Although the growth of microbes is inhibited at temperatures below freezing, they are not killed and can grow rapidly after thawing. The primary packaging material must prevent microbial contamination during handling at all stages of processing and distribution, and at the same time prevent loss of moisture and an exchange of gases.

Few poultry products are affected by color changes except when directly exposed to light and oxygen or air. The packages which protect products from dehydration and gas exchange, along with secondary packages, protect poultry from light and potential color change.

C. Protection from Odors, Microbes, Filth, and Insects

Generally speaking, the packages designed for physical protection and for prevention of moisture loss or gas exchange will protect poultry from microbes, insects, filth, and outside odors.

D. Adding Shape and Character to Products

Flexible films are used to package most retail units of uncooked poultry. These films may be either stretchable or shrinkable, in which case they follow the contour of the poultry product to provide a natural shape or character to the product. This package characteristic has little effect on microbial incidence on poultry but is essential for optimum merchandising of some products.

E. Consumer Assurance of Product Cleanliness

The package must not only protect poultry from contamination and other changes but must provide consumers the assurance that their food has been well protected and is free from tampering and physical, chemical, or microbiological changes.

F. Identification and Promotion

Unrelated to microbial invasion or growth, all poultry packages, especially at the retail level, should allow brand identification, product and price identification, inspection and grade labeling, and handling and cooking instructions. The transparent films developed and improved in recent years have allowed great advances in this area.

III. PACKAGE MATERIALS AND THEIR DEVELOPMENT

During the late nineteenth century and early twentieth century, major developments and advancements occurred in some packaging areas which contributed to a standardization of container forms. It was not until sometime later that poultry products received recognizable and identifiable standard packaging forms, and these have continued to change. During this package developmental period, advancements in color printing, graphic arts, and effective decoration made modern packaging possible.

New packaging materials including cellophane, aluminum foil, and glassine were introduced in the early 1900s and allowed the new flexible packaging industry to develop and thrive. Along with packaging machinery developments, the discovery of such products as polyethylene, polyester, polypropylene, stretched films, shrink films, metal foil, and improved laminated materials revolutionized the packaging industry since 1940.

The number of packaged poultry products on the market has never stopped growing in variety and volume. Poultry products that are processed-sliced, premixed, unitized, prepared ready for eating, protected, or made sanitized or sterile, safe, and properly labeled are now commonplace in the merchandising counters in ever-expanding and attractive displays. Much of this has lead to a new concept referred to as convenience packaging.

In this discussion, the primary function of an immediate container or package, or packaging procedure, is related to microbial control. Other functions of the package material must not be overlooked, such as ability of the package to allow U.S. Department of Agriculture (USDA) approved labeling. This includes label requirements for fresh, frozen, or cured poultry products, and the label must be readable, attractive, permanent under all storage conditions and must not interfere with the protective characteristics of the package.

IV. PACKAGING FILMS FOR POULTRY

All major films used for packaging poultry products provide adequate protection from microbial recontamination, and some of the newer films under development could offer an excellent means for inhibiting bacterial growth. At present, antioxidant-impregnated films are available, and it should be only a matter of time when approved bacterial inhibitors might be impregnated into films to help control growth of microorganisms in packaged unfrozen poultry. Characteristics required of a film include level of permeability of moisture, oxygen, and CO_2; transparency; sealability; label printabililty; toughness; shrinkability; stretchability; and resistance to temperature changes. In evaluating a film for a particular purpose, consideration must be given to all packaging requirements, including microbial control. A processor has access to many suppliers of packaging materials and packaging systems, and must be in constant contact for latest developments and technical improvements. Package materials must also be approved by USDA.

Some of the packaging materials currently in use for poultry products are included in the following list. This list is not complete, nor does it infer approval by the author or editors.

Cellophane: When cellophane became moisture proof and heat sealable, numerous food packaging opportunities became apparent. With newer technologies of coating and laminations it became possible to build in desired gas or moisture barrier properties. Cellophane is used for bags requiring a strong heat seal, for tray overwraps, and for processed poultry meats.

Nylon: Polyamide thermoplastic polymers are excellent barriers to oxygen, flavors, and aromas and are extremely tough at low temperatures. Nylon is frequently used in composite structures with low-density polyethylene (LDPE), ethylene vinyl acetate (EVA), and polyvinylidine chloride (PVDC). Nylon is used for fresh and processed poultry meats.

Polyester: This film requires a coating for effective heat sealing. The film is heat shrinkable and can be used for vacuum-processed poultry meat, boil-in-bag products, and overwraps for tray-packaged poultry.

Polyethylene (PE) and low- and high-density products (LDPE, HDPE). Polyethylene is a low-cost film and can be impregnated with an antioxidant inhibitor.

Polyvinyl chloride (PVC) is a stretchable film, used for many poultry products.

Polyvinylidene chloride (Saran-Cryovac) is a copolymer of vinylidine chloride with vinyl chloride. PVDC is a shrinkable film used for packaging a variety of poultry products. It can be heat sealed. This is an excellent barrier to microbes and oxygen or moisture exchanges.

Polystyrene (PS): Polystyrene is a colorless transparent thermoplastic product, is hard, and has a fairly high tensile strength. This product may be used for tray packaging.

Molded pulp trays may be used for packaging small cuts or combinations of cuts of poultry. These are usually overwrapped with a transparent plastic film.

Expanded polystyrene trays are increasing in use for tray packaging poultry. These are usually overwrapped with a transparent plastic film.

Packaging films are frequently used in combination with specific processes for increased protection of the poultry products from microbial entrance or growth. These necessarily must accompany approved sanitation procedures prior to packaging and temperature control after packing for optimum bacterial control.

V. PACKAGING AS IT AFFECTS MICROBIAL CONTROL ON POULTRY

Many poultry firms have switched to a chill-pack method for individual bird packaging and distribution to retailers. Denton and Gardner (1981) found no difference in bacterial counts (psychrophilic and mesophilic) on broilers from ice-packed and dry-chilled birds. Initial counts favored the ice-chilled birds; however after 12 days of retail storage, more bacteria were found on the ice-packed birds than on dry-chilled birds. Breast meat was also more heavily contaminated than leg samples.

A. Atmosphere

The in-package environment, controlled by process and package film, can significantly influence microbial quality of poultry. Sander and Soo (1978) evaluated growth of potential pathogens on chicken following bulk packaging in nylon–surlyn coextruded film and two levels of CO_2. At 1.1°C, growth of pathogens was negligible. CO_2 restricted growth of most objectionable putrefactive bacteria but not lactic acid organisms.

They recommended the use of a high CO_2 concentration for extending shelf life up to 27 days at 1.1°C.

Thomson (1970) found that packaging materials, storage atmosphere, and temperature all affected the microbial counts and rancidity of fresh fryer chickens. Chickens, packaged in either high (polyethylene) or low (vinylidene) gas permeability film, stored in a CO_2 atmosphere or air, at −0.5° or 6.0°C were evaluated after 3 and 6 days of storage for microbiological condition and oxidative deterioration (rancidity). Microbial counts were not significantly affected by the type of packaging film; however, they were significantly lower in CO_2 atmosphere storage than in air. Rancidity was significantly higher in polyethylene packaged birds than in vinylidene, but atmosphere or temperature of storage did not significantly affect rancidity development. The author further stated that rancidity development as well as microbiological factors should be considered in selecting storage conditions and packaging systems for chilled poultry.

A comparison of oxygen-permeable and -impermeable wrapping materials for the storage of chilled eviscerated poultry has been reported by Shrimpton and Barnes (1960). Polyethylene bags, vinylidene chloride–vinyl chloride copolymer bags, and a modified heat-shrinkable polyethylene film were all compared in this preliminary report. Changes during subsequent holding at 1°C in the concentrations of oxygen and CO_2 within the wraps and the total number of bacteria on the carcasses were reported. The most significant differences occurred in the copolymer bags in which oxygen concentration increased to 9–10%. Development of spoilage flora was slowed down, and off-odors did not develop until carcasses had been held for 16 days at 1°C.

The effect of freezing and packaging methods on the survival and biochemical activities of spoilage organisms on chicken was reported by Rey and Kraft (1971). Fluorescent pigment production, proteolytic, and lipolytic spoilage of chicken stored at 5°C was directly related to the availability of oxygen provided by the packaging procedures. Bacterial numbers paralleled increases in biochemical activities. Freezing followed by defrosting and refrigerated storage increased the proportion of biochemically active psychrophiles on the surface of the meat. Vacuum packaging generally limited the amount of spoilage. When samples were alternatively frozen and thawed, total aerobic bacterial counts were only slightly different from those on chicken frozen for 22–25 days.

The effects of packaging and storage on the quality of chicken frankfurters were evaluated by Baker et al. (1972). Frankfurters were made by two formulas (with and without a phosphate mixture plus sodium isoascorbate). They were stored at 2°C, vacuum-packaged and non-vacuum-packaged, for up to 24 days. Vacuum-packaged frankfurters with added phosphates and sodium isoascorbate showed little change in flavor, juiciness, or overall acceptability over the 24-day storage period. With no additives and non-vacuum packaging, quality differences showed up by 10 days. There was no problem with bacterial growth, although some mold growth occurred on non-vacuum-packaged frankfurters.

Some of the current films and packaging procedures may create an atmosphere which will support growth of anaerobic bacteria. Koutter et al. (1979) packaged several meat sandwiches in a nitrogen atmosphere. Hamburger and sausage became toxic (*Clostridium botulinum*) in 4–7 days at 25°C, but turkey did not. In a similar study using *Staphylococcus aureus*, Bennett and Amos (1979) found toxin in sausage in 2 days, hamburger in 4 days—but turkey did not support bacterial growth. No reasons have been given for these findings.

An atmosphere high in CO_2 affects growth of some bacteria. Bailey et al. (1979) reported that *Pseudomonas* was the primary spoilage organism on chicken packaged in a CO_2 environment. Chicken cubes were inoculated with *C. botulinum* spores and were held under a CO_2 environment (Silliker and Wolfe, 1979). Any hazard from *C. botulinum* was neither enhanced nor reduced by the CO_2. Arafa and Chen (1975) reported that vacuum packaging changed the microflora on poultry in refrigerated storage but did not protect fresh poultry during refrigerated storage. The change in microflora did change the aroma of the products as they spoiled.

B. Effect of Package Material on Microbial Control on Poultry

The effect of packaging materials and techniques on shelf life of fresh poultry meat has been reported by Wells et al. (1958). They showed that packaging materials, vinylidene chloride copolymer (VCP) and cellophane film did not exert a direct bacteriostatic effect on the spoilage microorganisms which developed in poultry meat. These researchers also reported that shelf life for birds wrapped in cellophane and VCP was the same under nonvacuum conditions, but when the air was evacuated, the shelf life increased 4 days beyond that obtained without vacuum. It was also reported that the type of film and packaging will influence the microflora and in turn will influence the type of off-odor which persists at spoilage.

A combination of vacuum packaging with heat-shrinkable package film or bag is effective in increasing shelf life and reducing microbial growth. Spencer et al. (1956) compared shelf life of poultry packaged in cellophane and polyethylene with that of poultry heat shrunk and vacuum packed using polyvinylidene and polyethylene. Additional shelf life of 2 days was obtained by use of vacuum plus a heat-shrinking film.

Handling poultry products during a postcook packaging operation can contribute more to bacterial spoilage than the packaging operation itself. Denton and Gardner (1981) reported that turkey product wrapping had only a minor effect on numbers of bacteria. The significant increase in tissue bacterial content of oven-roasted boneless breast turkey and weiners was due to the handling operation, probably from equipment surfaces.

Igbinedion et al. (1981) evaluated the effectiveness of several films in combination with vacuum packaging and holding conditions for controlling bacterial growth on poultry. Packaging films were identified as Barrier, Super L, and L. Film. Most effective bacterial control was obtained by vacuum packaging and use of the "Barrier" film. Holding products under artificial light resulted in increased aerobic and fluorescent pseudomonads on products packaged with Super L and L. Films.

A difference in bacterial growth on chicken attributed to package film treatment was reported by Debevere and Voets (1973). At 4°C holding, bacterial growth rate was less when poultry was packaged in heat-shrinkable polyvinyl chloride than stretchable polyvinylchloride. Use of unstretched and unshrunk film resulted in bacterial growth equal to that of unpackaged poultry.

Shrinkable films have a low oxygen permeability and envelopes the chicken tightly and smoothly so that little oxygen remains. Polyvinylchloride film increases in thickness when shrunk, and gas permeability decreases. Oxygen permeability of the stretchable polyvinylchloride film is much higher than that of shrinkable PVC, and considerable oxygen remains in the package to support aerobic bacterial growth.

VI. SUMMARY

A definite need exists for careful and possible aseptic handling and packaging of poultry meat, especially after cooking. Raw-product packaging results in minor bacterial contamination, whereas a significant increase in bacterial content of cooked product has been noted for oven-roasted boneless and other emulsion-type products as a result of handling associated with packaging. This recontamination can occur in spite of the use of disposable sterile gloves and periodic cleaning with soap and hot water solutions followed by sanitizing with iodine solutions for all contact work surfaces.

At the present time, poultry processors can obtain packaging materials for any specific product, process, and distribution method. Information on latest available packaging products can be obtained from suppliers of the films, trays, or other containers.

For maximum bacterial control, effective sanitary procedures must be used in processing and preparation of the products in order to start with products having low bacterial numbers. An appropriate film should be selected which will allow vacuum sealing and low moisture and/or gas transmission. Carbon dioxide flushing or nitrogen flushing will reduce available oxygen for bacterial growth and for other chemical changes. Effective low-temperature control is essential, along with adequate packaging, to assure minimum bacterial growth and maximum product quality.

REFERENCES

Arafa, A. S., and Chen, T. C. 1975. Effect of vacuum packaging on microorganisms on cut-up chickens and in chicken products. J. Food Sci. 40: 50.

Bailey, J. S., Reagen, J. O., Carpenter, J. A. Schuler, G. A., and Thompson, J. E. 1979. Types of bacteria and shelf-life of evacuated carbon dioxide injected and ice packed broilers. J. Food Protection. 42: 218.

Baker, R. C., Darfler, J., and Vadehra, D. V. 1972. Effect of storage on the quality of chicken frankfurters. Poultry Sci. 51: 1620.

Bennett, R. W., and Amos, W. T. 1979. Staphylococcal growth and toxin production in nitrogen-packed sandwiches. Presented at 39th Ann. Meet. of Inst. of Food Technologists, St. Louis, MO.

Debevere, J. M., and Voets, J. P. 1973. Influence of packaging materials on quality of fresh poultry. British Poultry Sci. 14: 17.

Denton, J. H., and Gardner, F. A. 1981. Effect of further processing systems on selected microbiological attributes of turkey meat products. J. Food Science 47: 214.

Igbinedion, J. E., Orr, H. L., Johnston, R. A., and Gray, J. I. 1981. The influence of packaging in flexible films and light on the shelf-life of fresh chicken broiler carcasses. Poultry Sci. 60: 950.

Ito, K., and Bee, G. R. 1980. Microbiological hazards associated with new packaging techniques. J. Food Technol. 34: 78.

Koutter, D. A., Lynt, R. K., Lilly, T. J., and Solomon, H. M. 1979. Evaluation of the botulism hazard from nitrogen-packed sandwiches. Presented at 39th Ann. Meet. Inst. of Food Technologists, St. Louis, MO.

Paine, F. A., and Paine, H. Y. 1983. "Handbook of Food Packaging." Leonard Hill, Pub.

Rey, C. R., and Kraft, A. A. 1971. Effect of freezing and packaging methods on survival and biochemical activity of spoilage organisms on chicken. J. Food Sci. 36: 454.

Sander, E. H., and Soo, H. M. 1978. Increasing shelf-life by carbon dioxide treatment and low temperature storage of bulk pack fresh chickens packaged in nylon surlyn film. J. Food Sci. 43: 1519.

Shrimpton, D. H., and Barnes, E. M. 1960. A comparison of oxygen permeable and impermeable wrapping materials for the storage of chilled eviscerated poultry. Chem. Ind., pp. 1492–1493.

Silliker, J. H., and Wolfe, S. L. 1979. Microbiological safety considerations in controlled atmosphere storage of meats. Food Technol. 34: 59.

Spencer, J. V., Eklund, M. W., Sauter, E. A., and Hard, M. M. 1956. The effects of different packaging materials on the shelf-life on antibiotic treated chicken fryers. Poultry Sci. 35: 1173.

Taylor, M. H., Helbacka, N. V., and Kotula, A. W. 1966. Evacuated packaging of fresh broiler chickens. Poultry Sci. 45: 1207.

Thomson, J. E. 1970. Microbial counts and rancidity of fresh fryer chickens as affected by packaging materials, storage atmosphere and temperature. Poultry Sci. 49: 1104.

Wells, F. E., Spencer, J. V., and Stadelman, W. J. 1958. Effect of packaging materials and techniques on shelf-life of fresh poultry meat. Food Technol. 12: 425.

CHAPTER 8

Radiation Preservation of Poultry Meat

Richard E. Faw and Tsing-Yuan Chang Mei

Department of Nuclear Engineering
Kansas State University
Manhattan, Kansas 66506

I. INTRODUCTION

Preservation of poultry meat by treatment with ionizing radiation is an internationally accepted procedure for reduction of the public health risks of foodborne illness as well as enhancement of wholesomeness and marketability of poultry products. This chapter provides a brief history of, and an introduction to, the technology of radiation food processing. Poultry meat irradiation is examined within the broad context of the many foods so treated. The extraordinary degree of international cooperation in the technology of radiation food processing is examined along with the international standards adopted for quality assurance in the trade of irradiated foods. Two types of poultry treatment are discussed—low-dose treatment for destruction of pathogens and extension of shelf life and high-dose treatment, or sterilization, permitting storage for years under ambient conditions. The chapter concludes with a discussion of irradiation facilities and economics of poultry irradiation.

II. A BRIEF HISTORY OF RADIATION FOOD PRESERVATION

Although the bacteriocidal action of X-rays was recognized very early (Minck, 1896), the first reported investigation of potentially commercial-scale radiation food processing for elimination of pathogens dates from 1943 (Procter *et al.*, 1943). In work sponsored by the U.S. Army Quartermaster Corps at the Massachusetts Institute of Technology, the preservation of hamburger meat by X-rays was studied. Slow but steady research progress was made in the United States through 1953 when, endorsed by the National Academy of Sciences, a National Food Irradiation Program was undertaken by the Department of the Army. Also in 1953, President Eisenhower brought before the United Nations his "Atoms for Peace" policy. This policy and passage of the 1954 Atomic Energy Act stimulated international interest and the establishment of many research programs in radiation food preservation.

The first public health clearance for radiation-processed food took place in the USSR in 1958 when approval was given for human consumption of potatoes treated for inhibition of sprouting. Use of X-rays for this purpose had been investigated in the USSR as early as 1936 (Matsuyama and Umeda, 1983). USSR clearance of grain irradiated for insect disinfestation was given in 1959. Studies of the use of X-rays for insect control had been reported as early as 1912 (Tilton and Burditt, 1983). Irradiated potatoes were cleared in Canada in 1960 and in the United States in 1964. Treated grain was cleared in the U.S. in 1963 and Canada in 1969. These early clearances have been followed by many others, for many foods, and in many countries. Irradiated poultry was cleared in the USSR in 1966, in the Netherlands in 1971, and for test marketing in Canada in 1973.

Evaluations of the wholesomeness of irradiated foods were taken up by individual countries—in the United States, for example, under supervision of the Surgeon General of the Army working in cooperation with the Food and Drug Administration (FDA), which is the organization responsible for final evaluation of research reports and petitions for clearance. International cooperation in wholesomeness studies was instituted in 1961 at a meeting of the International Atomic Energy Agency (IAEA), the Food and Agriculture Organization (FAO), and the World Health Organization (WHO). In the case of poultry, for example, favorable evaluation led to joint FAO/IAEA/WHO approval (FAO, 1977) of chicken irradiation for the purpose of reduction in pathogenic microorganisms and microbial spoilage.

III. RADIATION TECHNOLOGY

A. Sources

Irradiators used in food processing are of two types, isotopic and machine. Isotopic irradiators employ one or the other of the radi-

TABLE I
General Characteristics of Isotopic Sources Used in Irradiators for Food Preservation

	^{60}Co	$^{137}Cs-^{137m}Ba$	^{134}Cs
Half-life (years)	5.26	30.0	2.05
Principal gamma-ray energies (MeV) and frequencies (%)[a]	1.332 (100) 1.173 (100)	0.662 (85.0)	0.605 (96.9) 0.796 (86.7) 0.569 (13.2) 0.563 (8.5) 0.802 (7.9) 1.365 (3.5) 1.168 (2.0) 0.476 (1.6) 1.038 (1.0)
Usual form[b]	Metal	Chloride	Chloride
Density (g/cm³)	8.7	1.7–2.7[c]	1.7–2.7[c]
Melting point (°C)	1495	646	646
Typical specific activity (Ci/g)	25–100	25	

[a] Source: Chilton et al. (1984).
[b] Source: Jarrett (1982).
[c] As fabricated.

onuclides Cobalt-60 (^{60}Co) or Cesium-137 (^{137}Cs)[1] as sources of gamma rays for application to food. Machine irradiators use electron accelerators, usually for direct application of high-energy electrons to food, but in some cases for the production of high-energy X-rays for applicaton to food. ^{60}Co is produced in nuclear reactors by neutron-induced transmutation of naturally occurring ^{59}Co. ^{137}Cs is a fission product and is extracted from the by-products produced in nuclear reactor fuel elements. Also present as a fission product is ^{134}Cs, found to a certain degree in the cesium source material for irradiators. Properties of the cobalt and cesium radionuclides are given in Table I. Jarrett (1982) gives a comprehensive description of the preparation, fabrication, and testing of isotopic sources for irradiators.

The "strength" of an isotopic source is ordinarily expressed in terms of "activity," the rate of disintegration of a radionuclide. The traditional unit for activity is the curie (Ci) defined as 37 billion disintegrations per second. The Système International unit is the becquerel (Bq), defined as one disintegration per second. One Ci is thus equivalent to 37 gigabequerels (GBq). Activity alone is inadequate to characterize the utility of a given source or irradiator. Important also is the frequency of gamma-ray emission and the energy of the gamma ray. ^{60}Co, for example, emits two

[1] Actually, an equilibrium mixture of metastable barium-137, ^{137m}Ba, and ^{137}Cs.

gamma rays per disintegration while ^{137}Cs emits a gamma ray in only 85% of its disintegrations. Radioactive cobalt and cesium also emit beta particles (electrons), but these particles do not escape from the source material which is encapsulated in metal, usually stainless steel.

Another important characteristic of an isotopic source is the half-life, the length of time for the activity of a source to be halved as a result of decay. Were it not for other factors, ^{137}Cs would be chosen over ^{60}Co as an irradiator source because of its much longer half-life. The frequencies of source replacement and product flow rate adjustment would be significantly lower. Other factors (Jarrett, 1982) make cost comparison and source selection less straightforward.

1. *Source form.* ^{60}Co is usually in a highly insoluble metallic form, while ^{137}Cs is usually in the form CsCl, which is highly soluble in water and thus presents a hazard in the event of failure of source encapsulation.

2. *Efficiency.* If this term is defined as the fraction of the gamma-ray energy that is deposited productively in irradiator contents, then typical values are 15–25% for ^{137}Cs and 20–35% for ^{60}Co. Typically, to accomplish a given task in food irradiation, the required ^{137}Cs activity would be about 4–5 times the required ^{60}Co activity.

3. *Dose uniformity.* For irradiator contents up to 15 cm thick and densities up to 1.3 g/cm^3, there is typically little difference in dose uniformity between ^{60}Co and ^{137}Cs. At greater thicknesses or densities, ^{60}Co offers slightly improved uniformity.

4. *Shielding.* Because of the lower energy of the gamma ray, ^{137}Cs, as compared to ^{60}Co, requires subtantially less shielding for personnel protection.

Accelerators as sources of high-energy electron beams (EB devices) are of various types. For electron energies up to 4 MeV, direct-current accelerators are used. For higher-energy electrons, radio-frequency devices (Linacs) are called for. In either case, the capacity of the machine may be expressed as the beam power, the product of the average EB current and the energy of the accelerated electrons. Morganstern (1978) describes EB devices for food irradiation, ranging in power from 25 kW at 10 MeV to 150 kW at 4 MeV, the latter for use as an X-ray generator. He points out that although low conversion efficiencies from EB to X-rays are inhibiting, technological advances are making this option more promising. Ramler (1982) refers to developments which may lead to EB devices operating at megawatt powers.

EB devices are well suited to treatment of packaged products of limited thickness. Because of the limited penetration of electrons into mate-

8. Radiation Preservation of Poultry Meat

rial, multiple passage of products through EB irradiators, with irradiation from various directions, is the standard procedure, as is the case for isotopic irradiators. Figure 1 illustrates the absorbed dose as a function of depth of penetration of electrons in water.

B. Dosimetry

The traditional unit for specifying the "dose" given an irradiated product is the rad of absorbed dose. The absorbed dose is defined as the energy locally deposited along tracks of charged particles—in this context, electrons—either produced by gamma rays or X-rays or as a result of interactions of other electrons. The rad is defined as 100 ergs per gram. The corresponding SI unit is the gray (Gy), defined as 1 joule (J)

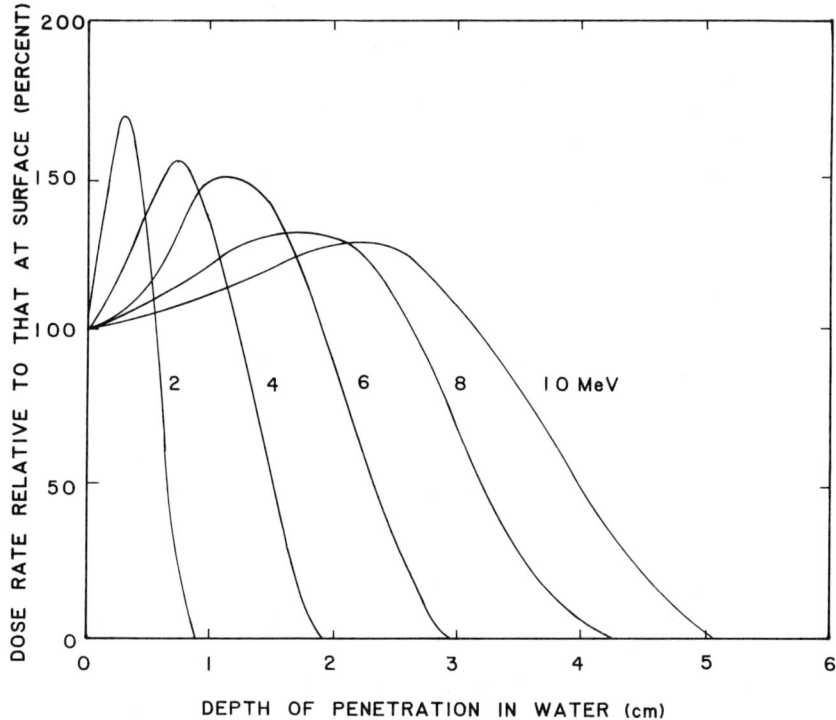

Fig. 1. Variation in dose with depth for electron beams normally incident on water (1 g/cm³). [From Brynjolfsson (1963).]

per kilogram—100 times greater than the rad. Multiples of the units are commonly used, and a useful conversion factor is 1 kGy = 0.1 Mrad. Careful measurement of absorbed doses in irradiated food is exceedingly important, not only in meeting legal requirements for quality assurance but also in adjusting product flow rates and minimizing wasteful excessive doses. The IAEA (1977) published a practical manual for dosimetry in food irradiation, and McLaughlin *et al.* (1982) give an exhaustive treatment of both the theoretical and practical aspects of radiation dosimetry in radiation food processing.

The range of doses used in food preservation is from about 0.1 kGy to 50 kGy at rates so high as to preclude use of conventional ionization chambers for measurement. Two primary, or reference, dosimetry systems are in common use. One is calorimetry, suitable for short durations of irradiation. Another is ceric sulfate dosimetry. Ceric ions in acidic aqueous solution are reduced by the action of ionizing radiation to cerous ions. The change in ceric ion concentration, proportional to the dose, may be measured readily by spectrophotometric methods. The method is suitable for doses in the range of 2–50 kGy, (McLaughlin *et al.*, 1982). Many secondary dosimetry systems are in use, calibrated against primary dosimeters. Most involve color changes in a wide variety of liquid and solid systems. One such system, well suited to use with food irradiation, involves darkening of polymethyl methacrylate (PMMA, lucite, or perspex). Spectrophotometric measurements can determine absorbed dose over the range of 1–50 kGy.

C. Protection

A radiation protection program for a food irradiation facility must deal routinely with radiation surveys, personnel monitoring, and record keeping to assure compliance with legal requirements and protection standards. It must have as a goal the minimization of personnel exposures. It must assure preparation for dealing with emergencies through plans and procedures, training, and maintenance of emergency supplies.

Standards for personnel exposure are expressed in terms of the dose equivalent, using the traditional rem unit or the SI unit sievert (Sv) unit. The conversion is 1 Sv = 100 rem. The dose equivalent is defined as the absorbed dose in tissue multiplied by the quality factor, a measure of the relative biological effectiveness of the type of radiation. For gamma rays, X-rays, and electrons, the quality factor is unity.

The International Commission on Radiological Protection (ICRP, 1977)

8. Radiation Preservation of Poultry Meat

has made recommendations for maximum dose equivalents in occupational exposure and in exposure of the public. These recommendations are taken into account by national protection boards and commissions and by regulatory agencies. For example, in the United States, national standards are recommended by the National Council on Radiation Protection and Measurement (NCRP, 1971), as summarized in Table II. The Environmental Protection Agency formally establishes standards, and federal agencies such as the Nuclear Regulatory Commission issue and enforce standards by way of regulations (e.g., Code of Federal Regulations, Title 10, Part 20). The interior of a food irradiation facility would be classified as a "high radiation area," one in which potentially lethal radiation exposures could take place. Extraordinary provisions for preventing access would thus be required. Requirements set forth in U.S. regulations are summarized as follows:

1. Redundant automatic controls permitting personnel entry only into regions in which a person could receive a dose equivalent no greater than 0.1 cSv in 1 hr.

TABLE II

General Provisions of the External Exposure Protection Standards of the U.S. National Council on Radiation Protection and Measurement[a]

Circumstances of exposure	Dose-equivalent limits (rem or cSv)
Routine occupational	
whole body[b]	5 per year
skin	15 per year
hands	75 per year
forearms	30 per year
Emergency occupational	
lifesaving	
Whole body	100
Hands and forearms	300
general	
whole body	25
hands and forearms	100
General public (whole body)	
individuals	0.5 per year
populations	0.17 per year

[a]From NCRP (1971).
[b]Special provisions are made for exposure of the fetus. See also (ICRP, 1977).

2. Visible and audible alarm signals alerting personnel to radiation hazards and functioning in such a way as to permit time for manual override in preventing commencement of source operation.
3. Various surveillance and administrative procedures further assuring personnel safety.

Guidance in establishing a radiation protection program for a food irradiation facility may be found in the following standards:

1. ICRP Publication 26, "Recommendations of the International Commission on Radiological Protection," Annals of the ICRP, Vol. 1, No. 3, 1977.
2. ICRP Publication 28, "The Principles and General Procedures for Handling Emergency and Accidental Exposures of Workers," Annals of the ICRP, Vol. 2, No. 1, 1978.
3. NCRP Report 51, "Radiation Protection Design Guidelines for 0.1–100 MeV Particle Accelerator Facilities," National Council on Radiation Protection and Measurements, Washington, D.C., 1977.
4. NCRP Report 57, "Instrumentation and Monitoring Methods for Radiation Protection," National Council on Radiation Protection and Measurements, Washington, D.C., 1978.
5. American National Standard ANSI N13.2—1969, "Guide for Administrative Practices in Radiation Monitoring."
6. American National Standard ANSI N13.6—1966 (R 1972), "Practice for Occupational Radiation Exposure Records System."

Protection of personnel from toxic gases is also a requirement for food irradiation facilities. Ozone is produced by interaction of ionizing radiation with oxygen. Production rates and ventilation requirements are discussed by Brynjolfsson and Martin (1971).

D. Radiation Shielding

A critical consideration in the design of an irradiator is the radiation shielding required for personnel protection. Design is complicated because of the need for access portals for product flow and for personnel movement for maintenance purposes. Massive concrete structures with labyrinth portals typify irradiators. In some cases, terrain features may permit natural earth and rock structures to serve as radiation shields.

Shielding design and fabrication is highly specialized and ordinarily would be supplied under a contract including the radiation source and ancillary equipment. Conservatism in design is called for not only for reasons of licensability of the facility but also in view of the potential need for increases in radiation source strength.

Approximate concrete shielding requirements for isotopic sources may be taken from Fig. 2. The abscissa is the concrete shield thickness (cm). The distance r (cm) is the linear distance from the source approximated as a point source and the outside of the concrete shield. The ordinate is r^2 times the dose equivalent rate H (cSv/hr) at the outside

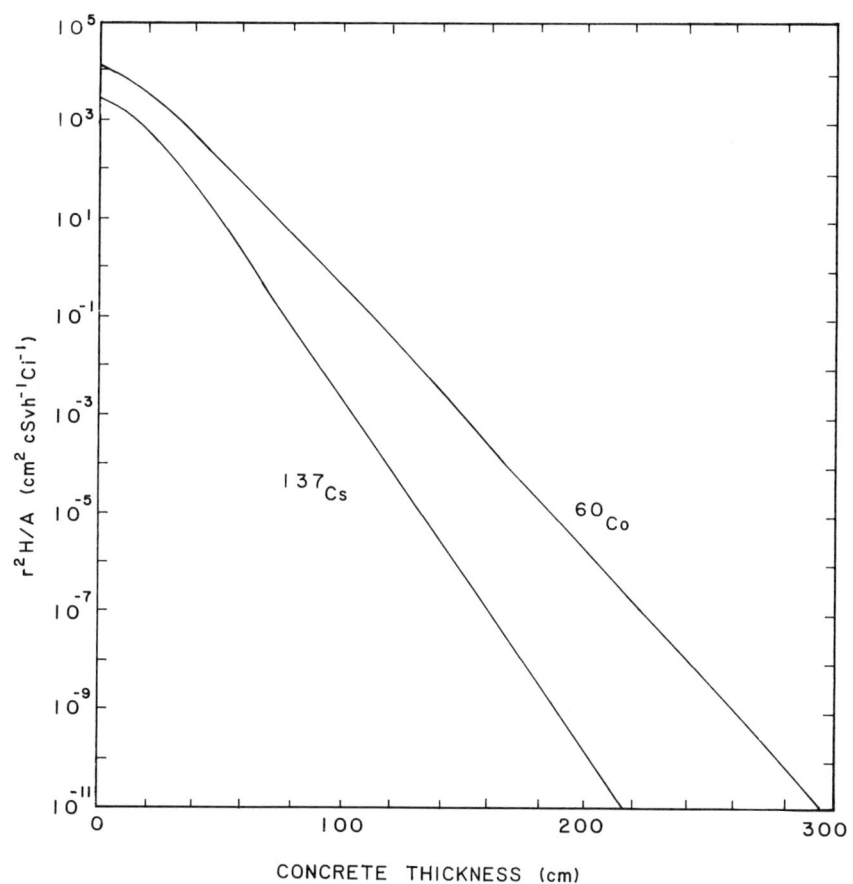

Fig. 2. Approximate variation of dose-equivalent rate with thickness of a concrete shield (2.35 g/cm^3) for ^{60}Co and ^{137}Cs sources of activity A (Ci). See test for explanation. [From Chilton et al. (1984).]

surface of the shield divided by the source activity A in curie units. Similarly, approximate concrete shielding requirements for EB devices used in the production of X-rays may be derived from Fig. 3. In this figure, the ordinate is normalized to the electron beam current, in mA units.

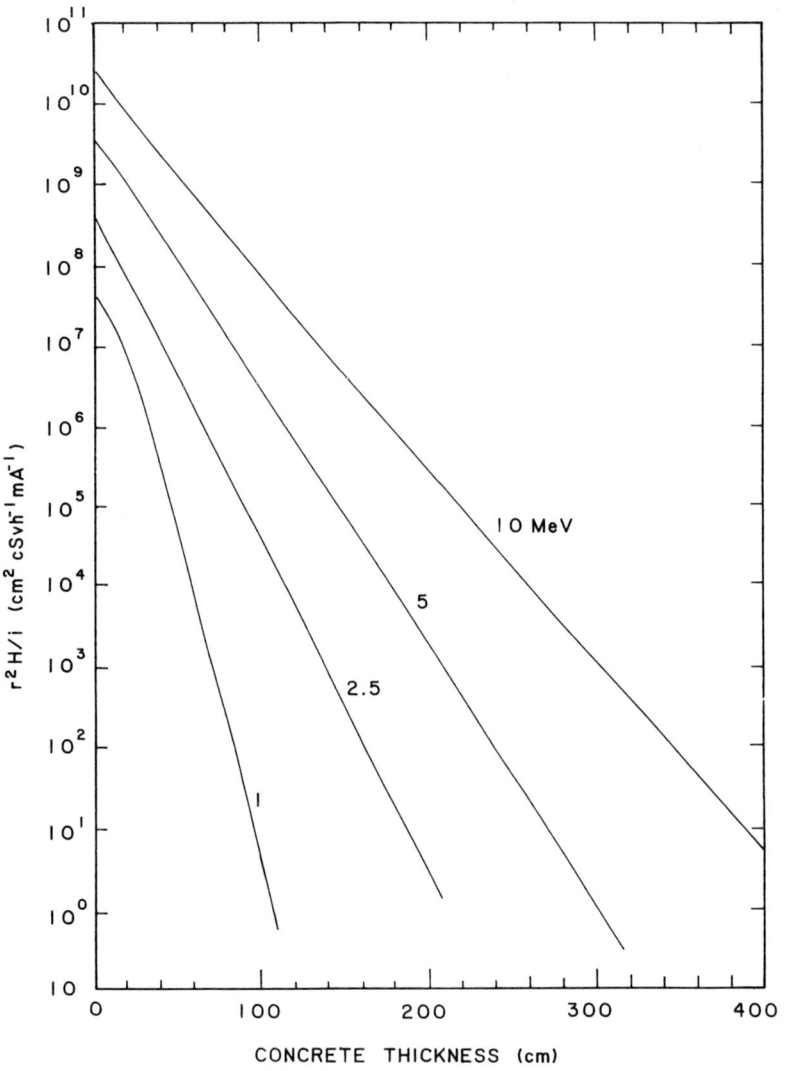

Fig. 3. Approximate variation of dose-equivalent rate with thickness of a concrete shield for electron beam devices of current i (mA). [From Brynjolfsson and Martin (1971).]

IV. CLASSIFICATION OF RADIATION TREATMENTS

Radiation treatments of food by ionizing radiation may be described by three broad categories: radappertization, radicidation, and radurization. These terms are defined as follows (FAO, 1977);

1. *Radappertization:* Treatment of food with a dose of ionizing radiation sufficient to reduce the number and/or activity of viable microorganisms to such a level that very few, if any, are detectable by any recognized bacteriological or mycological testing method applied to the treated food. The treatment must be such that no spoilage or toxicity of microbial origin is detectable, no matter how long or under what conditions the food is stored after treatment, provided it is not recontaminated.
2. *Radicidation:* Treatment of food with a dose of ionizing radiation sufficient to reduce the number of viable specific non-spore-forming pathogenic bacteria to such a level that none is detectable in the treated food when it is examined by any recognized bacteriological testing method.
3. *Radurization:* Treatment of food with a dose of ionizing radiation sufficient to enhance its keeping quality by causing a substantial reduction in the numbers of viable specific spoilage microorganisms.

Radappertization may require doses as high as 50–60 kGy. At the other extreme, radurization doses may be as low as 0.1 kGy, depending on the product. The distinction between categories is not always clear. Radicidation which, for a given product, requires a greater dose than that for radurization, may also enhance the keeping qualities of the product. On the other hand, the radicidation dose may be so high that the quality of the product is diminished.

An alternative categorization used by United Nations agencies is as follows (FAO, 1981):

High-dose applications (about 10–50 kGy)
  Sterilization for commercial purposes
  Elimination of viruses
Medium-dose applications (about 1–10 kGy)
  Reduction of microbial load
  Reduction in the number of non-sporing pathogenic microorganisms

Improvements in technological properties of food
Low-dose applications (up to about 1 kGy)
Inhibition of sprouting
Insect disinfestation
Delay of ripening

V. FOOD PRODUCT POTENTIALITIES FOR RADIATION TREATMENT

Table III lists food irradiation treatments, as of 1981, given unconditional clearance by the Expert Committee on the Wholesomeness of

TABLE III

Food Irradiation Treatments Unconditionally Approved by the Joint FAO/IAEA/WHO Expert Committee on the Wholesomeness of Irradiated Food[a]

Goal of treatment	Food	Maximum dose or dose range (kGy)
Sprouting inhibition	Potatoes	0.03–0.15
	Onions	0.15
Insect disinfestation	Wheat and ground products	1.0
	Rice	1.0
	Teleost fish and products (dried)	1.0
	Mangoes	1.0
	Papayas	1.0
	Pulses	1.0
	Spices and condiments	1.0
	Dates	1.0
	Cocoa beans	1.0
Delayed ripening	Papayas	0.5–1.0
	Mangoes	1.0
Reduction in microbial spoilage	Chicken[b]	2.0–7.0
	Strawberries	1.0–3.0
	Teleost fish and products[c]	2.2
	Cocoa beans	5.0
	Mangoes	1.0
	Spices and condiments	10.0
Reduction in pathogenic microorganisms	Chicken	5.0–7.0
	Teleost fish and products	2.2
	Spices and condiments	10.0

[a] Adapted from FAO (1977, 1981).
[b] Storage at less than 10°C.
[c] Storage in melting ice.

Irradiated Food of the WHO, FAO, and IAEA. Foods listed in the table are representative of the many under examination in some 43 national projects around the world (Vas, 1982).

Loaharanu and Urbain (1982) list a number of advantages of food irradiation other than those specifically identified in Table III. For example, irradiation can improve the baking quality of wheat; can increase juice yields of grapes, and can reduce cooking time, improve texture, increase chlorophyll retention, and reduce oligosaccharides in vegetables. When used as quarantine treatment of fruit, irradiation avoids not only the problems of dealing with fumigant residues but also the deleterious effects of thermal treatments. Irradiation as a quarantine measure is effective at all stages of ripeness. Because inspection and packaging can be done before treatment, wastage and the likelihood of reinfestation are reduced. Radiation treatment in general permits reduction in the use of postharvest insecticides, preservatives, and sterilizing agents. Use of sodium nitrite, as a *Clostridium botulinum* control, can be minimized by irradiation of certain canned meat products.

VI. INTERNATIONAL COOPERATION IN FOOD IRRADIATION

A statutory role of the FAO of the United Nations is promotion of action in the improvement of the nutrition, processing, marketing, and distribution of food. The IAEA, with its responsibility for peaceful applications of atomic energy, joined forces with the FAO in 1964 in establishing a "Joint Division of Atomic Energy in Food and Agriculture" with a Food Preservation Section. Between 1964 and 1977, this section promoted the award of 50 research contracts, with a value of over $650,000, to laboratories all over the world (Vas, 1982). The two goals of the section have been (1) examination of the technological and economic feasibility of food irradiation (see, for example, IAEA, 1975) and (2) achievement of global public health acceptance of food irradiation as a means of reducing food losses (see, for example, IAEA 1967, 1974). The second of these goals brought forth the cooperation of the WHO of the United Nations. In 1969, a Joint FAO/IAEA/WHO Expert Committee on the Wholesomeness of Irradiated Food gave clearance for 5 years for human consumption of irradiated wheat and potatoes (Goresline, 1982). The 1976 and 1980 meetings of the Expert Committee resulted in the clearances listed in Table III.

As a result of the 1969 Expert Committee action, an International Project in the Field of Food Irradiation (IFIP) was undertaken to study irradiated rice, fish, spices, mangoes, and dates, as well as wheat and potatoes. This project, in part, led to 1976 and 1980 clearances listed in Table III. A key conclusion of the 1980 meeting was ". . . that the irradiation of any food commodity up to an overall average dose of 10 kGy presents no toxicological hazard; hence, toxicological testing of foods so treated is no longer required" (FAO, 1981). This conclusion was based on the following three considerations:

1. All the toxicological studies carried out on a large number of individual foods (from almost every type of food commodity) have produced no evidence of adverse effects as a result of irradiation.
2. Radiation chemistry studies have now shown that the radiolytic products of major food components are identical, regardless of the food from which they are derived. Moreover, for major food components, most of these radiolytic products have also been identified in foods subjected to other, accepted types of food processing. Knowledge of the nature and concentration of these radiolytic products indicates that there is no evidence of a toxicological hazard.
3. Supporting evidence is provided by the absence of any adverse effects resulting from the feeding of irradiated diets to laboratory animals, the use of irradiated feeds in livestock production, and the practice of maintaining immunologically incompetent patients on irradiated diets.

Findings of the committee were submitted to the FAO/WHO-sponsored the Food Standards Program Codex Committee on Food Additives for referral to the Codex Alimentarius Commission (FAO 1980a, 1980b).

Two other notable international projects in food irradiation have been described by Vas (1982). Both have IAEA or FAO sponsorship. One is the International Facility for Food Irradiation Technology at Wageningen in the Netherlands. The other is the Asian Regional Project on Radiation Preservation of Fish and Fishery Products. Both projects deal with the technological and economic feasibility of food irradiation. Vas also addresses the question of why food irradiation has required such a high degree of institutional rather than private industry involvement. He gives the following reasons (1) uncertainties in public health acceptance and (2) fragmentation of the food industry (especially in developing nations), and (3) decentralization of marketing, with no concentrated private financial forces able to undertake technological development.

VII. U.S. REGULATION OF RADIATION FOOD PROCESSING

The key statute affecting the development of food irradiation in the United States has been the Food, Drug and Cosmetic Act of 1958, which legally defined ionizing radiation as a food additive rather than a process such as thermal canning or freeze-drying. This food-additive approach to irradiation has enormously complicated clearance of irradiated food for human consumption. It has been interpreted to mean that evaluation of the toxicological aspects of wholesomeness had to be based on concepts of an acceptable daily intake and safety factors, as with pesticide residues (FAO, 1977). While pesticide levels, for example, can be exaggerated in animal feeding studies, it is not practical to exaggerate the doses on the feeding levels of irradiated foods.

The FDA of the U.S. Department of Health and Human Services is the federal agency responsible for administration and enforcement of the Food, Drug and Cosmetic Act. It has been constrained by the food-additive approach and is not free to treat irradiation as a process, as is done by the FAO/IAEA/WHO Expert Committee.

In 1963, the FDA did give clearance to low-dose treatment of grain and high-dose treatment of canned bacon. In 1964, clearance was given to low-dose treatment of potatoes. In 1965, while considering a petition for high-dose treatment of canned ham, the FDA reversed its decision on bacon. Aside from clearance of irradiated packaging materials, no additional clearances had been granted by the FDA when, in 1979, an Irradiated Food Committee was established by the Bureau of Foods of the FDA. As reported by Goresline (1982), the findings of this committee (Brunetti 1980), along with those of the FAO/IAEA/WHO Expert Committee (FAO, 1981), led to FDA publication of proposed procedures for radiation food preservation (Federal Register, 1981). The Brunetti (1980) report concluded that food irradiated at doses not exceeding 1 kGy is safe for human consumption, and foods comprising no more than 0.01% of the daily diet and irradiated at 50 kGy or less could be safely irradiated without any specific toxological testing.

In 1984, FDA proposed regulations (Federal Register, 1984) which would permit the use of food irradiation at doses not exceeding 1 kGy for insect disinfestation and for inhibiting growth and maturation of fresh fruits and vegetables. FDA also proposed changes in criteria for establishing the safety of foods given medium-dose treatment (1–10 kGy).

In 1985, FDA approved irradiation at doses up to 10 kGy for control of insects and microorganisms in a wide variety of dried spices, herbs, and vegetable seasonings (Federal Register, 1985a, 1985b). Also approved was irradiation at doses from 0.3 to 1 kGy for control of *Trichinella spiralis* in pork (Federal Register, 1985c). In 1986, approval was given for irradiation of dried spices, herbs, and seasonings at doses up to 30 kGy and for irradiation of fresh foods at doses up to 1 kGy for the purpose of inhibiting growth and maturation as well as disinfestation of arthropod pests (Federal Register, 1986).

Two other federal agencies are involved in a major way with food irradiation regulation. The Nuclear Regulatory Commission is responsible for licensing of facilities using by-product isotope sources ^{60}Co and ^{137}Cs. In many states, so-called agreement states, the establishment and enforcement of radiation protection regulations has been delegated to state authorities. Once clearance has been given for a food irradiation procedure, the U.S. Department of Agriculture (USDA) has the responsibility for approval of procedures followed at specific food irradiation installations.

VIII. INCENTIVES FOR RADIATION PRESERVATION OF POULTRY MEAT

In terms of public health, the incentive for medium-dose radiation treatment of poultry meat is the elimination of salmonellae microorganisms in fresh, unfrozen carcasses, thereby preventing cross-infection of other foods. There is also a public health incentive for high-dose (sterilization) treatment, namely, elimination of *Clostridium botulinum* as well as less radioresistant microorganisms. There is also a nutritional incentive for the high-dose, or radappertization, treatment. In the many areas of the world in which refrigeration is nonexistent, radiation-sterilized chicken, shelf-stable indefinitely at ambient temperature, could be a new and important diet component. There are also special incentives for sterilization of poultry meat, for example, in military or space-travel rations and in diets for persons undergoing medical immunosuppressive treatment.

There are significant economic incentives in medium-dose treatment of fresh poultry meat. These have to do with food distribution and are discussed by Urbain (1983), who reminds us that fresh poultry is predominantly marketed as prepackaged items sold from a display case. Because of the short display-case life of a retail cut, final cutting and packaging is done at the retail market. Urbain argues that the extension of shelf life through the delay of microbial spoilage brought about by medium-dose radiative treatment will permit centralized final cutting and packaging of poultry meat. He cites the following cost-reduction consequences:

1. Labor savings through improved efficiency and utilization, i.e., increased productivity.
2. Reduced shipping costs, since only what is sold is shipped—trimmings, bones, etc., are retained at the point of origin.
3. Better utilization of by-product trimmings, etc., due to concentration of volume at a central location.
4. Reduced capital requirements for facilities—one central unit can produce the volume of a number of store units, thereby saving on equipment and building space.
5. Opportunities for better quality control in a centralized production system.
6. Increased merchandising flexibility. (Because the retail cuts will be available, a given store needs to stock only what its customers want. This avoids forced selling at that store of all the cuts in a side or other wholesale unit.)

IX. MEDIUM-DOSE TREATMENT OF POULTRY MEAT

A joint FAO/IAEA/WHO Expert Committee (FAO, 1977) concluded, after a detailed and critical review of the available information, that the microbiological aspects of food irradiation were fully comparable to those of conventional processes used in modern food technology. To reduce microbial spoilage and thereby to prolong storage life of eviscerated chicken stored below 10°C, 2–7 kGy irradiation dose was found to be satisfactory. These doses also reduce the number of pathogenic microorganisms. Gamma rays from ^{60}Co or ^{137}Cs, or fast electrons of up to 10 MeV energy, could be used as the radiation source. Insofar as its purpose is the extension of product life through reduction of microbial population, medium-dose treatment is called radurization. Radurization does not prevent ultimate microbial spoilage; it serves only to delay it, and the delay constitutes the product life extension. Insofar as its purpose is reduction in numbers of pathogenic microorganisms, the medium-dose treatment is called radicidation.

With poultry, interest in radurization is limited largely to chicken, especially to fresh eviscerated chicken. In the Unites States the majority of chickens are delivered to stores as ice-packed or dry-packed whole carcasses, where the carcasses are individually cut, packed, and sold from a display case. With good refrigeration, the display case life is only about 2 days (Mountney, 1976). This short shelf life of poultry requires that final cutting and packaging be done in the retail store, thereby minimizing microbial contamination as well as exposure to atmospheric oxygen.

A. Technology

Irradiation dose and storage temperature both affect the shelf life of poultry. In addition to microbial spoilage, there are some changes in poultry that occur in storage which adversely affect consumer acceptance. These changes include discoloration caused by oxidation of the pigment myoglobin, fluid exudate (drip) from cut surfaces, and off-flavor caused by oxidation of lipids. Many investigations of these types of deterioration of fresh poultry have been carried out and are described below.

Urbain and Giddings (1972) (Urbain, 1973) developed a procedure to control microbial contamination, discoloration, drip, and off-flavor of irradiated meat and poultry. For chicken, the procedure could be simplified to the following steps:[2]

1. Chicken carcasses are wrapped in oxygen-permeable, moisture-impermeable flexible film. A number of wrapped carcasses are placed in a bulk vacuumized container.

2. The bulk-packaged chicken is irradiated with gamma radiation using doses of 1–2 kGy at product temperature between 0 and 10°C.

3. After irradiation, the bulk package is stored and transported to the retail store at temperatures between 0 and 5°C.

4. At the retail store, the individually wrapped carcasses are removed from the bulk package about 30 min prior to displaying for sale.

5. The retail products are displayed at temperature between 0 and 5°C and are sold within 3 days.

[2]Urbain (1983) recommends the following preliminary step: to the cut of meat is added not more than 0.5% of a condensed phosphate (e.g., sodium tripolyphosphate). This may be done by dipping the meat in an aqueous solution or by other appropriate means.

B. Microbiological Aspects of Radurization and Radicidation

The principal microbial contaminants of poultry meat of public health significance are salmonellae, *Staphylococcus aureus*, *Clostridium botulinum*, and *C. perfringens*. Only the salmonellae and *S. aureus* are effectively destroyed by radicidation. Kampelmacher (1981) noted that, for poultry-related food poisoning outbreaks in the United States, 44% are due to the salmonellae, and 26% each are due to *C. perfringens* and *C. botulinum*. Principal microorganisms of spoilage are bacteria such as *Acinetobacter*, *Flavobacterium*, *Pseudomonas*, the lactobacilli group, and the *Achromobacter–Alcaligenes* group (Mountney, 1976; Mulder, 1982; Urbain, 1983). Destruction of these microorganisms is the goal of radurization.

Microbial examination typically involves colony counting methods, i.e., determination of total plate counts (TPC) per square centimeter of poultry skin surface (Mountney, 1976). Howker et al. (1976), for example, take swabs from a 6.3-cm^2 surface beneath the wing or at the junction of the leg and thigh.

C. Effects of Radiation on Spoilage Microorganisms

Table IV shows the effect of radiation dose on poultry spoilage microflora for freshly slaughtered chickens, wrapped in plastic film, and stored at 2°C after irradiation. From the results, it is clear that a great reduction occurs at doses between 2 and 5 kGy. After 20-day storage, the number of organisms surviving a 5 kGy dose had not yet reached the

TABLE IV

Effects of Irradiation Dose on Total Plate Counts[a] of Chicken Carcasses Stored at 2°C[b]

Days stored	Irradiation dose (kGy)				
	Control	2	3	4	5
1	8.2×10^5	$<10^2$	~50	<10	<10
6	1.2×10^8	1.7×10^4	~50	<10	<10
12	—	4.5×10^6	3×10^3	1×10^2	<10
20	—	—	5×10^6	2×10^5	5×10^3

[a] Total plate counts per square centimeter.
[b] From Niemand et al. (1977).

initial contamination level. These results confirm the earlier results reported in Table V. The results shown in Table VI illustrate the importance of storage at low temperature.

Kahan and Howker (1978) argue that a 2.5-kGy irradiation dose and storage at 1.6°C are adequate for the radurized chicken process. They note that the product is free from microbial spoilage and of excellent quality for at least 15 days and, in addition, is essentially free of salmonellae and other organisms of public health significance. They argue that the doses prescribed by the FAO/IAEA/WHO Expert Committee for chicken treatment are unnecessarily high and can result in degradation in quality without improvement in microbiological or enzymatic preservation. They recommend an upper dose limit of 3 kGy for irradiation of fresh poultry.

D. Effects of Radiation on Salmonellae

In most countries salmonellosis is the most important causal agent of foodborne disease. According to Kampelmacher (1981), 60–70% of poultry production worldwide may be contaminated. He estimates that there are some 2 million cases of salmonella poisoning annually in the United States. Ouwerkerk (1981) estimates that there are some 300,000 cases per year in Canada, growing at a 3% annual rate.

Mulder (1982) carried out an extensive review of known radiation sensitivities of salmonellae. Sensitivities are expressed in terms of a D

TABLE V

Organoleptic Spoilage of Retail Cuts of Poultry Stored at 5°C after Irradiation[a]

Dose (kGy)	Storage time (days)	Number of birds analyzed	Percentage spoiled
0	1	18	0
	6	18	33
	9	18	100
5	16	36	0
	20	36	16
	30	36	83
7	19	18	0
	24	54	22
	29	52	22
	37	36	66
	40	18	66

[a] From Idziak and Incze (1968).

TABLE VI

Effects of Irradiation Dose on Total Plate Counts[a] of Chicken Carcasses Stored at 1.6 and 4.4°C[b]

Days stored	Control		1.3 kGy		2.8 kGy		5.6 kGy
	1.6°C	4.4°C	1.6°C	4.4°C	1.6°C	4.4°C	4.4°C
1	2.0×10^4	1.2×10^3				$<1 \times 10^2$	$<1.1 \times 10^2$
4	7.6×10^4	6.5×10^4					
7	1.3×10^6					$<1 \times 10^2$	$<1.0 \times 10^2$
9		3.9×10^7		2.8×10^3			
11	7.8×10^6	4.1×10^8					
13							6.2×10^2
15	2.6×10^8		6.6×10^3	4.0×10^6		2.0×10^5	
17	2.6×10^8						
19							7.4×10^3
21				6.3×10^8	1.0×10^4	1.0×10^8	
23							
25			6.9×10^6			8.2×10^4	
27			5.6×10^6				
29						4.1×10^5	2.5×10^4

[a] Total plate counts per square centimeter.
[b] Adapted from Howker et al. (1976) and Kahan and Howker (1978).

value, defined as that dose resulting in a tenfold decrease in numbers of microorganisms. For 19 serotypes, he found from the literature D values ranging from 0.18 to 0.62 kGy at 0°C and from 0.30 to 0.72 kGy at 22°C. In his own work on artificially contaminated skin samples and broiler carcasses, he found the D values listed in Table VII. These results are consistent with the earlier results of Idziak and Incze (1968), who observed D values of 0.5 kGy for salmonellae and 0.45 kGy for *Staphylococcus aureus* strains. Mulder concluded that, for the Dutch industry, the

TABLE VII

D Values (kGy) of Salmonellae and *E. coli* on Artificially Contaminated Broiler Carcasses[a]

	Irradiation temperature	
Organism	+5°C	−18°C
E. coli (K12 NDA)	0.27–0.44	0.72
Salmonella niloese	0.54–0.68	0.88–1.25
Salmonella panama	0.67	1.25–1.32

[a] From Mulder (1982).

application of a dose of 2.5 kGy did not guarantee a *Salmonella*-free product, but based on a *D* value of 0.8 kGy, only at most 1 in 55 carcasses would be *Salmonella*-positive, a factor of 14 decrease due to irradiation. He also observed that microflora surviving the 2.5 kGy irradiation consisted of bacilli, micrococci, streptococci, yeasts, and molds at such low numbers that no dangerous effects could be expected.

E. Nutritional and Toxicological Aspects of Medium-Dose Treatment

From all available data for irradiations of all types of foods, there is no evidence of nutritional effects of medium-dose treatment, save for a slight loss in vitamin C activity in orange juice and a loss of thiamine in dried mackerel (Murray, 1983). Although concern has been expressed over thiamine loss in irradiated chicken (FAO, 1977), such losses are not significant (McGown *et al.*, 1979).

Elias (1983) has reviewed ongoing toxicological research in the following areas: long-term feeding and reproduction studies in dogs and mice; teratology studies in mice, rabbits, and hamsters; dominant lethal and heritable translocation tests in mice; Ames tests (reverse mutations in *Salmonella*); and *Drosophila* sex-linked recessive lethal tests. No adverse effects have been reported in these studies or in genetic toxicity investigations. In over 30 years of toxicity, carcinogenicity, and teratology study, covering every class of foods processed by radiation up to 10 kGy, there has been no evidence of any toxicologically significant compounds in the irradiated food.

Teufel (1983) has reviewed the microbiological safety of food irradiation under the FAO/IAEA/WHO guidelines (FAO, 1977). It has been shown that no new health hazards arise from naturally radiation-resistant microorganisms. Nor is there evidence of radiation-induced mutations or radiation-enhanced pathogenicity.

F. Organoleptic Acceptance

Although the results in Table VI show the advantage of higher doses, there are limitations on the radiation dose that can be used, because of deleterious effects on odor, appearance, and sensory preference of irradiated products. Harikumar *et al.* (1975) irradiated retail cuts of white and dark poultry meat at doses of 1.0 and 2.5 kGy. Postirradiation storage was at 0–4°C. Controls were acceptable organoleptically for 10 days. Irradiated cuts were acceptable for 20 days. Acceptability was better for

8. Radiation Preservation of Poultry Meat

VIII
of Nonirradiated and Irradiated White Chicken Stored at 1.6°C[a]

parameter	Four days			Eight days		
	Nonirradiated	2.5 kGy	5.0 kGy	Nonirradiated	2.5 kGy	5.0 kGy
ance	Fresh	Fresh	Slight odor	No	Fresh	Fresh
nce scores[b]	No discoloration	Slight pink	Salmon pink	Dull	Slight pink	Salmon pink
	7.0	6.6	6.4	6.2	7.0	6.6
on flavor	1.1	1.3	1.2	1.6	1.3	1.4
sity[c]						
ess inten-	1.5	1.3	1.3	1.3	1.4	1.1

m Howker et al. (1976).
ng a 9-point hedonic scale, a hedonic rating of 5 or above indicates an acceptable product.
nsity ratings: 1—none, 2—trace, 3—slight, . . . 9—extreme.

vacuum-packed rather than aerobically packed cuts and for the lower dose rather than the higher. Niemand et al. (1977) irradiated light and dark cuts at 3, 5, 7.5, and 10 kGy. Dishes were prepared in a variety of styles—roasted, baked, etc.—presumably shortly after irradiation. A panel of 10 could detect no significant differences in quality between control and irradiated products.

With the help of consumer-type panels, the data shown in Table VIII were obtained for physical and sensory evaluation (Howker et al., 1976).

Gamma Radiation on the Organoleptic Qualities of Boiled Chicken Pieces[a]

Organoleptic property[b]	Radiation dose (kGy)											
	Control		8				10				15	
	Breast	Leg	Breast		Leg		Breast		Leg		Breast		Leg
			A	B	A	B	A	B	A	B	A	B	A	B
Appearance	9.3	8.3	8.3	9.0	7.3	9.0	8.6	9.2	7.6	8.6	8.6	7.8	7.7	8.2
Odor	9.1	8.7	8.3	9.0	7.8	8.0	8.6	9.5	8.0	8.3	8.4	7.8	8.0	8.7
Flavor	9.2	8.4	8.7	8.9	7.8	8.1	8.5	9.2	7.8	8.1	8.5	9.1	7.9	8.8
Appearance	8.9	7.2	6.4	8.3	6.2	8.1	7.2	8.7	6.4	7.8	7.6	8.7	7.2	8.7
Odor	8.3	7.9	6.1	8.0	5.9	7.7	7.3	8.9	6.8	7.3	7.5	8.0	7.1	8.4
Flavor	8.5	7.7	6.5	8.1	6.4	7.5	7.7	8.9	6.7	7.8	8.1	8.8	7.3	8.3
Appearance	7.7	7.1	—	—	6.9	6.0	7.7	5.3	7.3	6.7	7.8	6.0	7.4
Odor	7.9	7.1	—	7.7	—	7.0	6.2	8.1	5.8	6.9	7.0	7.9	6.8	7.3
Flavor	7.8	7.7	—	7.3	—	7.1	6.3	7.5	5.8	7.1	6.9	7.7	6.7	7.3
Appearance	7.0	7.0	—	7.5	—	6.8	—	6.7	5.2	7.0	6.4	7.6	6.0	7.0
Odor	6.8	6.8	—	7.6	—	6.8	—	7.0	5.7	6.8	6.9	7.8	6.2	7.1
Flavor	7.2	7.1	—	7.4	—	7.0	—	6.8	5.8	7.1	6.9	7.5	6.5	7.2

A refers to unsalted chicken. Column B refers to chicken pieces immersed for 30 min in an aqueous solution of m chloride and 0.25% sodium pyrophosphate. Chicken pieces were irradiated frozen. Controls were kept frozen.
akeil et al. (1978).]
nt hedonic scale.

Fresh chicken carcasses were irradiated within 24 hr after slaughter. After irradiation the chickens were stored at a temperature of 1.6°C. For sensory evaluation, the whole chicken was wrapped in aluminum foil and oven-roasted at 177°C. Based on these evaluations, Howker et al. (1976) suggested that a 2.5-kGy irradiation dose and storage at 1.6°C were sufficient for a radurized chicken process. The resulting chicken was free from microbial spoilage and was of excellent quality for at least 15 days. The 2.5-kGy irradiation dose has the following advantages: sufficient microbial reduction, death or growth inhibition of coliform and fecal streptococci, salmonellae negativity, less irradiation odor and discoloration of carcasses, and lower costs.

Irradiation at subfreezing temperatures is very necessary for the radappertization of meats (see Section X), because the high dose requirements of this process generate a strong irradiation flavor if the irradiation is carried out at temperatures above freezing (Wierbicki et al. 1975). Theoretically this low-temperature technique can be used in radurization of poultry to allow use of doses above 2.5 kGy. The extra costs for freezing and subsequent thawing and the possible associated product quality loss make this approach less desirable. Also, Mercuri et al. (1966) reported that fresh chicken irradiated in the frozen state and stored thawed developed spoilage earlier than chicken irradiated unfrozen. Nevertheless, local conditions of climate and processing technology may be such that higher doses, under frozen condition, is desirable for public health reasons. Table IX shows the results for such a technique. Data are shown for salted, as well as unsalted, carcasses.

X. HIGH-DOSE TREATMENT OF POULTRY MEAT

Radappertization is a process applicable to enzyme-inactivated (precooked) foods in sealed containers and involves irradiation with sterilizing doses of 25–70 kGy at temperatures of −40 to −20°C. The radappertized products are free from organisms causing spoilage and public health hazard, comparable to the "commercial sterility" achieved in thermal canning (Josephson, 1983). The products can be stored indefinitely without refrigeration. Although radappertized products are ready-to-eat, they can also be warmed before serving.

The commercial feasibility of food irradiation depends on economic factors such as production costs and product appeal. Product appeal involves more than just convenience and taste attractiveness. There is also the element of consumer unease over the use of "radiation." It is well established that this unease is not justified, as will be seen shortly. Before commercial feasibility can be established, there are other criteria which must be met in achieving public clearance for unrestricted consumption of irradiated food. These are (1) absence of measurable induced radioactivity, (2) freedom from viable pathogens and their toxins, (3) avoidance of excessive loss in nutritive value, and (4) absence of toxic, mutagenic, or carcinogenic radiolysis products (Josephson, 1983). These criteria, along with technological and commercial aspects of radappertization, are considered in the following paragraphs.

A. Technology of High-Dose Treatment of Poultry

There are a number of requirements for production of a high-quality product through radappertization. It is essential to inactivate enzymes which catalyze spoilage. To minimize deterioration of texture, flavor, and odor, it is necessary to irradiate under vacuum at low temperature. The radiation treatment, of course, must eliminate both microorganisms of decay and pathogens. Finally, vacuum-tight, long-lived packaging must be used to prevent recontamination.

Either whole or cut-up chicken may be treated, through distribution and marketing considerations would no doubt favor the treatment of serving portions in individual packages. Minced or rolled products may also be treated, but final mixing should be done under vacuum to avoid oxygen retention (Wierbicki, 1980). Either metal cans or flexible pouches may be used for storage of the product.

Although radiation alone causes some inactivation, thermal treatment is necessary. Typical treatment is 2 hr at an external temperature of 177°C, resulting in internal temperatures of 75–80°C (Josephson, 1983), or an external temperatures of 70°C for 2 hr and 85°C for 5 hr, resulting also in internal temperatures of 75–80°C (Shults et al., 1977). The enzyme inactivation results in loss of some 35% of natural juices and water-soluble nutrients. Though this may be acceptable, losses can be reduced to 10–15%, with improvement in organoleptic quality, by pretreatment with aqueous sodium chloride (<0.75%) and sodium tripolyphosphate (0.25–0.50%) (Josephson, 1983). There is also an important side advantage of the heating used for enzyme inactivation. It also inactivates most foodborne viruses and vegetative bacteria (Grecz et al., 1983).

For metal cans, the required pressure before radiation treatment is 13.9 kPa. For flexible pouches designed for 100–125 g of meat, the headspace after sealing should be not more than 0.5 cm^3 (Wierbicki, 1980).

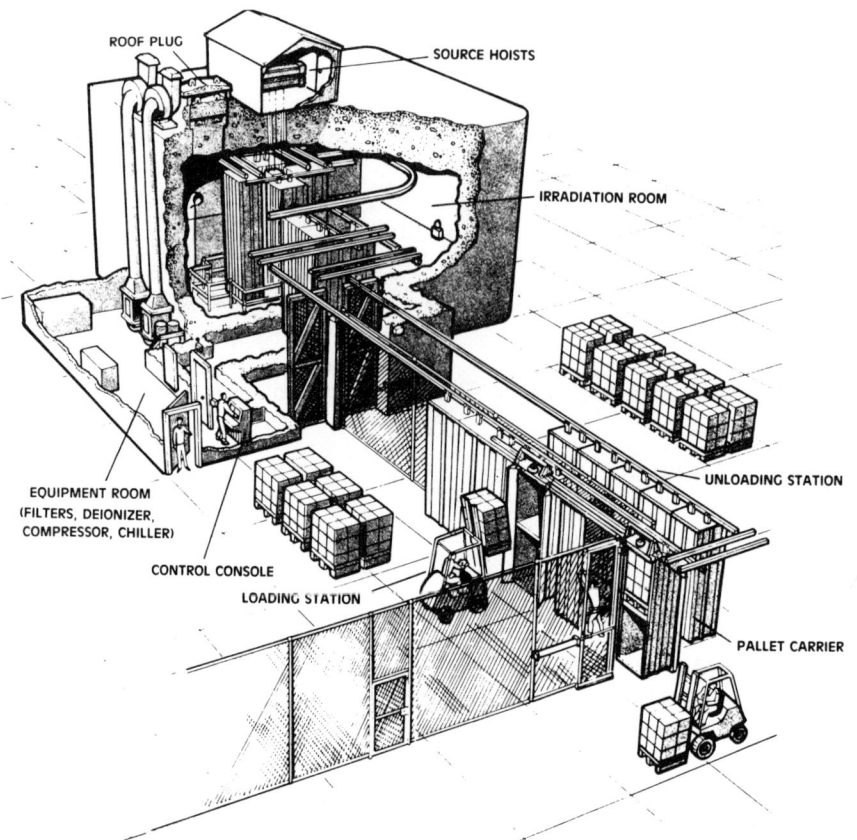

Fig. 4. Multipurpose automatic pallet irradiator suitable for uninterrupted processing of products requiring varying doses. [Illustration provided by courtesy of AECL Radiochemical Company.]

Although destruction of *Clostridium botulinum* is most effective at 0°C (Grecz et al., 1983), irradiation at lower temperatures reduces deterioration resulting from products of water radiolysis. Considering both phenomena and taking into account refrigeration costs, the optimum seems to be about −30°C on average, −40°C at the beginning of irradiation and no more than −20°C at the end (Josephson, 1983).

After radappertization, meat can be stored and distributed without refrigeration. Wierbicki (1980) notes that maximum storage duration depends on storage temperature and degree of enzyme inactivation. Precooking to 70–75°C and storage at 21°C or less permits storage for at least 2 years (Heiligman, 1965). Integrity of flexible pouches is guaranteed for 8 years (Welt, 1984).

B. Induced Radioactivity in Irradiated Meat

Becker (1983) has reviewed the theory and experimental data relating to induction of radioactivity by irradiation with gamma rays or electrons. Gamma rays from ^{60}Co or ^{137}Cs sources have insufficient energy to induce radioactivity through nuclear transmutations. Therefore, attention is centered on high-energy electrons and X-rays (up to 10 MeV) associated with electron-beam irradiators. There are two mechanisms for radioactivity induction: (1) direct transmutations resulting from nuclear interactions of electrons or X-rays and (2) indirect transmutations resulting from capture of neutrons produced by X-rays or electrons, especially through interactions with naturally occurring deuterium. All living matter contains naturally occurring radionuclides such as ^{14}C and ^{3}H. Once food is processed, this naturally occurring radioactivity decays away. Becker has shown that, as long as electron energies in irradiators do not exceed 10.5 MeV, the radioactivity levels produced in irradiated foods are indetectibly low. Furthermore, because processed food is inevitably stored before consumption, food processed by any means is likely to be less radioactive than food that is fresh.

C. Microbiological Considerations in High-Dose Treatment of Poultry Meat

Radappertization treatment must be such that no spoilage or toxicity of microbial origin is detectable no matter how long or under what condition the food is stored. Assurance of complete sterilization is based on the 12-D concept (FAO, 1965). Since *C. botulinum* spores are the most radiation-resistant organisms of concern in radappertization, a dose 12 times that required for 90% reduction would lead to a factor of 10^{12} reduction in the number of viable spores present in the irradiated food.

Grecz et al. (1983) have reviewed the action of radiation on bacteria and viruses. They note that second in importance to *C. botulinum* is *C. perfringens*. Of the six groups of *C. botulinum* (types A–F), types A and B are of main concern because of their greater radiation resistance. Type E, associated with marine products, is more sensitive to radiation but will grow and produce toxin at refrigeration temperatures. Types A and B are inactive below 10°C. *Clostridium perfringens* is classified into five types. Type A is associated more commonly with food poisoning. It is less resistant to radiation than *C. botulinum*, and its food poisoning is less severe.

Although it is not the purpose of radappertization to eliminate viral contamination, the enzyme deactivation thermal treatment does inactivate viruses that may occur in food processing. Furthermore, there is no evidence that viral mutants of public health significance may emerge from foodborne viruses in radiation sterilized foods (Grecz et al., 1983).

There is wide variability in sensitivity between strains of types A and B C. botulinum, with D-values ranging from 1.1 to 3.4 kGy (Grecz et al., 1983). In studies involving some 2000 cans of inoculated chicken treated with 0.75% sodium chloride and 0.3% sodium tripolyphosphate and radappertized at −30°C after enzyme inactivation to a final temperature of 73–75°C, Anellis et al. (1977) determined that the 12-D dose was 42.7 kGy. Sensitivity to radiation is greatest at 0°C for aqueous media and at 35°C in the dry state (Grecz et al., 1983), but for reasons of minimizing effects of water radiolysis, irradiation is done at −30°C.

In contrast to bacterial spores, the radiation resistance of vegetative cells is sensitive to pre- or postirradiation thermal treatment (Maxcy and Rowley, 1978). Both *Moraxella* and *Acinetobacter* are sensitive to irradiation temperature. At −30°C, the D-6 dose (10^{-6} survival in poultry meat) is about 70 kGy, much greater than that for C. botulinum. At +30°C, the D-6 dose is about 20 kGy. The thermal resistance is much less

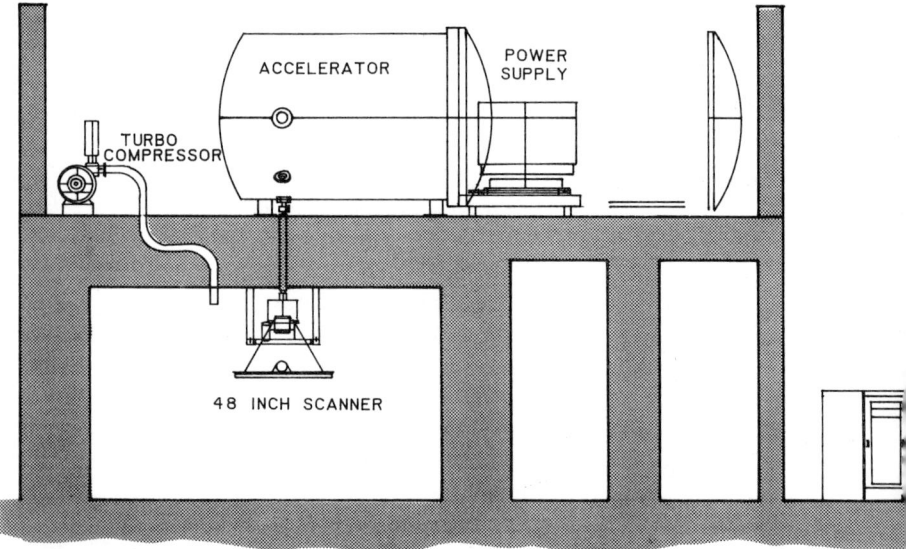

Fig. 5. Electron-beam processing system typical of that used in the elimination of pathogens in poultry feed. Electron energy: 1.5 MeV. Beam current: 75 mA. [Illustration provided by courtesy of High Voltage Engineering Corporation.]

than that for spores. At 70°C, there is a 90% reduction in number in about 7 min. Even if the microorganisms were to survive enzyme inactivation and radappertization, they could not multiply without addition of water.

D. Toxicity Considerations in High-Dose Treatment of Poultry Meat

Toxicity considerations for radappertization are not different in kind from those discussed for radurization and radicidation. They differ in degree only because of the greater radiation doses involved. Proof of wholesomeness in public health clearance involves demonstration that either the radiated food, as a whole, is nontoxic or that the radiolytic products produced during the treatment are individually nontoxic.

Elias (1983), in his review of toxicological aspects of food irradiation, reminds us that testing for toxicity of processed foods is quite different from testing for toxicity of a single, definable food additive. The principal difference rests in the difficulty in applying the traditional animal feeding studies. Many candidates for food irradiation constitute a substantial percentage of the human diet, and it is therefore not possible to exaggerate the feeding levels of irradiated food to test animals to the point of assuring adequate safety factors. This point has been stressed by the United States FDA (Federal Register, 1984) in their recommendation that scientists should focus on the safety of radiolytic products to evaluate the safety of irradiated food; in other words, a "chemiclearance" procedure (Brynjolfsson, 1981) should be followed. This is a very promising approach, because of the great similarity of radiolytic products in related foods; the uniformity of reaction to radiation of the constituent proteins, lipids, and carbohydrates; and the fact that few, if any, radiolytic products are not normal constituents (Elias and Cohen, 1977; Elias, 1983). Furthermore, it is known that the yield of radiolytic products is proportional to the radiation dose (Merritt et al., 1978b); thus extrapolation of experimental data is straightforward. These observations are widely supported (see, e.g., Taub et al., 1979a,b; Merritt et al., 1978a).

This chemiclearance approach has been accepted by the Joint FAO/IAEA/WHO Expert Committee (FAO, 1981) in their conclusion that irradiation of any food commodity up to a dose of 10 kGy is toxicologically safe and that testing is not required. Indeed, the same approach has been taken, though more cautiously, by the United States FDA (Federal

Register, 1984) in their proposed regulation that a dose up to 1 kGy may be applied to any fresh fruits or vegetables for ripening inhibition or to any foods for insect disinfestation. FDA clearance for food irradiation at higher doses has been deferred pending review of animal feeding studies. Some 300,000 lb of radappertized chicken have been fed over a 2-year period to numerous animal species. Results of the studies are under evaluation by the USDA (Thayer, 1984).

That clearance in the United States may ultimately be granted by the FDA may be anticipated. Their 1984 proposed regulations are based largely upon the report of Brunetti (1980), wherein it was accepted that doses of 1 kGy would result in about 30 parts per million (ppm) of radiolytic products, of which 90% are known natural components of food and 10% are, even if unique, similar to known food components. As more and more sophisticated analytical techniques have been applied, it has been found that in irradiation of chicken to doses as great as 91 kGy, no unique radiolytic products are formed (Welt, 1984). At a sterilizing dose of 43 kGy, the total yield of radiolytic products would be only about 10 ppm—lower even than the "acceptable" 30 ppm for a 1-kGy dose.

E. Organoleptic Considerations in High-Dose Treatment of Poultry Meat

Having established the required dose for sterilization, the necessary thermal treatment for enzyme inactivation, and the appropriate vacuum-sealed packaging materials for long-term storage, one must then establish the optimum irradiation temperature and salt treatment leading to high-quality, shelf-stable products appealing to consumers. There are certain clear advantages of sterilization by irradiation over sterilization by thermal retorting. Irradiation does not appreciably alter the water retention of meat. Nor does irradiation cause texture losses comparable to those caused by heat. Furthermore, energy requirements for meat sterilization by irradiation are less than those for purely thermal treatment. There are certain quality problems associated with radiation treatment of meat, namely, discoloration, off-flavors and off-odors, a distinctive "irradiation flavor," and a certain "mushiness" (overtenderization of texture). These factors are considered in the test results described below. In the test results, intensity ratings are based on a scale of 1 to 9, with 1 denoting "none" and 9 denoting "extreme." Preference ratings are based on a scale of 1 to 9, with 1 denoting "dislike extremely" and 9 denoting "like extremely." A preference rating of 5, denoting "neither like nor dislike," is the baseline for product acceptability. Readers are referred to the original literature for test details and statistical analyses.

By marinating poultry meat in aqueous sodium chloride and one of various sodium phosphates, distinct advantages can be gained in product flavor, texture, juiciness, and overall consumer acceptance (Wierbicki, 1981). Shults *et al.* (1977) and Schults and Wierbicki (1973) investigated various concentrations of sodium chloride in combination with sodium tripolyphosphate (NaTPP), pyrophosphate, and hexametaphosphate. They found that, in terms of color, texture, and sensory qualities, a marinade of 1% NaCl and 0.5% NaTPP led to significant improvements in irradiated products. No significant effects of additives on sensory qualities were found in nonirradiated controls. Nor were significant differences in sensory qualities found between controls without additives and irradiated products with additives, stored up to 3 months.

F. Effects of Irradiation Temperature and Storage Duration

Many of the deleterious effects of irradiation on organoleptic quality arise from chemical products of the radiolysis of water. These effects are minimized when water is in the frozen state. Ideally, irradiation should be done at as low a temperature as possible, with the penalty of high refrigeration costs. It has been found, though, that improvements are marginal at temperatures below about $-30°C$, and that significant ill effects occur at temperatures above about $-20°C$ (see Merritt *et al.*, 1978b).

Table X illustrates the effects of irradiation temperature on chicken breasts marinated in 1.5% NaCl + 0.5% NaTPP and enzyme inactivated prior to vacuum packing and irradiation. For testing, specimens were deep-fat fried. Color, flavor, odor, and texture ratings are the preference scores of a 12-person technological panel. Because of the quality decrease at $-20°C$ irradiation temperature, Wierbicki (1981) has recommended that surface temperature during treatment be maintained at $-40 \pm 5°C$, with the temperature at the center of the product after irradiation not exceeding $-20°C$.

Table XI illustrates effects of storage time and temperature on chicken rolls made from boneless, skinless breast meat irradiated at $-30°C$ to

TABLE X
Effects of Irradiation Temperature on Sensory Qualities of Chicken Breasts[a]

Irradiation conditions (kGy at °C)	Conditions of storage	Ratings				
		Color	Odor	Flavor	Texture	Preference
Control	−29°C	7.1	7.4	7.2	6.9	7.1
41 at −20°C	10 Days at 21°C	7.0	6.6	6.5	7.0	6.1[b]
43 at −40°C	10 Days at 21°C	6.9	6.3	6.1	6.6	6.5
45 at −60°C	10 Days at 21°C	7.0	6.8	5.7	7.0	6.7
45 at −30°C	3 Years at 20–32°C	6.3	6.0	6.2	6.3	—

[a] From Wierbicki (1980, 1981).
[b] Significantly less preferred.

TABLE XI
Effects of Storage Time and Temperature on Organoleptic Quality of Irradiation-Sterilized Chicken Rolls[a]

Storage time (months)	Storage conditions (°C)	Ratings				
		Discoloration	Off-odor	Irradiation flavor	Off-flavor	Mushiness
0	21	1.4	2.6	2.3	1.9	2.0
	38	1.4	1.9	1.7	2.0	2.1
	Control	1.6	1.3	1.0	1.4	1.7
1	21	1.7	2.3	3.1	1.7	2.1
	38	1.7	1.7	1.9	1.7	2.3
	Control	1.1	1.6	1.3	1.6	1.9
3	21	1.6	1.4	1.6	1.6	2.1
	38	1.9	1.4	2.0	1.4	2.3
	Control	2.1	1.1	1.1	2.4	1.7
6	21	1.9	1.4	2.1	2.1	1.9
	38	2.4	1.4	1.6	2.0	2.0
	Control	1.9	1.1	1.0	1.7	1.7
12	21	2.2	1.9	2.2	2.1	2.4
	38	2.4	1.9	1.9	2.6	2.7
	Control	1.6	1.1	1.0	1.5	2.0
18	21	1.6	2.1	3.1	1.4	2.1
	Control	1.4	1.1	1.0	1.6	2.0
25	21	1.6	1.9	2.9	1.9	1.7
	Control	1.4	1.1	1.0	1.4	1.4

[a] From Shults et al. (1977).

doses of 45–56 kGy. Pretreatment included marination in 1% NaCl and 0.5% NaTPP, enzyme inactivation, and vacuum canning. Ratings are intensity scores given by 8-person technological panels. Controls were kept frozen prior to evaluation, Investigators concluded that there were no significant differences in sensory-characteristic ratings of samples stored at 21°C and 38°C and control samples for up to 12 months. At 21°C, the samples were shelf-stable for at least 25 months.

G. Nutritional Considerations in High-Dose Treatment of Poultry Meat

Beginning in 1970, an International Project in the Field of Food Irradiation undertook a very broad program of investigation into the wholesomeness of irradiated food. Some 23 nations took part in the project, carried out under the auspices of the Organization of Economic cooperation and Development, the Nuclear Energy Agency, and the Joint FAO/IAEA Division of Atomic Energy in Food and Agriculture. Elias and Cohen (1977) have reviewed the major findings of the project. Conclusions related to nutritional aspects of food irradiation are as follows.

1. *Lipids:* In both whole food and model systems (fatty acids, triglycerides, etc.), chemical changes induced by radiation are similar to those induced in heat treatments.

TABLE XII

Vitamin Losses in Enzyme-Inactivated Beef (Irradiation 47–71 kGy, −40 to −20°C)[a]

		Vitamin content (ppm)			
Vitamin	Storage (months)	Freezing	Heat sterilization	Gamma-ray irradiation	Electron beam irradiation
Thiamin	0	0.97	0.63	0.83	0.77
	15	0.68	0.14	0.21	0.26
Riboflavin	0	2.80	2.63	2.83	2.60
	15	1.69	2.60	2.60	1.46
Niacin	0	48.6	48.1	48.8	46.8
	15	57.2	54.9	50.1	44.5
Vitamin B_6	0	2.50	2.13	3.93	5.20
	15	0.97	0.57	0.35	0.42

[a] From Josephson (1983).

2. *Proteins:* Regardless of type or food origin, proteins respond to irradiation in a reasonably uniform manner. Substances formed in irradiated meats are also obtained in irradiated amino acids, polypeptides, and pure protein.

3. *Vitamins:* Responses to irradiation vary. Vitamin E is the most sensitive of the fat-soluble vitamins. Vitamin B_1 is the most sensitive of the water-soluble vitamins. Degradation products are formed to a greater extent in model systems than in whole foods, and many vitamins are no more susceptible to radiation than they are to the action of light, oxygen, and heat.

Josephson (1983) has reported that irradiation of lipids results in oxidation, degradation, and decarboxylation, with saturated fatty acids being less readily oxidized than unsaturated. At radappertization doses, fat digestibility is unaffected. Furthermore, at those doses, deleterious effects on protein are minimal, and nutritional losses are no more significant than those incurred in preservation by freezing or thermal sterilization. Effects on vitamins in irradiated beef are illustrated in Table XII. Information is not as well established on effects of irradiation on lesser-known vitamins. In general, when food is preserved by radappertization, nutritional losses are no greater than those which occur in more conventional methods of preservation.

XI. ECONOMICS OF POULTRY IRRADIATION

A. Plant Design

The selection of an irradiator for material processing is governed to a large extent by the load, or throughput, and the dose required. Generally an electron beam or machine irradiator is favored for very high loads of low-density material irradiated to high doses. Both machine and isotopic irradiation facilities are highly capital-intensive; thus continuous operation minimizes unit costs. Other considerations in selection of an irradiator are product densities, container dimensions, and seasonal variation in load.

A multipurpose isotopic irradiator is illustrated in Fig. 4. Though the irradiator shown is designed to be more efficient in food processing, it may also be used for sterilization of medical products, for example. When loaded with 2 MCi of ^{60}Co, this irradiator can sterilize up to 750 kg/hr of low-density medical products (Fraser, 1983). When loaded with 1 MCi of ^{60}Co, it can process up to 7500 kg/hr of poultry to a dose of 3–5 kGy (Ouwerkerk, 1981). It will accommodate pallets up to 1.0 m × 1.2 m × 1.95 m high. The RT4101 series of multipurpose isotopic irradiators offered by Radiation Technology, Inc. (Welt, 1983) can process up to fifty 1.0 m × 1.2 m × 2.44 m high pallets per hour (up to 57 tonne per hour). Maximum isotope loading is 3 MCi of ^{60}Co.

Figure 5 illustrates an electron beam irradiator used in the processing of poultry feed. The irradiation chamber is 20 ft × 20 ft × 12 ft high. Shielding walls are concrete, 4 ft thick.

B. Cost of Radiation Processing

Ouwerkerk (1981) reports the following figures for costs of poultry irradiation for salmonella control. The irradiator is designed for treatment of 2500 kg/hr to a dose of 3.0 kGy. The irradiator is supplied with 500 kCi of ^{60}Co. Total initial investment required is $1.2–1.5 million, including the cost of the ^{60}Co. Annual operating costs, mostly fixed, for 4000 hr/year (two-shift) operation are estimated to be $300–400 thousand. Thus, the unit cost for irradiation is $0.03–0.04 per kg. The costs given by Ouwerkerk are comparable to those given by Welt (1984), for poultry irradiation for salmonella control, namely, $0.01–0.03 per pound. Costs for high dose treatment are, of course, a good deal greater, not only because of the higher dose, but also because of refrigeration and packaging costs. Welt (1984) gives the following figures for the cost of sterilization of poultry meat: $0.12–0.30 per pound for irradiation, $0.03 per pound for refrigeration and related costs, $0.03–0.05 per serving container (single-layer pouch for up to 6-month storage), or $0.08–0.12 per serving container (three-layer pouch for up to 8-year storage).

Morganstern (1978) has analyzed the cost of poultry irradiation for salmonella control by use of an electron beam machine, a 4 MeV, 150 kW device used as an X-ray generator. The irradiator could process 2800 kg per hr to a dose of 5 kGy. Total capital cost is estimated to be $1.3 million. Unit costs would be $0.02–0.04 per kg, depending on the operating schedule. A 10 MeV, 2.0 mA electron linear accelerator for use in salmonella treatment of poultry and capable of treating about 6000 kg/hr to 5 kGy would have a capital cost of about $2.5 million, and estimated operating costs (exclusive of amortization of capital costs) in the range of $32–40 per hour (Bly, 1984). Total unit costs would be upward of $0.012 per kg. Operating costs for electron beam machines is very sensitive to

TABLE XIII

Energy Usage in Food Processing (kJ/kg)[a]

Low-dose radiation pasteurization (2.5 kGy)	21
High-dose radiation sterilization (30 kGy)	157
Thermal sterilization (retorting)	918
Blast freezing (4.4°C to −23.3°C)	7552
Refrigeration at −25°C for 3.5 weeks	5149
Refrigeration at 0°C for 5.5 days	318
Cooking whole, thawed chicken	2558

[a]From Brynjolfsson (1978).

the cost of electrical power. Morganstern's estimates were based on $0.07 per kWh and a 50% power efficiency. Likewise, operating costs for isotope irradiators are sensitive to the cost of the source. Ouwerkerk's estimates were based on a ^{60}Co cost of $0.65–0.75 per Ci.

C. Energy and Food Irradiation

The production, processing and distribution of food is very demanding of energy resources. Processing of food by radiation can effect considerable savings in energy, permitting reallocation of scarce resources. Brynjolfsson (1978) has made a thorough analysis of energy use in conventional food processing and distribution, taking into account transportation, refrigeration, and preparation uses as well as processing uses. Similar analysis for radiation treatment includes, as well, energy use in irradiator construction and energy use in the production of radiation sources. Components of energy usage are listed in Table XIII. Energy

TABLE XIV

Total Energy Usage from Poultry Slaughter through Preparation (kJ/kg of Edible Portions)[a]

Refrigerated, raw, cut-up chicken	17,760
Refrigerated and radiation-pasteurized, raw, cut-up chicken	17,860
Frozen, raw, cut-up chicken	46,600
Frozen, cooked long chicken rolls	27,550
Retorted, canned chicken meat	20,180
Radiation-sterilized, cooked long chicken rolls	14,260
Radiation-sterilized, cooked individual servings	15,460

[a]From Brynjolfsson (1978).

usage in various processing methods are listed in Table XIV. Radiation sterilization is significantly less demanding of energy than thermal sterilization and, as well, offers quality improvements. Radiation pasteurization of fresh chicken adds only marginally to energy usage, but offers considerable advantages through reduced spoilage, reduced salmonella hazard, and added flexibility in distribution.

REFERENCES

Anellis, A., Shattuck, E., Morin, M., Srisara, B., Qvale, S., Rowley, D. B., and Ross, E. W., Jr. 1977. Cryogenic gamma irradiation of prototype pork and chicken and antagonistic effect between Clostridium botulinum types A and B. Applied and Environmental Microbiology. 34: 823.

Becker, R. L. 1983. Absence of radioactivity in irradiated food. In "Recent Advances in Food Irradiation." Elias, P. S., and Cohen, E. J. eds. Amsterdam: p. 285. Elsevier Biomedical Press.

Bly, J. H. 1984. Radiation Dynamics, Inc., Private Communication, 17 Aug 1984.

Brunetti, A. P. 1980. Recommendation for evaluating the safety of irradiated foods. Final Report Prepared for the Director, Bureau of Foods, US Food and Drug Administration.

Brynjolfsson, A. 1963. Three dimensional dose distribution in samples irradiated by electron beams. In "Radiation Research," p. 116. Proceedings of an International Conference, U.S. Army Natick Laboratory, Natick, Massachusetts.

Brynjolfsson, A. 1978. Energy and food irradiation. In "Food Preservation by Radiation," Vol. II, p. 285. Vienna: International Atomic Energy Agency (STI/PUB/470).

Brynjolfsson, A. 1981. Chemiclearance of food irradiation process: its scientific basis. In "Combination Processes in Food Irradiation." p. 367. Vienna: International Atomic Energy Agency (STI/PUB/568).

Brynjolfsson, A., and Martin, T. G., III. 1971. Bremsstrahlung production and shielding of static and linear electron accelerators below 50 MeV. Int. J. Applied Radiation & Isotopes 22: 29.

Chilton, A. B., Shultis, J. K., and Faw, R. E. 1984. "Principles of Radiation Shielding." Englewood Cliffs, N.J.: Prentice-Hall, Inc.

Elias, P. S., and Cohen, A. J. (eds.) 1977. "Radiation Chemistry of Major Food Components." Elsevier/North-Holland, Inc., New York.

Elias, P. S. 1983. Toxicological aspects of food irradiation. In "Recent Advances in Food Irradiation." Elias, P. S., and Cohen, A. J. eds. P. 235. Amsterdam: Elsevier Biomedical Press.

El-Wakeil, F. A., El-Magoli, S. B. M., and Salama, N. A. M. 1978. Effect of radurization on the chemical, microbiological and organoleptic characteristics of poultry meat. In "Food Preservation by Radiation," Vol. I, p. 467. Vienna: International Atomic Energy Agency (STI/PUB/470).

FAO. 1965. Technical Basis for Legislation on Irradiated Food. Report of the FAO/IAEA/WHO Expert Committee. World Health Organization Technical Report Series No. 316. Geneva: World Health Organization.

FAO. 1977. Wholesomeness of Irradiated Food. Report of the Joint FAO/IAEA/WHO Expert Committee. Report No. 6 of the FAO Food and Nutrition Series, Food and Agriculture Organization of the United Nations, Rome.

FAO. 1980a. Recommended international general standard for irradiated foods. Report CAC/RS 106–1979, Codex Alimentarius Commission, Food and Agriculture Organization of the United Nations, Rome.
FAO. 1980b. Recommended international code of practice for the operation of radiation facilities for the treatment of foods. Report CAC/RCP 19–1979. Codex Alimentarius Commission, Food and Agriculture Organization of the United Nations, Rome.
FAO. 1981. Wholesomeness of Irradiated Food. Report of a Joint FAO/IAEA/WHO Expert Committee. World Health Organization Technical Report Series No. 659. Geneva: World Health Organization.
Federal Register. 1981. Policy for irradiated foods 46FR59, p. 18992, 27 Mar 1981. Advance notice of proposed procedures for the regulation of irradiated foods for human consumption.
Federal Register. 1984. Irradiation in the production, processing, and handling of food. 49FR31, p. 5714, 14 Feb 84.
Federal Register. 1985a. Irradiation in the production, processing and handling of food. 50FR75, p. 15415, 18 Apr 1985.
Federal Register. 1985b. Irradiation in the production, processing and handling of food. 50FR111, p. 24190, 10 Jun 1985.
Federal Register. 1985c. Irradiation in the production, processing and handling of food. 50FR140, p. 29658, 22 Jul 1985.
Federal Register. 1986. Irradiation in the production, processing and handling of food. 51FR75, p. 13376, 18 Apr 1986.
Fraser, F. M. 1983. Gamma radiation processing equipment and associated energy requirements in food irradiation. In "Preservation of Food by Ionizing Radiation," Vol. III. Josephson, E. S. and Peterson, M. S. eds., p. 253. Boca Raton, Florida: CRC Press.
Goresline, H. E. 1982. Historical aspects of the radiation preservation of Food. In "Preservation of food by Ionizing Radiation," Vol. I. Josephson, E. S., and Peterson, M. S. eds., p. 1. Boca Raton, Florida: CRC Press, Inc.
Grecz, N., Rowley, D. B., and Mataurama, A. 1983. The action of radiation on bacteria and viruses. In "Preservation of Food by Ionizing Radiation," Vol. II. Josephson, E. S., and Peterson, M. S. eds., p. 167. Boca Raton, Florida: CRC Press, Inc.
Harikumar, P., Ninjoor, B., and Kumta, U. S. 1975. Studies in radurization of chicken meat. Symposium Proceedings: "Use of Radiation and Radioisotopes in Studies of Animal Production," Indian Veterinary Research Institute, Izatnagar.
Heiligman, F. 1965. Storage stability of irradiated meats. Food Technol. 19: 114.
Howker, J. J., Kahan, F. S., and Wierbicki, E. 1976. Radurization of Fresh Poultry. Report FEL-61, US Army Natick R&D Laboratories, Natick, MA.
IAEA. 1967. Microbial Problems in Food Preservation by Radiation. Vienna: International Atomic Energy Agency (STI/PUB/370).
IAEA. 1974. Improvement of Food Quality by Irradiation. Vienna: International Atomic Energy Agency (STI/PUB/370).
IAEA. 1975. Requirements for the Irradiation of Food on a Commercial Scale. Vienna: International Atomic Energy Agency (STI/PUB/394).
IAEA. 1977. Manual of Food Irradiation Dosimetry. Vienna: International Atomic Energy Agency (STI/DOC/10/178).
ICRP. 1977. Recommendations of the International Commission on Radiological Protection. ICRP Publication 26, Annals of the ICRP, Volume 1 No. 3.
Idziak, E. S., and Incze, K. 1968. Radiation treatment of foods. Radurization of fresh eviscerated poultry. Appl. Microbio. 16: 1061.
Jarrett, R. D., Sr. 1982. Isotope (gamma) radiation sources. In "Preservation of Food by Ionizing Radiation," Vol. I. Josephson, E. S., and Peterson, M. S. eds., p. 137. Boca Raton, Florida: CRC Press, Inc.

Josephson, E. S. 1983. Radappertization of meat, poultry, finfish, shellfish, and special diets. In "Preservation of Food by Ionizing Radiation," Vol. III. Josephson, E. S., and Peterson, M. S. eds., p. 231. Boca Raton, Florida: CRC Press.

Kahan, R. S., and Howker, J. J. 1978. Low-dose irradiation of fresh, nonfrozen chicken and other preservation methods for shelf life extension and for improving its public-health quality. Food Preservation by Radiation, Vol. II. International Atomic Energy Agency, Vienna, p. 221.

Kampelmacher, E. H. 1981. Prospects of eliminating pathogens by the process of food irradiation. In "Combination Processes in Food Irradiation." p. 265. Vienna: International Atomic Energy Agency (STI/PUB/568).

Loaharanu, P., and Urbain, W. M. 1982. Certain utilization aspects of food irradiation. In "Preservation of Food by Ionizing Radiation," Vol I. Josephson, E. S., and Peterson, M. S. eds., p. 65. Boca Raton, Florida: CRC Press, Inc.

Matsuyama, A., and Umeda, K. 1983. Sprout inhibition in tubers and bulbs. In "Preservation of Food by Ionizing Radiation," Vol. III. Josephson, E. S., and Peterson, M. S. eds., p. 159. Boca Raton, Florida: CRC Press, Inc.

Maxcy, R. B., and Rowley, D. B. 1978. Radiation-resistant vegetative bacteria in a proposed system of radappertized meats. In "Food Preservation by Radiation," Vol. I, p. 347. Vienna: International Atomic Energy Agency (STI/PUB/470).

McGown, E. L., Lewis, C. M., and Waring, P. P. 1979. Investigation of possible antithiamine properties in irradiation sterilized chicken. Report 72, Letterman Army Institute of Research, San Francisco, CA.

McLaughlin, W. L., Jarrett, R. D., Sr., and Olejnik, T. A. 1982. Dosimetry. In "Preservation of Food by Ionizing Radiation," Vol. I. Josephson, E. S., and Peterson, M. S. eds., P. 189. Boca Raton, Florida: CRC Press, Inc.

Mercuri, A. J., Kotula, A. W., and Sanders, D. H. 1966. Low dose ionization radiation of tray-packed, cut-up fryer chickens. Poultry Sci. 45: 1105.

Merritt, C., Jr., Angelini, P., and Nawar, W. W. 1987a. Chemical analysis of radiolysis products relating to the wholesomeness of irradiated food. In "Food Preservation by Irradiation," Vol. II, International Atomic Energy Agency, Vienna, p. 97.

Merritt, C., Jr., Angelini, P. and Graham, R. A. 1978b. Effect of radiation parameters on the formation of radiolysis products in meat and meat substances. J. Agr. Food Chem. 26: 29.

Minck, F. 1896. On the question of the effect of Roentgen rays on bacteria and the possibility of their eventual application. Munch. Med. Wochensch. 5: 101.

Morganstern, K. H. 1978. Economics of electron accelerators in the preservation of food by irradiation. In "Food Preservation by Radiation," Vol II, P. 267. Vienna: International Atomic Energy Authority (STI/PUB/470).

Mountney, F. J. 1976. "Poultry Products Technology," 2d ed. Westport, Connecticut: The Avi Publishing Co., Inc.

Mulder, R. W. A. W. 1982. "Salmonella Radicidation of Poultry Research," Beekbergen, The Netherlands.

Murray, T. K. 1983. Nutritional aspects of food irradiation. In "Recent Advances in Food Irradiation." Elias, P. S., and Cohen, A. J. eds. p. 203. Amsterdam: Elsevier Biomedical Press.

NCRP. 1971. Basic radiation protection criteria. NCRP Report 39, National Council on Radiation Protection and Measurements, Washington, D.C.

Niemand, J. G., Haauser, G. A. M., Clarke, I. R., and Thomas, A. C. 1977. Radiation Processing of Poultry. Report PER-16, Atomic Energy Board, Pelindaba, South Africa.

Ouwerkerk, T. 1981. Salmonella control in poultry through the use of gamma irradiation. In "Combination Processes in Food Irradiation." p. 335. Vienna: International Atomic Energy Agency (STI/PUB/678).

Procter, B. E., Van de Graff, R. J., and Fram, H. 1943. Effect of x-ray irradiation on the bacterial count of ground meat. Research Reports on the U.S. Army Quartermaster Contract Projects, Food Technology Laboratories, Massachusetts Institute of Technology.

Ramler, W. J. 1982. Machine sources. In "Preservation of Food by Ionizing Radiation," Vol. I. Josephson, E. S., and Peterson, M. S. eds., p. 165. Boca Raton, Florida: CRC Press, Inc.

Regulla, D. F., and Deffner, U. 1983. A system of transfer dosimetry in radiation processing. Radiat. Phys. Chem. 22:305.

Shults, G. W., and Wierbicki, E. 1973. Effects of NaCl and condensed phosphates on the water-holding capacity, pH and swelling of chicken muscle. J. Food Sci. 38: 991.

Shults, G. W., Cohen, J. S., Mason, V. C., and Wierbicki, E. 1977. Flavor and textural changes in radappertized chicken as affected by irradiation temperature, NaCl and phosphate additions. J. Food Sci. 42: 885.

Taub, I. A., Kapreilian, R. A., Halliday, J. W., Walker, J. E., Angelini, P., and Merrit, C., Jr. 1979a. Factors affecting radiolytic effects in food. Radiat. Phys. Chem. 14: 639.

Taub, I. A., Robbins, F. M., Simic, M. G., Walker, J. E., and Wierbicki, E. 1979b. Irradiation on meat proteins. Food Technol. 33(5): 184.

Teufel, P. 1983. Microbiological aspects of food irradiation. In "Recent Advances in Food Irradiation." Elias, P. S., and Cohen, A. J. eds. p. 217. Amsterdam: Elsevier Biomedical Press.

Thayer, Donald W. 1984. Summary of supporting documents for wholesomeness studies of precooked (enzyme inactivated) chicken products in vacuum sealed containers exposed to doses of ionizing radiation sufficient to achieve "commercial sterility." Report PB84-186980, Eastern Regional Research Center, U.S. Department of Agriculture, Philadelphia.

Tilton, E. W., and Burditt, A. K., Jr. 1983. Insect disinfestation in grain and fruit. In "Preservation of Food by Ionizing Radiation," Vol. III. Josephson, E. S., and Peterson, M. S. eds., p. 215. Boca Raton, Florida: CRC Press, Inc.

Urbain, W. M. 1973. The low-dose radiation preservation of retail cuts of meat. In "Proceedings of the Bombay Symposium on Radiation Preservation of Food," p. 505. Vienna: International Atomic Energy Agency.

Urbain, W. M. 1983. Radurization and radicidation: meat and poultry. In "Preservation of Food by Ionizing Radiation," Vol. III. Josephson, E. S., and Peterson, M. S. eds., p. 1. Boca Raton, Florida: CRC Press, Inc.

Urbain, W. M., and Giddings, G. G. 1972. Factors related to market life extension of low dose irradiated fresh meat and poultry. Radiat. Res. Rev. 3: 389.

Vas, K. 1982. National and international programs. In "Preservation of Food by Ionizing Radiation," Vol. I. Josephson, E. S., and Peterson, M. S. eds., p. 47. Boca Raton, Florida: CRC Press, Inc.

Welt, M. A. 1983. Status of commercial food irradiation in the United States. Radiation Phys. Chem. 22:215.

Welt, M. 1984. Radiation Technology, Inc. Private communication, 9 Aug 1984.

Wierbicki, E. 1980. Technology of irradiation preserved meats. Proc. 26th European Meeting of Meat Research Workers, Colorado Springs, CO, 31 Aug–5 Sep 1980.

Wierbicki, E. 1981. Technological feasibility of preserving meat, poultry and fish products by using a combination of conventional additives, mild heat treatment and irradiation. In "Combination Processes in Food Irradiation." p. 181. Vienna: International Atomic Energy Agency (STI/PUB/568).

Wierbicki, E., Brynjolfsson, A., Johnson, H. C., and Rowley, D. B. 1975. Preservation of meats by ionizing radiation. European Meeting Meat Research Workers, Berne, Switzerland, Paper No. 14.

CHAPTER 9

Methods of Preservation of Poultry Products

F. E. Cunningham

Department of Animal Sciences
Kansas State University
Manhattan, Kansas 66502

I. INTRODUCTION

Poultry meat was responsible for 12% of foodborne disease outbreaks in the United States during 1966 to 1974, according to reports submitted to the Centers for Disease Control, Atlanta, Georgia (Horwitz and Gangarosa, 1976). In 1978, 4656 cases of foodborne bacterial diseases were reported to the Centers for Disease Control (Centers for Disease Control, 1979). Of those, 1181 (25%) were traced to poultry meat. After reviewing reports from 20 countries on poultry-related foodborne disease, Todd (1980) indicated that salmonella infections were still a major worldwide public health problem.

Control of microorganisms on poultry meat products is the major concern in preparation of high-quality foods. This chapter reviews those techniques which have been reported to reduce microbial load and prolong shelf life of poultry meat items.

II. HEAT TREATMENTS

Surface-pasteurized fryers were compared with untreated and chlorotetracycline- (CTC-)treated fryers by Dawson *et al.* (1963), who studied bacterial population and odor of flourescence during storage at 1.1°C. They found that partial cooking resulted when pasteurized at a temperature higher than 60°C for longer than 1 min. They also found that CTC treatments, with proper pasteurization, extended acceptability about 1 week longer than that of controls.

Scholtyssek et al. (1969) found that a combination of scalding carcasses at 50°C and 52°C, then chilling in a stream of cold air increased shelf life by 9–13 days over scalding followed by spin-chilling, indicating that a combination of treatments can be effective.

The use of subatmospheric pressure steam not exceeding 75°C for 4 min by Klose et al. (1971) reduced external and internal bacterial loads by 1,000- to 5,000-fold on whole, ready-to-cook chicken carcasses. Klose et al. reported that optimum results were obtained when the steam diffuser was placed directly into the body cavity.

Kyle (1971) pasteurized skin surfaces of poultry meat by using condensing vapors from boiling liquids. The vapors from boiling solutions of actone–water, isopropyl alcohol, trichloroethylene–water, and water under reduced pressure reduced bacteria 10^3- to 10^5-fold in 4- to 8-min exposures. Kyle found that trace amounts of the liquids were difficult to remove from the cooked product. The shelf life exceeded the controls by 15 days in many cases.

Young and Lyon (1973) heated mechanically deboned chicken meat to 65°C and found that they could use 30% of the heated meat in the preparation of chicken frankfurters.

Cox et al. (1974) studied the effects of hot water treatments on the quality of broiler carcasses. Whole eviscerated broiler carcasses were immersed in water, with or without agitation and at different temperatures, for either 1 or 3 min. The carcasses were then bagged, rechilled, and stored at 4°C. Those researchers found that shelf life was not significantly extended with treatments of hot water alone.

Mast and MacNeil (1975) pasteurized mechanically deboned poultry meat for up to 6 min at temperatures of 59, 62, 65, 68, or 71°C. They found that the shelf life of the meat (as measured by total bacterial numbers) was extended by heat pasteurization, since the initial level of bacteria was greatly reduced.

Cunningham and Albright (1977) reported that total bacteria were substantially reduced after exposure of fresh chicken to subcooking microwave energy. Later, Cunningham (1978) found that briefly exposing fresh chicken patties to microwaves reduced total counts from 100,000 cells/cm^2 to less than 1,000 cells/cm^2.

Pure cultures of bacteria in nutrient broth at 10^7 or 10^8 organisms/ml and various raw meat products were exposed to either 915- or 2450-MHz microwaves by Cunningham (1980). After various timed exposures, temperature increases were noted and samples removed for plate-count determination of survivors. In nutrient broth, all psychrotrophic bacteria counts were dramatically reduced by short exposures (5–20 sec) to microwave radiation. In some instances, initial counts of 10^7–10^8 per milliliter were reduced to zero after 15 sec of microwave exposure and a temperature of 60°C. *Moraxella acinetobacter* (MA-3), *Serratia marcescens*, *Pseudomonas synthaxa*, and *Alcalligenes viscolactis* were extremely susceptible to the microwave treatments. When raw poultry parts with initial counts of 10^4/cm^2 of surface area were exposed to microwave radiation, organisms were reduced by 1 log cycle in 20 sec and by 2 log cycles in 40 sec. Total counts on microwave-treated chicken parts indicated that such treatment could substantially increase shelf life.

III. USE OF ANTIBIOTICS

The literature is full of research indicating that use of antibiotics in chill water or on poultry carcasses is very effective in prolonging shelf life. However, since those chemicals are not now in use they will not be discussed here. Antibiotics were in commercial use in the United States until Food and Drug Administration (FDA) approval was withdrawn in 1967 (Potter, 1973).

IV. USE OF SANITIZERS

Patterson (1968) examined the bacterial flora of chicken carcasses treated with high concentration of chlorine. It was found that shelf life was extended by approximately 20% when carcasses were immersed in chilled water containing 200 or 400 ppm of free residual chlorine (Cl_2) for 4 hr. Similar results were reported by Ranken et al. (1965). Patterson found that concentrations of available chlorine of 500 ppm or more resulted in "tainted" carcasses.

Dixon and Pooley (1961) suggested that chlorination with 200 ppm for 10 min be incorporated into the routine processing of poultry carcasses. Their suggestion was based on the fact that in their experiments that treated carcasses with 200 ppm of chlorine for 10 min, subsequent recovery of salmonellae (when fewer that 1000 organisms had been inoculated) was usually prevented.

Lillard (1979) also found that 20 ppm Cl^2 and 3 ppm chlorine dioxide (ClO^2) introduced into the chiller with the fresh water input were equally effective, and that using $\frac{1}{17}$ as much ClO^2 as Cl^2 would be less corrosive to processing equipment.

Baran et al. (1973) studied the influence of ClO_2 water treatment on numbers of bacteria associated with processed turkey. Comparing ClO_2-treated birds to birds chilled using the existing in-plant chlorination program, they reported that the ClO_2 treatment was most effective in reducing total and coliform counts of surface skin samples when contact times of 20–72 hr were used. However, holding times of 6 hr showed no differences in bacterial numbers between the two treatments. Total and coliform counts were very low with no detection of *Salmonella* in any samples.

Chlorination could be used to eliminate both fecal and spoilage bacteria from water used for chilling eviscerated poultry carcasses (Mead and Thomas, 1973a). In the later report, Mead and Thomas (1973b) determined the quantitative changes in the flora of chicken carcasses after passage through a series of three separate spin-chillers. Organisms were eliminated from the chill water by using 1.7 liters of water per carcass and 45–50 ppm of total residual chlorine in the first two chillers and 1.0 liters of water per carcass and 25–30 ppm of residual chlorine in the third chiller. It was concluded that the main effect of chlorination in the chillers was to destroy organisms washed from the carcass, thus avoiding recontamination.

Thiessen et al. (1985) treated chill water in a poultry processing plant with various levels of ClO_2 and found that increasing the concentration from 0 to 1.39 mg/liter resulted in a reduction of the bacterial counts to the point where salmonellae could not be isolated from the chill water or the chilled broiler carcasses. In addition, coliform, psychrotroph, and aerobic plate counts were all greatly reduced. Sensory panelists reported no off-flavors for any ClO_2 concentration.

A study on the effect of chlorine, antibiotics, β-propiolactone, acids, and washing on *Salmonella typhimurium* on eviscerated fryer chickens was conducted by Thomson et al. (1967). Breast skin areas were inoculated with *S. typhimurium* and were then sprayed with one of seven sterile aqueous solutions of chlorine, aureomycin, β-propiolactone, succinic acid, citric acid, neomycin, distilled water, or for controls, left unsprayed. A significant reduction in salmonellae counts was effected by spraying all treatment solutions.

Jones et al. (1968) found that hypochlorite solutions with the addition of ethylene glycol to prevent freezing were effective against *Bacillus subtilis* var. *niger* spores from 0 to 40°C.

Hayes et al. (1967) found that successive treatments with low levels of iodophor (6 ppm) followed by hypochlorite (12–25 ppm) resulted in a high level of destruction of all test organisms.

9. Methods of Preservation of Poultry Products

Further work done by Spencer et al. (1968) showed that carcasses chilled 4 hr in solutions containing 50, 100, and 200 ppm iodophor had total plate counts reduced by 50, 66, and 72% respectively. In a second experiment, iodophor extended shelf life by 1–2 days at 100 ppm, while at 200 ppm, shelf life was extended 2–3 days. A dark yellow color was observed about the feather follicles, but this dissipated with storage, and no differences in odor and taste were noted after 3 days of storage.

Ozone treatments were used by Yang and Chen (1979) to extend the shelf life of poultry parts. They found that ozone-treated parts had consistently lower microbial counts than the control parts during the entire refrigerated observation period and that the shelf life had been extended 2.4 days.

Quanternary ammonium compounds and hypochlorite sanitizers were used by Scheusner and Harmon (1971) to evaluate their effects on *Staphylococcus aureus, Streptococcus faecalis,* and several strains of *Esherichia coli*. Sanitizer concentration appeared to be the main factor in the course of injury and death of these organisms. The authors also emphasized that the inability or failure of common selective media to detect injured bacteria in food could have serious public health consequences.

Olson et al. (1981) investigated the effectiveness of a five-cycle, rapid freeze–rapid thaw process in conjunction with and without the chemicals chlorine, potassium sorbate, lactic acid, calcium propionate in reducing the numbers of *Salmonella typhimurium* cells inoculated on chicken wings. They also studied the effectiveness of the chemicals without a freeze–thaw process.

Their results are shown in Table I. They concluded that a five-cycle,

TABLE I

Percentage of Reduction of Numbers of *Salmonella typhimurium* Inoculated on Chicken Wings Treated with Chemicals and on Chemically Treated Wings Subjected to a Five-Cycle, Rapid Freeze–Rapid Thaw Process[a]

Treatment	Without freeze-thaw (%)	With freeze-thaw (%)
Control	0	95
Calcium propionate, 5%	90	96
Potassium sorbate, 5%	85	96
Lactic acid, 5%	82	98
Chlorine, 20 ppm	46	95

[a] From Olsen et al. (1981).

rapid freeze-rapid thaw process could be used in commercial poultry processing plants to reduce numbers of *S. typhimurium* cells. A chemical used in conjunction with the freeze-thaw process may extend the shelf life of poultry meat.

V. USE OF FOOD ACIDS

Many acids have been tested as aids in reducing bacterial numbers on fresh poultry. Murphy and Murphy (1962) reported that chill water with 500–2500 ppm of lactic, citric, or hydrochloric acid resulted in lower counts in the chill water and on the surface of the carcasses.

Lactic acid starter culture of *Pediococcus cerevisiae* and *Lactobacillus* were inoculated in cooked, mechanically deboned poultry meat, by Raccach and Baker (1978), to suppress growth of three species of *Pseudomonas*, *Salmonella typhimurium*, and *Staphylococcus aureus*. A 50–50 mixture of the two starter cultures was the most effective treatment, by delaying the time necessary for the three *Pseudomonas* spp. to attain 10^7 cells/g, which is associated with spoilage, and by totally suppressing the growth of *S. typhimurium* and *S. aureus*.

Perry et al. (1964) found that shelf life of eviscerated, refrigerated, cut-up, "dry-pack" poultry was extended by the synergistic effect of a two-step process consisting of acidic hydration and sorbic acid application. Treated parts did not spoil until after 18 days at 7.2°C. Development of putrid odor was also markedly suppressed, and flavor was not impaired. Their studies also indicated that sorbic acid applied to skin surfaces was absorbed into underlying muscle tissue.

Juven et al. (1974) tested hot succinic acid for destruction of *Salmonella* and found that a treatment of 85°C for 2–3 min was required to achieve elimination.

Mountney and O'Malley (1965) added the following 10 acids to adjust the chill water to pH 2.5: acetic, adipic, citric, fumaric, hydrochloric, lactic, malonic, phosphoric, succinic, and sorbic. Acetic, adipic, and succinic acids prolonged the shelf life by 6 days; however, acetic acid was considered unacceptable because of its pungent odor and adverse effect of making skin hard and leathery. Cox et al. (1974) extended the shelf life of broiler legs stored at 4.4°C for 6.5 days by treatment with hot (60°C) sucinic acid (30,000 ppm). Dipping broiler parts into 10,000 ppm ascorbic acid for 3 min retarded microbial growth and increased the shelf life by 7 days (Arafa and Chen, 1977).

Lin and Chen (1981) researched the use of sorbic acid–ice for its effect in controlling bacterial populations on cut-up broiler parts. The chickens were ice-packed and held at 2–4°C with crushed ice which was made with either tap water, water containing 0.75% sorbic acid, or water containing 0.10% potassium sorbate. They found that the shelf life of broiler parts ice-packed with 0.75% sorbic acid–ice and 0.10% potassium sorbate–ice was 11.5 and 3.6 days longer than the ice-packed controls.

Both sorbic acid–ice and potassium sorbate–ice decreased the incidence of gram-negative rod-type organisms and increased incidence of gram-positive cocci on broiler parts after 4 days of storage at 2–4°C.

VI. USE OF CONTROLLED ATMOSPHERE

Carbon dioxide was shown to inhibit the growth of slime-forming bacteria and to extend the shelf life of poultry (Oglivy and Ayres, 1951). CO_2 preservation of fresh poultry was also studied by Wabeck et al. (1968). Administration of CO_2 throughout storage, at 1°C, had an inhibitory effect on part of the bacterial population (Table II). Coliform populations were relatively stable during storage, due to the low tempera-

TABLE II

Logarithmic Bacterial Counts of Drip from Birds Stored at 1°C in Air and 10% and 20% Carbon Dioxide[a]

Storage time period (days)	Bacterial count (log average)[b]								
	Total			Psychrophilic			Coliform		
	Air	10% CO_2	20% CO_2	Air	10% CO_2	20% CO_2	Air	10% CO_2	20% CO_2
0	5.26	5.55	5.29	3.27	3.26	3.02	2.78	3.12	3.25
2	5.23	5.53	5.27	3.05	3.43	2.77	2.29	3.07	2.72
4	5.13	5.97	5.12	3.47	4.12	3.87	2.56	2.31	2.47
6	4.93	5.18	4.86	4.82	4.76	4.68	2.33	2.50	2.84
8	5.96	5.46	5.95	4.79	5.06	4.52	2.30	2.74	2.73
10	6.33	6.08	5.62	7.21	5.68	4.90	2.15	3.33	2.75
12	7.94	6.60	5.85	8.10	7.41	5.84	2.72	3.98	3.40
14	9.17	7.70	5.65	9.21	8.50	5.67	2.80	4.89	3.44
16	9.40	8.17	6.14	10.08	9.15	6.20	3.28	5.52	3.21

[a] From Wabeck et al. (1968).
[b] Average of four samples.

ture, but some lipolytic and proteolytic psychrophiles were suppressed by CO_2. Differences in pH of the skin were insignificant, although values for birds in CO_2 were lower than values for those in air. Growth patterns of organisms on birds in air were normal, whereas those on birds in CO_2 atmospheres indicated selective inhibition. Flourescent bacteria were not observed on birds in the CO_2 atmosphere.

Further studies were conducted by Jurdi et al. (1980) using 100% CO_2, 30% CO_2, N_2, or air as storage atmospheres. Samples stored at 5°C and 100% CO_2 showed repression of total aerobic bacterial numbers, significantly ($P < .05$) lower than the samples exposed to either nitrogen or air.

Vacuum packaging encouraged the growth of *Enterobacter* over the aerobic pseudomonads but had little effect on shelf life (Arafa and Chen, 1975).

Bailey et al. (1979) found that significantly lower microbial counts and extended shelf life occurred when storage was at 2°C instead of 5°C, after 12 days of storage when low-permeable packaging film was used instead of high-permeable film, and after 15 days of storage, when 65% CO_2 was used in packaging instead of 20%.

Carbon dioxide snow is also an effective means of preserving poultry meat (Reddy and Kraft, 1980). Storage was for 48 hr for different ratios of CO_2 to chicken. In general, CO_2 snow storage decreased the fluorescent *Pseudomonas* counts to a greater extent than psychrotrophic counts but had no effect in reducing mesophilic counts.

Gray et al. (1984) inoculated fresh chicken thighs with *Salmonella enterditis* and *Staphylococcus aureus* prior to dipping them in 1.0, 2.5, or 5.0% potassium sorbate solutions adjusted to pH 6.0. The treated thighs were packaged in Nylon/Plexar/Surrllyn bags under air; vacuum; or 20%, 60%, or 100% CO_2 atmospheres and were stored at 10°C for 10 days. They found that *S. aureus* growth on poultry was more effectively inhibited by exposure to high levels of CO_2. Increased concentrations of sorbate dip solutions in combination with higher concentrations of CO_2 in the package environment provided a more effective inhibitory system against growth of both pathogens on fresh poultry.

VII. USE OF SALTS

The effect of pH levels and sodium chloride concentrations, alone and in combination, on the enumeration of unstressed and heat-stressed cells of three strains of *Staphylococcus aureus* was determined by Smolka et al. (1974). The researchers reported that as the concentration of NaCl increased, the optimum pH for enumeration became more narrowly defined for both types of cells. Heat-stressed cells showed a marked sensitivity to NaCl concentrations of 4% and above. Because of this sensitivity, recoveries were poorer than with unstressed cells, which showed no marked sensitivity.

The effect of potassium sorbate on salmonellae, *Staphylococcus aureus*, *Clostridium perfringens*, and *C. botulinum* in skinless, precooked, uncured sausage links was studied by Tompkin et al. (1974). Growth of salmonellae was markedly retarded by sorbate. Growth of *S. aureus* was delayed 1 day in the presence of sorbate, after which growth occurred at the same level as in product without sorbate. *Clostridium perfringens* declined to below detectable levels within the first day in products with and without sorbate.

Labbe and Duncan (1970) investigated the method by which sodium nitrite acts to prevent germination or outgrowth, or both, of heat-injured spores of *C. perfringens* in canned, cured meats. These authors reported that concentrations of 0.02 and 0.01% sodium nitrite prevented outgrowth of heat-sensitive and heat-resistant spores, respectively.

Duncan and Foster (1968) studied the effects of meat-curing agents on germination and outgrowth of spores of a putrefactive anaerobe. They found that in the presence of 3–6% NaCl, most of the spores germinated and produced vegetative cells, but cell division was often blocked.

Bowen et al. (1974) reported that toxin production by *C. botulinum* in weiners was inhibited by sodium nitrite levels above 50 µg/g.

A reduction in bacterial growth (4–10 days) was demonstrated with 100 and 150 ppm nitrite in chicken white-meat patties. Four hundred parts per million of nitrite reduced bacterial growth in dark-meat patties for 6 days. In white-meat patties, the use of sorbate (0.26%) in combination with nitrite (40 ppm) was as effective in reducing bacterial growth as higher concentrations of nitrite, salt (2.5%) alone or in combination with sorbate (0.26%), or nitrite (40 ppm) during the first few days of storage.

Robach and Ivey (1978) reported that a 5% potassium sorbate dip significantly reduced the total plate count of chicken. They also observed that a 5% dip in potassium sorbate reduced the growth of salmonellae inoculated onto the surface of the chicken.

In 1979, Cunningham determined shelf life and quality of poultry parts after dipping in potassium sorbate. Fresh broiler parts were dipped 30–60 sec in solutions of potassium sorbate. Drumsticks were stored at refrigerator temperature and evaluated daily for off-odor development and total counts. Thighs and breasts were baked to internal breast temperatures of 85°C, cooled, then evaluated for shear press values, moisture retention, and sensory properties. Drumsticks dipped in potassium sorbate developed off-odors later than controls. Those parts dipped in 10% potassium sorbate had a 20-day shelf life at 4°C, based on off-odor development and time for total counts to reach 10^7. Parts dipped in potassium sorbate for 30 sec, drained, and then baked had similar shear press (tenderness) and Carver press (moisture retention) values regardless of the sorbate concentration used. Taste panel members could not distinguish between those parts dipped in 5 or 10% potassium sorbate and those parts dipped in distilled water. Sensory evaluations were significantly lower for those parts dipped in 15% potassium sorbate solutions for 1 min. Parts dipped in 5 or 10% solutions for 1 min were indistinguishable from the controls.

To and Robach (1980) found that whole broilers treated with potassium sorbate showed no spoilage until the fourteenth or fifteenth day. The object of that study was to determine the antimicrobial efficacy of a potassium sorbate dip on fresh broiler drumsticks.

Cunningham (1981) used fresh broiler drumsticks to study the effectiveness of a potassium sorbate dip to control bacterial growth. Some pieces were inoculated with *Salmonella typhimurium* before being dipped. Other pieces, used to determine total counts, were not inoculated with salmonellae. Treated drumstricks were held at 4, 10, or 22°C. When stored at 4°C, the untreated pieces had counts of 10^6 in about 6 days, but parts dipped in 10% potassium sorbate took nearly 10 days to reach the same total counts. At 22°C, potassium sorbate extended the shelf life only about 1 day. Potassium sorbate solutions of 5% strength were nearly as effective as the 10% solutions on drumsticks held at 4°C, but not at 22°C. The 10% dip (for only 30 sec) effectively reduced numbers of salmonellae after 2 days of storage at 22°C.

Ikeme and Stadelman (1981) have also studied the effects of citric acid, potassium sorbate, and modified corn starch on chicken. Solutions 1 and 2 contained all of the above, at different levels. The pH of solution 1 was 3.2, and solution 2 was 2.4. Initial tests showed that untreated samples remained edible for 7 days at 4°C and for 14 days at −2°C. Those chicken parts stored at 15°C were considered spoiled by the second day. After these initial determinations, samples were dipped for 10 sec in boiling water, then dipped in solutions 1 or 2 for 60 sec, drained, and packaged. Samples were stored at 15, 25, and 4°C. A shelf life of over 28 days was observed at the cooler temperature. Throughout the entire storage period, solution 2–dipped samples had lower microbial counts than samples dipped in solution 1.

VIII. DISINFECTANTS

Gluteraldehyde (1,5-pentanedial) has for years been used as a disinfectant in clinical applications. It has excellent disinfectant properties, is not inactivated by organic matter, and is not corrosive to metals. One of the first studies in which gluteraldehyde was used in contact with poultry was reported by Bailey et al. (1977). They found that *Salmonella typhimurium* (200–300 cells per carcass) was effectively eliminated when carcasses were held for 30 min in chill water containing 5000 ppm gluteraldehyde. Passage of salmonellae from inoculated to uninoculated carcasses was prevented by 100 ppm gluteraldehyde for 10 min. When carcasses were stored at 2°C, their shelf life was extended 6 days over control carcasses.

In 1978, Mast and MacNeil determined the efficacy of gluteraldehyde as a difinectant when used in simulated commercial chilling. Poultry carcasses were held in chill waters (3°C) consisting of either 50 ppm gluteraldehyde, 50 ppm chlorine, or water (control) for 45 min. Chilling reduced the number of bacteria on all carcasses: those chilled in gluteraldehyde retained lower counts throughout 10 days of storage at 3°C. Gluteraldehyde-chilled carcasses had a shelf life 2 days longer than chlorine-chilled carcasses and 3 days longer than controls. There were no significant ($P < .05$) differences in moisture uptake, moisture retention, or color among carcasses chilled in the three solutions.

Poly(hexamethylenebiguanide) (PHMB) is also an excellent disinfectant and shows great promise in poultry processing, although it has not yet received approval for application in foods. Thomson et al. (1980) found that drumsticks, inoculated with 50 or 10^6 cells of marker *S. typhimurium* and immersed for 10 min in a 50-, 100-, or 200-ppm solution of PHMB, had no detectable organisms remaining following the treatment. It was also found that low levels (50 ppm) of PHMB would prevent cross-contamination in the commercial chilling process, with low levels of salmonellae present. Further studies by Thomson et al. (1981) showed that with higher levels of salmonellae (60,000 cells), 10 ppm PHMB in chill water did not prevent cross-contamination, but 25 ppm in chill water would.

Islam and Islam (1979a) report that PHMB in indeed effective at improving the shelf life of fresh poultry. Birds treated with 200, 300, and 400 ppm PHMB had average shelf lives of 22.9, 25.9, and 26.0 days, respectively, compared to 10.5 days of shelf life for water-treated controls.

Furthermore, Islam and Islam (1979b) compared the effectiveness of three potential poultry preservatives: gluteraldehyde, iodacetamide, and PHMB. It was found that PHMB had the best preservative effect by increasing the shelf life of fresh poultry 17 days over control birds, followed by iodacetamide (11 days), and gluteraldehyde at 5 days longer than control birds.

IX. COATINGS

The use of sorbic acid in an edible coating was reported by Meyer et al. (1959). Sorbic acid was combined with agar or lipid coating resulting in an extended shelf life. The preservation of food by alginate coating was studied by the A. E. Staley Manufacturing Co. Poultry was preserved by coating with a dispersion containing 2.5–20 parts water-soluble alginate to 97.5–80.0 parts carbohydrate mass (dextrose) in pure drinking water. The coating was coagulated with an aqueous Ca^{+2} solution, without imparting a bitter taste to the food. The alginate can be removed by washing with a complexing solution, preferably alkali polyphosphate, without affecting the properties of the food.

X. POLYPHOSPHATES

Polyphosphate inhibition of growth of pseudomonads from poultry meat was reported by Elliot et al. (1964). Both commercial and chemically pure polyphosphates inhibited the growth of nonfluorescent pseudomonads in a synthetic medium. Fluorescent strains grew after a short lag. Inhibition was not caused by high pH, but rather by chelation of metal ions essential to the growth of the bacteria. Chilling chicken carcasses overnight in slush-ice containing polyphosphates lengthened subsequent shelf life by 17–25%. Carcasses in continuous contact with polyphosphates during storage at 2.2°C kept 17–67% longer. Only fluorescent strains developed in the presence of these compounds. Chickens held in antiseptic ice containing polyphosphates kept 60% longer than those in water-ice.

The effect of food-grade polyphosphates on the microbial population of chicken meat has also been reported by Steinhauer and Banwart (1964). The use of 8% concentrations of two commercial blends of polyphosphates in the cooling water resulted in lower average bacterial counts on treated carcasses as compared to controls. Differences in total and proteolytic counts were noted for treated carcasses in one experiment. Polyphosphates appeared to inhibit odor production, slime formation, and discoloration in carcasses but did not appear to alter the type of microorganism on carcasses held at 5°C.

XI. MISCELLANEOUS

The effect of ethanol on shelf life and flavor of chicken meat was reported by Hall and Spencer (1964). The authors found that a 70% ethanol dip increased the shelf life of tray-packaged chicken halves by 3–

TABLE III

Chemicals Tested for Antimicrobial Activity against Poultry Spoilage Organisms[a]

1. Acetamide	27. p-Hydroxybenzoic acid
2. Acetic acid	28. Iodoacetmide
3. Acetoin	29. Maleic acid
4. Acetyl glycine	30. Malonic acid
5. p-Amino benzoic acid	31. Methoxyacetic acid
6. Ammonium citrate, dibasic	32. Methylmalonic acid
7. 1,3-Butanediol	33. Methylsuccinic acid
8. 1,4-Butanediol	34. Monomethyl adipate
9. 2,3-Butanediol	35. Mucic acid
10. Chloroacetamide	36. 2-Oxobutyric acid
11. Chloroacetic acid	37. Pyruvic acid
12. p-Chlorobenzoic acid	38. Salicylic acid
13. a-Chlorobenzoic acid	39. Sodium acetate
14. Cis-3-Chlorocrotonic acid	40. Sodium chloroacetate
15. 3-Chloropropionic acid	41. Sodium citrate
16. Citric acid	42. Sodium glycolate
17. 2,3-Dibromosuccinic acid	43. Sodium iodoacetate
18. 2,2-Dichloroacetic acid	44. Sodium mercaptoacetate
19. 2,2-Dichloroacetamide	45. Sodium tartrate
20. Diglycolic acid	46. Succinamide
21. Dimethylfumarate	47. Succinic acid
22. Dimethylsuccinate	48. Succinimide
23. N,N-Dimethylsuccinamic acid	49. Succinonitrile
24. Diphenylacetic acid	50. Sulfanilic acid
25. Disodium ethylenediamine tetraacetate (EDTA)	51. Sulfosalicylic acid
	52. Tetrahydroxy succinic acid
26. Fumaric acid	53. p-Toluene sulfonic acid

[a] From Islam et al. (1978).

TABLE IV

Chemicals Exhibiting Antimicrobial Activity at Neutral pH (6.5 to 7.5)[a]

Chemical	pH at inhibitory concentration	Inhibitory concentration (ppm)
Iodoacetamide	7.1	30
2-Chloroacetamide	7.3	800
2-Oxobutyric acid	6.5	800
p-Aminobenzoic acid	6.8	1,100
3-Chloropropionic acid	6.5	1,200
Succinic acid	6.8	1,200
Sodium iodoacetate	7.2	1,900
Sodium acetate	7.3	2,200
Sodium mercaptoacetate	6.7	10,000
Sodium chloroacetate	6.8	11,000
2,2-Dichloroacetamide	7.2	16,700

[a] From Islam et al. (1978).

5 days. Control halves were characterized as having more-typical chicken flavor than ethanol-dipped halves when comparisons of either skin, white meat, or dark meat were made. Ehtanol-dipped halves were unacceptable more frequently than control halves.

Islam et al. (1978) screened 53 chemicals for their antimicrobial activity around neutral pH against five cultures isolated from spoiled poultry and three reference organisms (Table III). Eleven of those agents were found to be effective (Table IV). Of the 11 chemicals, iodacetamide and chloroacetamide were found to inhibit the growth of spoilage organisms at concentrations of 30 and 80 ppm, respectively. Freshly processed broilers were immersed in solutions of these two chemicals, drained, packed in polyethylene bags, and held at 5°C until spoilage ensued. Birds treated with solutions of iodacetamide and chloroacetamide at 5000 ppm were found to have average shelf-lives of about 11 and 10 days, respectively, compared to a shelf-life of 7.5 days for control samples.

REFERENCES

Arafa, A. S., and Chen, T. C. 1975. Effect of vacuum packaging on microorganisms on cut-up chickens and in chicken products. J. Food Sci. 40: 50.

Arafa, A. S., and Chen, T. C. 1977. Ascorbic acid dipping as a means of extending shelf life

and improving microbiological quality of cut-up broiler parts. Poultry Sci. 56: 1345. S.A.A.S. Abstr.
Bailey, J. S., Thompson, J. E., and Cox, N. A. 1977. Elimination of *Salmonella* and extension of shelf-life with a gluteraldehyde product on broiler carcasses. Poultry Sci. 56: 1346 S.A.A.S. Abstr.
Bailey, J. S., Reagan, J. O., Carpenter, J. A., and Schuler, G. A. 1979. Microbiological conditions of broilers as influenced by vacuum and carbon dioxide in bulk shipping packs. J. Food Sci. 44: 134.
Baran, W. L., Dawson, L. E., and Lechowich, R. V. 1973. Influence of chlorine dioxide water treatment on numbers of bacteria associated with processed turkey. Poultry Sci. 52: 1053.
Barnes, E. M. 1974. Microbiological aspects of poultry chilling. Pages 549–552. in Proc. XV World's Poultry Congr., New Orleans, LA.
Bowen, V. G., Cerveny, J. G., and Deibel, R. H. 1974. Effect of sodium ascorbate and sodium nitrite on toxin formation of *Clostridium botulinum* in weiners. Applied Micro. 27: 605.
Boyd, J., Weisner, H. H., and Winter, A. R. 1960. Influence of high level antibiotic rations with terephthalic acid on the antibiotic content and keeping quality of poultry meat. Poultry Sci. 39: 1067.
Bushway, A. A., Ficker, N., and Jen, C. W. 1982. Effect of nitrite and sorbate on total number of aerobic microorganisms in chicken white and dark meat patties. J. Food Sci. 47: 858.
Centers for Disease Control. 1979. Foodborne disease. U.S. Dept. Health, Educ., and Welfare, Publ. Health Serv., Ctr. Dis. Contr., Atlanta, GA.
Coleby, B., Imgram, M. and Shepard, H. J. 1960. Treatment of meats with ionizing radiations. III. Radiation pasteurization of whole eviscerated chicken carcasses. J. Sci. Food Agr. 11: 61.
Corey, R. R., and Byrnes, J. M. 1963. Oxytetracycline-resistant coliforms in commercial poultry products. Applied Micro. 11: 481.
Cox, N. A., Mercuri, A. J., Juven, B. J., Thomson, J. E., and Chow, V. 1974. Evaluation of succinic acid and heat to improve the microbiological quality of poultry meat. J. Food Sci. 39: 985.
Cunningham, F. E. 1978. The effect of brief microwave treatment on numbers of bacteria in fresh chicken patties. Poultry Sci. 57: 296.
Cunningham, F. E. 1979. Shelf-life and quality characteristics of poultry parts dipped in potassium sorbate. J. Food Sci. 44: 863.
Cunningham, F. E. 1980. Influence of microwave radiation on psychotrophic bacteria. J. Food Protection. 43: 651.
Cunningham, F. E. 1981. Microbiology of poultry parts dipped in potassium sorbate. Poultry Sci. 60: 969.
Cunningham, F. E., and Albright, K. M. 1977. Using microwaves to reduce bacterial numbers on fresh poultry. Microwave Energy Appl. Newsletter, Vol. 10(4): 3.
Dawson, L. E., Mallman, W. L., Bigbee, D. G., Walker, R., and Zabik, M. E. 1963. Influence of surface pasteurization and chlorotetracycline on bacterial incidence on fryers. Food Tech. 17: 218.
Dixon, J. M. S., and Pooley, F. E. 1961. The effect of chlorination on chicken carcasses infected with *Salmonellae*. J. Hyg. (Cambridge) 59: 343.
Duncan, C. L., and Foster, E. M. 1968. Effect of sodium nitrite, sodium chloride, and sodium nitrate on germination and outgrowth of anaerobic spores. Applied Micro. 16: 406.
Eklund, M. W., Spencer, J. V., Sauter, E. A., and George, M. H. 1961. The effect of

different methods of chlortetracycline application on the shelf-life of chicken fryers. Poultry Sci. 43: 81.
Elliot, R. P., Straka, R. P., and Garibaldi, J. A. 1964. Polyphosphate inhibition of growth of pseudomonads from poultry meat. Applied Micro. 12: 517.
El-Wakeil, F. A., El-Magoli, S. B., and Salama, N. A. M. 1978. Effect of radurization on the chemical, microbiological and organoleptic characteristics of poultry meat. Dept. of Food Sci. & Tech; Fac. of Agric; Cairo Univ, Cairo, Egypt.
Essary, E. O., Dawson, L. E., and Mallman, W. L. 1960. Effect of different length chill periods with chlortetracycline and different holding conditions on the shelf-life of dressed fryers. Food Tech. 14: 286.
Gray, R. J. H., Elliott, P. H., and Tomlins, R. I. 1984. Control of two major pathogens on fresh poultry using a combination of potassium sorbate/carbon dioxide packaging treatment. J. Food Sci. 49: 142.
Hall, K. N., and Spencer, J. V. 1964. The effect of ethanol on shelf-life and flavor of chicken meat. Poultry Sci. 43: 573.
Hays, H., Elliker, P. R., and Sandline, W. E. 1967. Microbial destruction by low concentrations of hypochlorite and iodophor germicides in alkaline and acidified water. Applied Micro. 15: 575.
Horwitz, M. A., and Gangarosa, E. J. 1976. Foodborne disease outbreaks traced to poultry. United States, 1966–1974. J. Milk Food Technol. 39: 859.
Idziak, E. S., and Incze, K. 1968. Radiation treatment of foods. I. Radurization of fresh eviscerated poultry. Applied Micro. 16: 1061.
Ikeme, A. I., and Stadelman, W. J. 1981. Extending the shelf-life of broiler meat. Poultry Sci. 60: 1673.
Islam, M. N., and Islam, N. B. 1979a. Extension of poultry shelf-life by poly(hexamethylenebiguanide hydrochloride). J. Food Protection. 42: 416.
Islam, M. N., and Islam, N. B. 1979b. Comparison of the efficacy of three potential poultry preservatives; gluteraldehyde, iodacetamide, and poly(hexamethylenebiguanidehydrochloride). J. Food Process. and Preser. 3: 1.
Islam, M. N., Gray, R. J. H., and Geiser, J. N. 1978. Development of antimicrobial agents for the extension of poultry shelf-life. Poultry Sci. 57: 1266.
Jones, L. A., Jr., Hoffman, R. K., and Phillips, C. R. 1968. Sporicidal activity of sodium hypochlorite at subzero temperatures. Applied Micro. 16: 787.
Jurdi, D., Mast, M. G., and McNeil, J. H. 1980. Effects of carbon dioxide and nitrogen atmospheres on the quality of mechanically deboned chicken meat during frozen and non-frozen storage. J. Food Sci. 45: 641.
Juven, B. J., Cox, N. A., Mercuri, A. J., and Thomson, J. E. 1974. A hot acid treatment of eliminating salmonella from chicken meat. J. Milk and Food Tech. 37: 237.
Kim, T. K., and Chipley, J. R. 1974. Effect of salts on penicillinase release by *Staphylococcus aureus*. Microbios 10A: 55.
Klose, A. A., Kaufman, V. F., Baune, H. G., and Pool, M. F. 1971. Pasteurization of poultry meat by steam under reduced pressure. Poultry Sci. 50: 1156.
Labbe, R. G., and Duncan, C. L. 1970. Growth from spores of *Clostridium perfringens* in the presence of sodium nitrite. Applied Micro. 19: 353.
Licciardello, J. J., Nicherson, J. T. R., Goldblith, S. A., Shannon, C. A., and Bishop, W. W. 1969. Development of radiation resistance in *Salmonella* cultures. Applied Micro. 18: 24.
Lillard, H. S. 1979. Levels of chlorine and chlorine dioxide of equivalent bactericidal effect in poultry processing water. J. Food Sci. 44: 1594.
Lin, Y. L., and Chen, T. C. 1981. Effect of sorbic acid-ice on microorganisms of broiler meat. Poultry Sci. 60: 1684 (abstract).

Mallmann, W. L., Dawson, L. E., and Jones, E. 1959. The influence of relatively high chlorine concentrations in chill tanks in shelf-life of fryers. Poultry Sci. 38: 1225.

Mast, M. G., and MacNeil, J. H. 1975. Heat pasteurization of mechanically deboned poultry meat. Poultry Sci. 54: 1024.

Mast, M. G., and MacNeil, J. H. 1978. Use of gluteraldehyde as a disinfectant in immersion chilling of poultry. Poultry Sci. 57: 681.

Mead, G. C., and Thomas, N. L. 1973a. Factors affecting the use of chlorine in the spin-chilling of eviscerated poultry. Brit. Poultry Sci. 14: 99.

Mead, G. C., and Thomas, N. L. 1973b. The bacteriological condition of eviscerated chickens processed under controlled conditions in a spin-chilling system and sampled by two different methods. Brit. Poultry Sci. 14: 413.

Mercuri, A. J., Kotula, A. W., and Sanders, D. H. 1967. Low-dose irradiation of try-packed cut-up chickens. Food Tech. 21: 1509.

Meyer, R. C., Winter, A. R., and Weiser, H. H. 1959. Edible protective coatings for extending the shelf life of poultry carcasses. Food Tech. 13: 146.

Mountney, G. J., and O'Malley, J. E. 1965. Acids as poultry meat preservatives. Poultry Sci. 44: 896.

Murphy, J. F., and Murphy, R. E. 1962. Method of treating poultry. U.S. Patent 3,025,170.

Nilsson, T., and Regner, B. 1963. The effect of chlorine in the chilling water on salmonellae in dressed chickens. Acta. Vet. Scand. 4: 307.

Oglivy, W. S., and Ayres, J. C. 1951. Post-mortem changes in stored meats. II. The effect of atmospheres containing carbon dioxide in prolonging the storage life of cut-up chicken. Food Technol. 5: 97.

Olson, V. M., Swaminathan, B., Pratt, D. E., and Stadelman, W. J. 1981. Effect of five cycle rapid freeze-thaw treatment in conjunction with various chemicals for the reduction of Salmonella typhiumurium. Poultry Sci. 60: 1822.

Patterson, J. T., 1968. Bacterial flora of chicken carcasses treated with high concentrations of chlorine. J. Applied Bacteriol. 31: 544.

Perry, G. A., Lawrence, R. L., and Melnick, D. 1964. Extension of poultry shelf-life by processing with sorbic acid. Food Technol. 18: 891.

Potter, N. N. 1973. Page 151 in "Food Science." 2nd ed. Avi Publishing Co., Inc., Westport, CT.

Raccach, M., and Baker, R. C. 1978. Lactic-acid bacteria as an anti-spoilage and safety factor in cooked mechanically deboned poultry meat. J. Food Prot. 41: 703.

Raevuori, M., and Genigeorgis, C. 1975. Effect of pH and sodium chloride on growth of *Baccilus cereus* in laboratory media and certain foods. Applied Micro. 29: 68.

Ranken, M. D., Clewlow, G., Shrimpton, D. H., and Stevens, B. J. H. 1965. Chlorination in poultry processing. British Poultry Sci. 6:331.

Reddy, K. V., and Kraft, A. A. 1980. Effect of carbon dioxide snow on shelf life of packaged chicken. J. Food Sci. 45: 1436.

Rhodes, D. N. 1965. The radiation pasteurization of broiler chicken carcasses. Brit. Poultry Sci. 6: 331.

Robach, M. S., and Ivey, F. J. 1978. Antimicrobial efficacy of a potassium sorbate dip on freshly processed poultry. J. Food Prot. 41: 284.

Sanders, D. H., and Blackshear, C. D. 1971. Effect of chlorination in the final washer on bacterial counts of broiler chicken carcasses. Poultry Sci. 50: 215.

Scheusner, D. L., and Harmon, L. G. 1971. Temperature range for production of four different enterotoxins by *Staphlococcus aureus* in BHT broth. Bacteriol. Proc. p. 18.

Scholtyssek, S., Heimbach, P., and Berner, H. 1969. Investigations into a new method of chilling poultry. III. Hygenic aspects of the process. Fleischwirtschaft. 49: 1617.

Smolka, L. R., Nelson, F. E., and Kelley, L. M. 1974. Interaction of pH and NaCl on enumeraton of heat-stressed *Staphylococcus aureus*. Appl. Micro. 27: 443.

Spencer, J. V., Streeter, E. M., and George, M. H. 1968. Effect of an iodophor on shelf life of poultry. Poultry Sci. 47: 1721. (An Abstract).

Staley, A. E. Mfg. Co. Food Research Inc, 1969. Preservation of food by alginate coating. Ueberzuege Auf Lebensmittel. West German Patent Application.

Steinhauer, J. E., and Banwart, G. J. 1964. The effect of food grade polyphosphates on the microbial population of chicken meat. Poultry Sci. 43: 1736.

Thatcher, F. S., and Loit, A. 1961. Comparative microflora of chlortetracycline-treated and nontreated poultry with special reference to public health aspects. Applied Micro. 9: 39.

Thiessen, G. P., Usborne, W. R., and Orr, H. L. 1985. The efficacy of chlorine dioxide in controlling *Salmonella* contamination and its effect on product quality of chicken broiler carcasses. Poultry Sci. 63: 647.

Thomson, J. E., Banwart, G. J., Sanders, D. H., and Mercuri, A. J. 1967. Effect of chlorine, antibiotics, β-propiolactone, acids and washing on *Salmonella typhimurium* on eviscerated fryer chickens. Poultry Sci. 46: 146.

Thomson, J. E., Cox, N. A., Bailey, J. S., and Islam, M. N. 1980. Elimination of *Salmonella* from processed broiler meat with Poly(hexamethylenebiguanide hydrochloride). Poultry Sci. 59: 1573.

Thomson, J. E., Cox, N. A., Bailey, J. S., and Islam, M. N. 1981. Minimizing *Salmonella* contamination on broiler carcasses with poly(hexamethylenebiguanide hydrochloride). J. Food Protection. 44: 440.

To, E. C., and Robach, M. C. 1980. Potassium sorbate dip as a method of extending shelf life and inhibiting the growth of *Salmonella* and *Staphylococcus aureus* on fresh, whole broilers. Poultry Sci. 59: 726.

Todd, E. C. D. 1980. Poultry-associated foodborne disease—its occurrence, cost, sources, and prevention. J. Food Prot. 43: 129.

Tompkin, R. B., Christiansen, L. N., Shaparis, A. B., and Bolin, H. 1974. Effect of potassium sorbate on salmonellae, *Staphylococcus aureus*, *Colstridium perfringens*, and *Clostridium botulinum* in cooked, uncured sausage. Appl. Microbiol. 28: 262.

Wabeck, C. J., Parmlee, C. E., and Stadelman, W. J. 1968. Carbon dioxide preservation of fresh poultry. Poultry Sci. 47: 468.

Yang, P. P. N., and Chen, T. C. 1979. Effects of ozone treatment on microflora of poultry meat. J. Food Processing Preserv. 3: 177.

Young, L. L., and Lyon, B. G. 1973. The use of heat treated meat in chicken frankfurters. Poultry Sci. 52: 1868.

Ziegler, F., and Stadelman, W. J. 1955. Increasing shelf-life of fresh chicken meat by using chlorination. Poultry Sci. 34: 1389.

CHAPTER **10**

Pathogens Associated with Processed Poultry

N. A. Cox
J. S. Bailey

U. S. Department of Agriculture
Agricultural Research Service
Richard B. Russell Agricultural Research Center
Athens, Georgia 30613

I. INTRODUCTION

In poultry processing, soil and fecal material on the feet and feathers of birds and their intestinal contents are the main sources of bacterial contamination (Walker and Ayres, 1956). Other sources of contamination may be the water supply, ice, air, humans, equipment, packaging material, birds, rodents, and insects. Many types of bacteria are present on processed poultry and are disseminated from carcass to carcass and throughout the plant as the carcasses move through the processing line. In 1971, a list was compiled, comprising 24 different genera of bacteria (Table I) that had been isolated and identified from poultry (Kraft, 1971). In addition to these 24 genera, the following have been added to the growing list: *Acinetobacter* (Barnes and Shrimpton, 1968), *Bacteroides* (Barnes and Impey, 1972), *Bifidobacterium* (Barnes and Impey, 1972), *Campylobacter* (Smith and Muldoon, 1974), *Chromobacterium* (Cox, 1975), *Citrobacter* (Cox et al., 1979), *Clostridium* (Barnes and Shrimpton, 1957), *Cytophaga* (Barnes and Shrimpton, 1968), *Eubacterium* and *Fusobacterium* (Barnes and Impey, 1972), *Klebsiella* (Cox et al., 1975), *Peptostreptococcus* (Barnes and Impey, 1972), *Propionibacterium* (Barnes and Impey, 1972),

TABLE I

Genera of Bacteria Isolated from Poultry[a]

Achromobacter	Lactobacillus
Actinomyces	Microbacterium
Aerobacter	Micrococcus
Alcaligenes	Neisseria
Arthrobacter	Paracolobacterum
Bacillus	Proteus
Brevibacterium	Pseudomonas
Corynebacterium	Salmonella
Escherichia	Sarcina
Flavobacterium	Staphylococcus
Gaffkya	Streptococcus
Haemophilis	Streptomyces

[a]From Kraft (1971).

Providencia (Cox et al., 1975), *Serratia* (Cox and Bailey, 1985), and *Yersinia* (Leistner et al., 1975).

The various bacteria present on poultry can be divided into two groups: (1) pathogenic (those capable of producing disease in humans) and (2) nonpathogenic (those not previously associated with a recognized disease). Many of the microorganisms in the nonpathogenic group are of concern because they can cause food spoilage. In this chapter, the following foodborne pathogens will be discussed in detail: *Campylobacter, Clostridium, Salmonella, Staphylococcus,* and *Yersinia.* A detailed description of their characteristics and symptoms of gastroenteritis they produce is given in Table II.

II. CAMPYLOBACTER

McFadyean and Stockman (1913) were the first to recognize the disease potential of microaerophilic vibrios, when they reported that these organisms were associated with abortion in cattle and sheep. Their findings were later confirmed by Smith (1918), who isolated similar organisms from aborted bovine fetuses. Smith and Taylor (1919) characterized these organisms and named them *Vibrio fetus.* Jones and Little (1931a,b) associated these organisms with winter dysentery in cattle. Then, Jones et al. (1931) were able to reproduce the disease in healthy

TABLE II

Characteristics and Symptoms of the Major Gastroenteritis-Producing Foodborne Bacteria

Organism	Type of food poisoning	Incubation period	Duration	Major gastroenteritis symptoms
Campylobacter jejuni	Infection, invasive enterotoxin possibly involved	2–10 days	2–7 days	Diarrhea, fever, cramping, bloody stool
Clostridium perfringens	Intoxication, toxin produced during sporulation of vegetative cells entering the intestine	9–15 hr	24–48 hr	Diarrhea, cramps, loose watery stool (no blood)
Salmonella	Infection, invasive enterotoxin possibly involved	6–48 hr (usually 12–24 hr)	1–7 days	Nausea, diarrhea, fever, cramping, vomiting
Staphylococcus aureus	Intoxication, toxin preformed in food	30 min to 8 hr (usually 2–4 hr)	6–24 hr	Nausea, explosive diarrhea, vomiting, cramping
Yersinia enterocolitica	Infection, invasive	48–96 hr (sometimes longer)	2 weeks or longer	Abdominal pain, fever, diarrhea, nausea and vomiting

cattle after feeding them a pure culture of vibrios that had been isolated from diseased cattle. They observed that the jejunum was the first site of the intestinal tract to be infected, and consequently, they proposed the name *Vibrio jejuni*. Doyle (1944) proposed that swine dysentery was caused by microaerophilic vibrios which he had isolated from the mucosa of the colon of hogs with dysentery. He later named the organism *Vibrio coli* (Doyle, 1948). Vinzent et al. (1947) were first to report a human infection caused by a microaerophilic vibrio. They isolated *V. fetus* from two blood cultures from a pregnant woman, who prematurely delivered a stillborn infant; *V. fetus* was also isolated from the placenta.

In 1956, Hofstad was the first to associate vibrios with hepatitis in poultry, when he isolated a vibrio from the liver of diseased adult chickens. In 1958, Peckman confirmed Hofstad's findings by reproducing hepatitis in healthy chickens after challenging them with a vibrio that was originally isolated from the livers of diseased birds. He proposed that the disease be called avian vibrionic hepatitis.

The term "campylobacter" was proposed by Sebald and Veron (1963) as the generic name for the microaerophilic vibrios, because these organisms differed from the classical cholera and halophilic groups in certain fundamental respects. They proposed that *V. fetus* should be removed from the genus *Vibrio* and be reclassified as *Campylobacter*, with *Campylobucter fetus* being the type species of the genus.

Campylobacters are gram-negative, very slender curved- to spiral-shaped rods that are 0.2–0.4 μm in width and 1.5–3.5 μm long. The individual cells have one curve or twist and appear vibriod but may also be S-shaped or gull-shaped (Smibert, 1978). They are motile by a single polar flagellum although occassionally they have a polar flagellum at each end of the cell. For the most part, campylobacters require a microaerophilic atmosphere for growth. A concentration of 5% oxygen and 10% carbon dioxide has been determined to be optimal for growth (Kiggings and Plastridge, 1956).

There are now about a dozen or so species and subspecies recognized in the genus *Campylobacter*. These may be separated into two broad groups using the catalase test. A list of the species and subspecies belonging to each of these groups is shown in Table III. *Campylobacter jejuni* is the species of greatest concern to the food microbiologist. The reservoir for *C. jejuni* in nature is enormous. The organism has been isolated from cattle, sheep, swine, fowl, dogs, cats, and monkeys, and also from fresh water and seawater (Karmali and Fleming, 1979). Less than 10 years ago, organisms belonging to the genus *Campylobacter* were rarely associated with human infection. However, today not only is *C. jejuni* recognized as an agent of foodborne disease, but *Campylobacter* enteritis

TABLE III

Campylobacter Species and Subspecies[a]

Catalase-positive	Catalase-negative
C. fetus subsp. fetus	C. sputorum subsp. sputorum
C. fetus subsp. venerealis	C. sputorum subsp. bubulus
C. jejuni	C. sputorum subsp. mucosalis
C. coli	EMC. concisus
NARTC group	
C. fecalis	
Areotolerant Campylobacter sp.	
Nitrogen-fixing Campylobacter sp.	
Free-living Campylobacter sp.	

[a] From Karmali and Skirrow (1984).

is among the leading causes of acute bacterial gastroenteritis in humans. In some of these cases, food and water have been implicated as being the vehicle responsible for transmitting *C. jejuni* to susceptible individuals. Foods of particular concern are raw milk, beef, and poultry, primarily because these foods have been implicated in cases of human *Campylobacter* enteritis. Commercial poultry, both edible and inedible parts, has been shown to be a major source of *C. jejuni* (Table IV). In addition to the examples shown in the table, Svedhem and Kaijser (1980) reported that the feces of 18 of 50 (36%) chickens sampled in Sweden had *C. jejuni*, while only 6 of 100 hens were positive. Parks and Stankiewicz (1981) sampled 100 fresh eviscerated whole market chickens purchased 1 per week over a 5-week period from each of 20 different food stores in the Ontario and Ohio regions. *Campylobacter jejuni* was recovered from 62 and 54% of the samples in Ontario and Ohio, respectively. Shanker *et al.* (1982) isolated *C. jejuni* from 18 of 40 broiler carcasses obtained from four processing plants in Sydney, Australia. The average level encountered was 8.6×10^4 cells of *C. jejuni* per contaminated carcass. In 1983, Wempe *et al.* isolated *C. jejuni* from 67, 69, 57% of the chicken hearts, livers, and wings, respectively, sampled in two California poultry processing plants. Sun (1984) discussed the findings of a recent study sponsored by the Food and Drug Administration in which poultry products were found to be contaminated with *C. jejuni* four times more frequently than with *Salmonella*.

Kinde *et al.* (1984) surveyed the incidence of *C. jejuni* on chicken wings at several retail sources. The authors observed that 82.9% of fresh chicken wing packages were positive for the organism. However, after 3–6

TABLE IV
Isolation of *Campylobacter* from Commercial Poultry[a]

Country	Animal	Stage of preparation	No. sampled	% Positive	Reference
United States	Chicken	Parts after freezing 3 weeks	165	2	Smith and Muldoon (1974)
England	Chicken	Cecal contents	167	68	Bruce et al. (1977)
England	Chicken	Intestinal contents	34	91	Ribeiro (1978)
England	Chicken	Eviscerated	50	72	Simmons and Gibbs (1979)
		Eviscerated, water-chilled	25	80	Simmons and Gibbs (1979)
		Eviscerated, air-chilled	10	80	Simmons and Gibbs (1979)
	Turkey	Eviscerated, water-chilled	6	83	Simmons and Gibbs (1979)
		Eviscerated, air-chilled	5	100	Simmons and Gibbs (1979)
United States	Chicken	Intestinal contents, retail	46	83	Grant et al. (1980)
Netherlands	Chicken	Intestinal contents	239	29	Goren and de Jong (1980)
		Carcasses after freezing	750	0	Goren and de Jong (1980)
United States	Chicken	Intestinal contents	62	50	Eiden and Dalton (1980)
		Carcasses after freezing	23	100	Eiden and Dalton (1980)
United States	Chicken	Whole chickens, retail	50	54	Lovett et al. (1981)
United States	Turkey	Freshly killed, cecal contents	600	100	Leuchtefeld and Wang (1981)
		Eviscerated carcass	33	94	Leuchtefeld and Wang (1981)
		Eviscerated carcass, water-chilled	83	94	Leuchtefeld and Wang (1981)
		Viscera	24	33	Leuchtefeld and Wang (1981)
Canada	Duck	Feces	94	88	Prescott and Bruin-Mosch (1981)

[a] From Blaser (1982).

days of storage of the product at the supermarket, the incidence of the organism dropped to 15.5%.

In the Centers for Disease Control's (CDC) 1981 Annual Summary of Foodborne Disease, there were 10 outbreaks caused by *C. jejuni* involving 487 cases; however, poultry was not implicated whatsoever (Centers for Disease Control, 1983). It should be mentioned that it was not until January 1, 1983, that *C. jejuni* became a reportable disease and was officially included in the surveillance program of the CDC (G. K. Morris, personal communication).

Gill and Harris (1984) found that in artificially contaminated hamburgers, minimal cooking eliminated *C. jejuni*. No organisms were detected in 50 samples of commercial ground meat. They concluded that the limited circumstances under which cooked poultry meat is likely to carry *C. jejuni* in significant numbers suggest a need for caution in ascribing outbreaks of *Campylobacter* enteritis to consumption of poultry.

III. CLOSTRIDIUM

Clostridium perfringens-type gastrointestinal disturbances have been reported in the literature since 1895, but the illness was not well authenticated in the United States until McClung (1945) described four outbreaks of food poisoning after consumption of chicken cooked the previous day. Eight years later, Hobbs *et al.* (1953) reported an outbreak of food poisoning with similar etiology in the United Kingdom. Since that time numerous documented outbreaks due to this organism have appeared in the literature. Every year since 1953, *C. perfringens* has been in the top three groups of organisms, along with *Salmonella* and *Staphylococcus*, associated with incidents of food poisoning in the United States.

Clostridium perfringens is an anaerobic gram-positive, spore forming, straight rod, 0.9–1.3 by 3.0–9.0 μm. Spores are oval, subterminal, and rare on usual laboratory media. *Clostridium perfringens* is differentiated into five types on the basis of production of major lethal toxins, types A, B, C, D, and E. Both enterotoxin types A and C have been associated with gastroenteritis, but type A is the most common cause of foodborne illness in humans (Walker, 1975). Dische and Elek (1957) showed that large numbers of cells of *C. perfringens* must be ingested to cause food poisoning. Hauschild *et al.* (1968) found that about 10^6 cells per gram of meat or milliliter of gravy might result in food poisoning.

Clostridium perfringens is commonly found in the environment and is a normal inhabitant of the intestinal tract of humans and poultry. Poultry carcasses can become contaminated from soil, feces, or worker's hands during slaughtering and processing. Mead and Impey (1970) studied the distribution of clostridia in a number of chicken and turkey processing plants and found that the organisms could readily be isolated from carcasses at each stage of the processing line. Of the *Clostridium* isolated, 91% were *C. perfringens* in the chicken plants, compared with 56% from the turkey plants. Birds entering the processing plant carry a relatively large number of *C. perfringens* in the breast feathers and feet (Table V). After scalding and plucking, very few organisms remain on the carcass and can only be recovered in significant numbers from the cecum (Barnes, 1960; Lillard, 1971). Lillard (1973) examined giblets and found that after scalding, *C. perfringens* were recovered from 18% of hearts and 50% of livers and gizzards.

Bryan and McKinley (1974) recovered *C. perfringens* from surfaces of equipment contacting raw turkey and from 3 of 13 cooked turkeys. Zottola and Busta (1971) found *C. perfringens* on 7 of 35 raw turkey products and 6 of 35 cooked turkey products examined. *Clostridium perfringens* was isolated from 52% of freshly ground turkey samples

TABLE V

Number of *C. perfringens* in Broiler Processing Plants[a]

	Count/ml, or cm²	
Source of sample	Plant I (range)	Plant II (range)
Feet (swab)[b,c]	10–11,000	14–790
Breast feathers[d]	2,700–100,000	390–3,700
Cecum[d]	<10–6,400	<10–23,000
Vent area (swab)[b,c]	<10–500	<10
Neck skin[d]	<10	<10–27
Scald water[d]	<10–170	<10–16
Chill water[e]	<10	<10
Breast surface after final washer[c]	<10	<10
Breast surface after chill tank[c]	<10	<10

[a] From Lillard (1971).
[b] Area swabbed is approximate.
[c] Number of organisms per square centimeter.
[d] Number of organisms per gram.
[e] Number of organisms per milliliter.

(Guthertz et al., 1976). Bryan and McKinley (1974) reviewed ways to thaw, cook, chill, and reheat turkeys in order to prevent foodborne illness in turkeys. Most organisms that compete with *C. perfringens* are killed when meat and poultry are cooked, but not *C. perfringens* spores. These spores are heat-shocked during cooking, so more of them germinate when meats and gravies cool while being held at room temperature, in warming devices, or in refrigerators. Cooked meat can also be contaminated by workers during boning, slicing, grinding, mixing, or handling if any of this equipment is contaminated with *C. perfringens*. Since *C. perfringens* has a short lag period and can double its numbers every 8–12 min under optimal conditions, room-temperature or warm-storage conditions will result in the proliferation of *C. perfringens*. Also, if foods are stored in large containers in refrigerators, *C. perfringens* spores may germinate, and vegetative cells may multiply to quantities that can cause illness unless foods are reheated to temperatures sufficient to kill vegetative forms of *C. perfringens*. Therefore, when meat- or poultry-borne outbreaks occur, one or more of the following events usually occurred: (1) improper cooling, (2) improper hot holding, (3) food was prepared a day or more before serving, or (4) inadequate reheating (Bryan, 1978, 1980).

In 1981, the latest year that data are available from the CDC, there were 28 reported foodborne outbreaks of *C. perfringens*, with a total of 1162 cases. Four of these outbreaks involving 73 cases were from chicken, and 4 outbreaks involving 380 cases were from turkey (Centers for Disease Control, 1983).

IV. SALMONELLA

Salmonella is a gram-negative, facultatively anaerobic nonsporing rod, usually motile by peritrichous flagella; however, nonmotile mutants can occur, and *S. gallinarum* and *S. pullorum* are always nonmotile. The generic term *Salmonella* was given to the microorganisms in 1900 in honor of Dr. D. E. Salmon, who with Theobald Smith (Salmon and Smith, 1885) were the first to discover and describe the microorganism known as *Salmonella cholerae-suis*. This generic term of *Salmonella* was officially adopted and has been universally accepted since 1933 (Edwards and Ewing, 1972). In the classification system for the family Enterobacteriaceae in use at the CDC, *S. cholerae-suis*, *S. enteritidis*, and *S. typhi* are the only three species of *Salmonella* recognized (Brenner et al.,

1977). The species of *S. cholerae-suis* and *S. typhi* each consist of a single serotype, while *S. enteritidis* consists of more than 2000 antigenically distinct serotypes. All *Salmonella* species are pathogenic to humans, animals, or both. Salmonellae, with few exceptions, are not host-adapted and may inhabit the intestinal tract of any warm-blooded animal and many species of cold-blooded animals (Bryan, 1968b).

Salmonellosis is a disease caused by the ingestion of salmonellae by humans or animals. Following ingestion, symptoms may or may not develop, depending on the dose level, serotype, and virulence of the infecting strain and host resistance (Bryan, 1968b). Clinical manifestations may occur as acute gastroenteritis, bacteremia (with or without focal extraintestinal infection), and enteric fevers (typhoidal syndromes), or the host can remain healthy but can excrete salmonellae and in this instance would be referred to as an asymptomatic carrier state (Dowell, 1982). Of the various clinical manifestations, acute gastroenteritis is the most common and can be caused by almost all serotypes of *Salmonella*.

The first described and laboratory-confirmed outbreak of *Salmonella* occurred in 1888 in the village of Frankenhausen, in Saxony, Germany. Gartner (1888) reported that 57 persons who ate meat from a sick cow became suddenly ill with symptoms of acute gastroenteritis. One of the individuals, a young man, who had eaten a large portion of the raw meat, died approximately 35 hr after consumption. Gartner isolated a microorganism he named *Bacillus enteritidis* from the blood vessels and meat of the diseased cow and also from the organs of the dead man.

Raw foods and particularly those of animal origin have long been known to be important vehicles of foodborne salmonellosis. Probably the major source of the salmonellosis problem in humans is foods of animal origin such as poultry, beef, and pork.

The primary source of salmonellae with poultry is contaminated feed or feed ingredients. Research has shown that small numbers of salmonellae in feed can infect young chicks (Gordon and Tucker, 1965; Murphy, 1969; Schleifer *et al.*, 1984). Erwin (1955) was the first to recover viable *Salmonella* from commercial poultry feed. Since then, many researchers have isolated salmonellae from poultry feed. The percentages of poultry mash (unpelleted poultry feed) found to be contaminated with salmonellae were 4% (Isa *et al.*, 1963), 5% (Morehouse and Wedman, 1961; Allred *et al.*, 1967), 9% (Hobbs, 1961), 21.5% (Morris *et al.*, 1969), 43.5% (MacKenzie and Bains, 1976), and 68% (Cox *et al.*, 1983). The percentages of pelleted poultry feed contaminated with salmonella were 0.7% (Allred *et al.*, 1967), 1.7% (Hobbs, 1961), 4.3% (Hacking *et al.*, 1978), and 23% (Cox *et al.*, 1983). The higher percentages reported in the more recent studies would seem to indicate a worsening of the problem or advancements in the methodology of isolating salmonellae from feeds, or a combination of both. Even though commercial feed is frequently contaminated, surveys have shown that the percentages of chickens arriving at the processing plant with salmonellae in their intestinal tracts are low, approximately 2%, and slightly higher with turkeys, 2–5% (Sadler et al., 1961; Sadler and Corstvet, 1965). Once inside the processing plant, there is widespread dissemination of microorganisms, including salmonellae, during processing, which frequently results in contamination of the final product. May (1974) listed 57 potential points in a poultry processing plant at which cross-contamination could occur. This results in a higher incidence of salmonellae-contaminated carcasses leaving the processing plants and in the retail market. There has been much published information on the incidence of salmonellae-contaminated poultry sampled from processing plants or retail outlets: 1.7% (Magwood et al., 1967), 1–10% (Felsenfeld et al., 1950), 8.3% (Ladiges and Foster, 1974), 10% (Dixon and Pooley, 1961), 12.9% (Morris and Wells, 1970), 7–14% (Morris and Ayres, 1960), 17% (Wilson et al., 1961), 20.5% (Surkiewicz et al., 1969), 21.5% (McBride et al., 1980), 27% (Woodburn, 1964), 34.8% (Duitschaever, 1977), 44.5% (Cox et al., 1978), 46% (Cox and Blankenship, 1975), 47% (Dougherty, 1974), and 50.2% (Wilder and MacCready, 1966). While the percentages of contaminated carcasses are frequently high, the level of salmonellae on *Salmonella*-positive carcasses is rather low: 1–30 cells per carcass (Surkiewicz et al., 1969). Seventeen cells per 100 g of skin (Notermans et al., 1975), 100 cells per 100 g of skin (Mulder et al., 1977). In other studies involving the incidence of contaminated carcasses, Glezen et al. (1966) reported 33% salmonellae contamination in one plant and 0.5% in another; Wilder and MacCready (1966) found 48.7% and 28.8% in two different processing plants, and Cox and Blankenship (1975) reported 80, 65, 35, and 5% in four different processing plants. Johnston (1982) presented data from paired surveys of 1967 and 1979 using the same methodology and the same 15 federally inspected poultry processing plants when possible. With approximately 600 samples taken in each study, the incidences of processed poultry contaminated with salmonellae was 28.6% in 1967 and 36.9% in 1979.

Poultry is therefore a major source of salmonellae, and it is obvious that these organisms, at least small numbers of them, enter the kitchens of most, if not all, homes, institutions, or food service establishments at one time or another. It is at this point that a potential problem exists. Bryan (1978) listed many factors that contributed to outbreaks of salmonellosis, and some of the more important ones are inadequate cooling of foods, ingestion of contaminated raw foods or ingredients, inadequate time and/or temperature during cooking or heat processing of foods, cross-contamination from raw foods (such as raw poultry and raw meat) to cooked foods, lapse of a day or more between preparing and serving, and inadequate cleaning of equipment and kitchen surfaces.

From the number of reported outbreaks and cases attributed to *Salmonella* it is obvious that salmonellosis is an important public health problem both in the United States and in the world. However, the problem may be far greater than most of us realize. Gangarosa (1978) estimated that only 1% of the cases of salmonellosis in the United States are reported and that the actual number is probably $2\frac{1}{2}$ million cases per year, resulting in the hospitalization of 500,000 persons and the death of 9000 persons annually. According to his estimates, U.S. citizens pay 1.2 billion dollars per annum in medical costs alone as a result of salmonellosis.

Meat-related salmonellosis is both a serious problem and one that will be difficult to solve. An example of the difficulty can be seen in the consensus statements endorsed by an International Symposium on *Salmonella* held in July, 1984, in New Orleans. Scientists actively involved in *Salmonella* research from around the world met, and at the close of the meeting, the following statements were endorsed by consensus (Anonymous, 1984):

1. The eradication of *Salmonella* in domestic animals is not attainable at this time except for specific infections such as *S. pullorum* and *S. gallinarum*. Serious efforts should be made to control and reduce the incidence of *Salmonella* infection in domestic animals.

2. Methods are available to reduce *Salmonella* contamination of processed feeds to safe levels, but feed contamination remains a widespread and serious problem. Further improvement is necessary.

3. It should be candidly recognized that raw foods of animal origin are frequently contaminated by *Salmonella* and that such contamination levels can not be expected to change greatly in the near future.

4. Trade barriers should not be erected and import requirements in respect to *Salmonella* should not exceed standards attained in the importing country.

5. There should be continuing programs to provide education and information on proper food handling to the consumer and to all those who handle foods.

In the Centers for Disease Control's 1981 Annual Summary of Foodborne Disease, there were 66 outbreaks in which chicken was implicated, involving 122 cases, and 5 outbreaks related to turkey or turkey products, involving 465 cases.

V. STAPHYLOCOCCUS

Bergdoll (1979) stated that illness similar to staphylococcal food poisoning was reported as early as 1830, although the organisms themselves were not observed until 1878 by Koch and 1890 by Pasteur. Although Ogston is credited with applying the name "staphylococcus" to these organisms in 1881 because of the grapelike clusters of cocci he observed in cultures, it was Rosenboch who in 1884 first observed pure cultures of the organism on solid media (Bergdoll, 1979).

Staphylococcus is a gram-positive, nonmotile, asporogenous organism. These spherical cells occur singly, in pairs, in tetrads, and in irregular clusters resembling bunches of grapes. The species of major concern to the food industry is *Staphylococcus aureus*, both the pigmented and nonpigmented varieties. The ability to produce coagulase (coagulase-positive), to ferment mannitol, and to produce a heat-stable endonuclease differentiates *S. aureus* from the other species of *Staphylococcus* (*S. epidermidis* and *S. saprophyticus*) (Baird-Parker, 1974).

The toxic agent which produces staphylococcal food poisoning is an enterotoxin. Five enterotoxins have been identified: A, B, C, D, and E. Two enterotoxin C's with differing isoelectric points (8.6 for C_1 and 7.0 for C_2) and slightly differing immunological reactions have been purified.

Not all strains of *S. aureus* will produce enterotoxins, but all strains should be regarded as potential food poisoning pathogens, until demonstrated to be enterotoxin-negative. The development of phage typing has made it possible to differentiate human strains of *S. aureus* from poultry strains (Gibbs *et al.*, 1978a). Gibbs and Patterson (1978) found 62% of strains of *S. aureus* isolated from poultry to be enterotoxigenic, the majority producing enterotoxin A. Harvey *et al.* (1982) found that strains of *S. aureus* isolated from poultry but of human origin produced entertoxin A, C, and D, while poultry strains produced only enterotoxin D.

Strains of staphylococci are frequently carried by live chicken or turkey and enter the processing plant in the liver, on the skin, and in the nasal cavity of many birds and can subsequently be found in low numbers throughout the plant (Barnes, 1972). Healthy poultry tissues do not support prolonged growth of staphylococci, but bruised tissues will allow persistance of these organisms (Bryan, 1976). Hobbs (1971) suggested that birds acquire their staphylococci from human sources, but subsequent research with phage typing has shown that both human-strain contaminants and strains indigenous to poultry can be isolated.

The degree of colonization and the numbers of *S. aureus* present in the intestinal tract are low during the first few weeks of a chick's life but increase as the chick gets older (Thompson *et al.*, 1980). Notermans *et al.* (1982) found that *S. aureus* was present in small numbers (about 10 cells/g) on the skin of broilers before processing. During processing, contamination of carcasses with this organism increased to >10^3 cells/g of skin. Based on phage typing the increase in contamination was shown to be due to a strain of *S. aureus* which was indigenous to the processing plant, and that picking and evisceration were the main points at which contamination of the carcasses occurred. Patterson (1969) found broiler carcasses to normally yield fewer than 50–100 cells of *S. aureus* per 16 cm^2 of breast skin, although counts as high as 750 cells/cm^2 were reported. Counts of *S. aureus* on older hens and turkeys were shown to be higher (>10^3 cells/gm^2) than on broiler carcasses (Gibbs *et al.*, 1978b; Thompson *et al.*, 1980).

The distribution of staphylococci in five turkey processing plants was studied by Walker and Ayres (1959), who found 30% of the isolates to be coagulase-positive strains. Genigeorgis and Sadler (1966) surveyed eight processing plants and examined 2513 poultry livers and found that 40 (1.59%) were contaminated with coagulase-positive *S. aureus*. Salzer *et al.* (1964) recovered coagulase-positive staphylococci from 62 of 360 (17%) turkey livers before washing, but from only three (<1%) after washing. Bryan and McKinley (1974) recovered *S. aureus* from 20% of raw turkey carcasses but did not recover *S. aureus* from any cooked turkeys.

Surkiewicz *et al.* (1969) found *S. aureus* in low numbers on eviscerated broiler carcasses and from chill tank water but showed that the washing action of continuous overflow chillers reduced the numbers of *S. aureus* on the birds. In 1967, da Silva sampled meat, equipment, and personnel in turkey processing plants producing rolls and roasts. Coagulase-positive staphylococci were isolated from several locations in the plants and from uncooked final products, but cooked rolls did not yield *S. aureus*. Hands of workers, defeathering operations, and chill tank water contributed to increases in incidence of *S. aureus* on the raw poultry meat. Notermans *et al.* (1982) found plucking and evisceration to be the main processing steps at which contamination of carcasses with *S. aureus* occurred.

Poultry products can become vehicles of staphylococcal enterotoxins when one of the following circumstances occur: The poultry meat is cooked, killing any staphylococci which might be present, but after cooking the poultry is touched during boning, slicing, grinding, mixing, or other handling by contaminated equipment or by a person who is harboring staphylococci. The food is then either left unrefrigerated for several hours or placed in too large of a container and takes too long to cool in the refrigerator, during which time the *S. aureus* multiplies and produces enterotoxin. Epidemiological data show that improper cooling, preparing foods a day or more before serving, and cooked foods being handled by infected persons are the most frequent contributing factors that lead to outbreaks of staphylococcal food poisoning (Bryan, 1980). Control of staphylococcal intoxication must be emphasized at places where foods are prepared (food processing plants), food service establishments, and homes (Bryan, 1968a).

In the 1981 Centers for Disease Control's Annual Summary of Foodborne Diseases there were a total of 44 outbreaks, with 2934 cases of *S. aureus* food poisoning reported. In 4 of these outbreaks, with 492 cases, turkey was the implicated food, and in 3 of the outbreaks, with 55 cases, chicken was the implicated food (Centers for Disease Control, 1983).

VI. YERSINIA

The name *Yersinia* was first established in 1944 by Vanlogham in honor of A. J. E. Yersin who first isolated the causative agent of plague in 1894. The genus *Yersinia* is comprised of seven species formerly classified in the *Pasteurella* family (Mollaret and Thal, 1974) and includes *Y. pestis, Y. pseudotuberculosis, Y. ruckerii, Y. enterocolitica, Y. intermedia, Y. frederiksenii,* and *Y. kristensenii*.

From the food microbiologists point of view, the main concern with this genus is with the organism *Y. enterocolitica. Yersinia enterocolitica* is a gram-negative, facultative anaerobic rod, 1.0–3.5 by 0.5–1.5 μm in size, arranged singly or in short chains or heaps, peritrichously flagellated at 25°C, but not at 37°C (Nilehn, 1969); temperature-dependent motility as well other temperature-related phenomena are characteristic of this microbial species. It is one of the few human pathogens that can grow in properly refrigerated foods (0–5°C) (Stern and Oblinger, 1980). The epidemiological knowledge concerning *Y. enterocolitica* infections in humans is still incomplete; however clinical data suggest that in the majority of cases the infection occurs via the digestive tract (Brenner *et al.*, 1977; Hurvell *et al.*, 1980). Therefore, the psychrotrophic nature of this microorganism presents a unique problem to the food industry in maintaining food safety. The presence of this bacterium in foodstuffs is potentially hazardous to human health (Feeley et al., 1976).

Yersinia enterocolitica has been isolated from a variety of animals, among the most common ones are swine, which harbor the same serotype (0:3) as the most common clinical strain associated with human disease in Europe and Japan. For this reason, swine serves as a reservoir and can be considered a source of yersiniosis (Toma and Diedrick, 1975; Christensen, 1980; Schiemann, 1980; Stern, 1982). This organism has also been isolated from sick cats (Toma, 1973; Rakvosky et al., 1973), cattle (Zen-Yoji et al., 1974; Inove and Kurose, 1975; Wooley et al., 1980), beaver, Canadian geese, racoons, Peking robins (Hacking and Sileo, 1974), hares (Rakvosky et al., 1973), chinchillas (Toma, 1973; Vandepitte et al., 1973), snails (Boltzer et al., 1976), deer, monkeys, (Otsuki et al., 1973), and chickens (Leistner et al., 1975; Norberg, 1981; De Boer et al., 1982; Bailey et al., 1984).

Leistner et al. (1975) isolated Y. enterocolitica from 29% of 121 samples of chicken meat. From 82 samples of frozen chicken obtained from retail stores in Sweden, Norberg (1981) recovered Y. enterocolitica, Campylobacter, and Salmonella from 20, 18, and 1 of the samples, respectively. De Boer et al. (1982) analyzed 108 poultry samples for the presence of Y. enterocolitica. They sampled chicken carcasses; chicken parts such as livers, stomachs, legs, wings, necks, and breasts; and also processed chicken products such as sausages, rolled roasts, and "chicken burgers." Of the 108 samples, 73 (68%) were positive for Y. enterocolitica. For 14 of the samples, they enumerated the Y. enterocolitica present with a three-tube most probable number procedure and found the level of Y. enterocolitica varied from 30 to 11,000 cells per gram, with a mean number of 1800 cells per gram of sample. Bailey et al. (1984) recovered Yersinia from 34 of 60 (57%) broiler carcasses sampled. More than one species of Yersinia were isolated from 11 carcasses, and 9 of 60 (15%) were contaminated with Y. enterocolitica.

Although yersiniosis is frequently suspected to be a foodborne disease, the linkage of Y. enterocolitica in foods and human infection is not clear. There have been only three major foodborne outbreaks of yersiniosis reported and confirmed in the United States. The first outbreak occurred in September 1976, in Oneida County, New York, affecting 220 children attending five schools. The symptoms of yersiniosis can, at times, mimic those of appendicitis, and in this outbreak, of the 36 children hospitalized, 16 had appendectomies. Chocolate milk was determined to be the vehicle of transmission (Black et al., 1978). The second major outbreak occurred in a summer camp in Liberty, Sullivan County,

New York, in July 1981, where 53% of the 455 campers and staff members had symptoms of abdominal pain and other gastrointestinal disorders. Appendectomies were performed on 5 of the 7 hospitalized camp members. The implicated organism was confirmed to be *Y. enterocolitica* serotype 0:8, and the incriminated foods were powdered milk and turkey chow mein (Shayegani et al., 1983). The third major outbreak occurred in June 1982, in three adjacent southern states, Arkansas, Mississippi, and Tennessee, in which it was estimated that several thousand people had yersiniosis. The implicated food was pasteurized white milk, and the implicated organism was *Y. entercolitica* serotype 0:13. This constituted the largest outbreak of this disease in the United States and also the first known yersiniosis outbreak in humans in the United States caused by a non–serotype 0:8 strain of *Y. enterocolitica*.

Yersinia enterocolitica is one of the two bacteria, the other being *Campylobacter jejuni*, that have been accorded the dubious distinction of being termed emerging pathogens or pathogenic bacteria of the 1980s (Swaminathan et al., 1982). There are several reasons why *Y. enterocolitica* was not previously incriminated in human gastroenteritis: (1) the lack of awareness of the presence of *Y. enterocolitica* in foods, (2) taxonomic confusion, (3) the use of inadequate media for the isolation of pathogenic serotypes (when this organism was first realized to be a foodborne pathogen, the media used for its isolation were those that had been originally developed for the isolation of *Salmonella* and *Shigella*), and (4) the inability to determine virulence.

Foods may contain both environmental (avirulent) and potentially pathogenic (clinical) strains of *Y. enterocolitica*. The effects of consuming the environmental (avirulent) strains of *Y. enterocolitica* are probably as minimal as those caused by consuming nonpathogenic coliform bacteria (Stern, 1982), while consumption of a virulent strain may cause yersiniosis, resulting in clinical symptoms of gastroenteritis. Therefore, the virulence potential of a *Y. enterocolitica* isolate should be tested. In the recent past, several tests have been suggested for the characterization of virulence of *Y. enterocolitica*. Some of these tests are (1) enterotoxin presence (Pai and Mors, 1978), (2) HeLa cell invasion (Lee et al., 1977), (3) plasmid analysis (Zink et al., 1980), (4) calcium dependency (Gemski et al., 1980), (5) autoagglutination test (Laird and Cavanaugh, 1980), (6) biochemical tests (Schiemann et al., 1981), and (7) a serological procedure (Doyle et al., 1982).

In the Centers for Disease Control's 1981 Annual Summary of Foodborne Disease, there were two outbreaks caused by *Y. enterocolitica* involving 326 cases; however, poultry was not implicated in either outbreak (Centers for Disease Control, 1983).

REFERENCES

Allred, J. N., Walker, J. W., Beal, V. C., Jr., and Germaine, F. W. 1967. A survey to determine the *Salmonella* contamination rate in livestock and poultry feeds. J. Am. Vet. Med. Assoc. 151: 1857.
Anonymous. 1984. "Proceedings of the International Symposium on Salmonella," July 19–20, New Orleans. Published by the American Association of Avian Pathologists, Univ. PA., New Bolton Center, Kenneth Square, PA.
Bailey, J. S., Cox, N. A., Del Corral, F., Thomson, J. E., and Reagan, J. O. 1984. Isolation and classification of *Yersinia* from broiler carcasses. Proceedings 81st Annual Meeting, Southern Association of Agricultural Scientists (Food Science Section) 21: 10.
Baird-Parker, A. C. 1974. The basis for the present classification of staphylococci and micrococci. Ann. N.Y. Acad. Sc: 236: 7.
Barnes, E. M. 1960. Bacteriological problems in broiler preparation and storage. Royal Soc. Health J. 80: 145.
Barnes, E. M. 1972. Food poisoning and spoilage bacteria in poultry processing. The Veterinary Record, June 24, pg. 720.
Barnes, E. M., and Impey, C. S. 1972. Some properties of the nonsporing anaerobes from poultry caeca. J. Appl. Bact. 35: 241.
Barnes, E. M., and Shrimpton, D. H. 1957. Causes of greening of uneviscerated poultry carcasses during storage. J. Appl Bact. 20: 273.
Barnes, E. M., and Shrimpton, D. H. 1968. The effect of processing and marketing procedures on the bacteriological condition and shelf life of eviscerated turkeys. Brit. Poultry Sci. 9: 243.
Bergdoll, M. S. 1979. Staphylococcal Intoxications. *In* "Food-borne Infections and Intoxications," 2nd Ed. Academic Press pp. 444.
Black, R. E., Jackson, R. J., Tsai, T., Medvesky, M., Shayegani, M., Feely, J. C., Macleod, K. E. E., and Walker, A. M. 1978. Epidemic *Yersinia enterocolitica* infection due to contaminated chocolate milk. New England J. Med. 298: 76.
Blaser, M. J. 1982. *Campylobacter jejuni* and Food. Food Technol. 36: 89.
Boltzer, R. G., Wetzler, F. T., Cowan, A. B., and Quan, T. J. 1976. Yersiniae in pond water and snails. J. Wildlife Dis. 12: 492.
Brenner, D. J., Farmer, J. J., III, Hicman, F. W., Asbury, M. A., and Steigewalt, A. G. 1977. Taxonomic and nomenclature change in *Enterobacteriaceae*. U.S. Dept. of Health and Human Services, Public Health Service Center for Disease Control, Atlanta, GA.
Bruce, D., Zochowski, W., and Ferguson, I. R. 1977. *Campylobacter* enteritis. Brit. Med J. 2: 1219.
Bryan, F. L. 1968a. What the sanitarian should know about staphylococci and salmonellae in non-dairy products. I. Staphylococci. J. Milk Food Technol. 31: 110.
Bryan, F. L. 1968b. What the sanitarian should know about staphylococci and salmonellae in non-dairy products. J. Milk Food Technol. 31: 131.
Bryan, F. L. 1976. *Staphylococcus aureus*. *In* DeFigeiredo, M. P., and Splittstoesser, D. F. (Eds.) "Food Microbiology: Public Health and Spoilage Aspects." Avi Publishing Co., Westport, CT. pp. 12.
Bryan, F. L. 1978. Factors that contribute to outbreaks of foodborne disease. J. Food Prot. 41: 816.
Bryan, F. L. 1980. Foodborne diseases in the United States associated with meat and poultry. J. Food Prot. 43: 140.
Bryan, F. L., and McKinley, T. W. 1974. Prevention of foodborne illness by time-tem-

perature. Control of thawing, cooking, chilling, and reheating turkey in school lunch kitchens. J. Milk Technol. 37: 420.
Centers for Disease Control. 1983. *Salmonella* surveilance. Annual Summary, 1984. U.S. Dept. of Health, Education and Welfare Center for Disease Control, Atlanta, GA.
Christensen, S. G. 1980. *Yersinia enterocolitica* in Danish pigs. J. Appl. Bact. 48: 377.
Cox, N. A. 1975. Isolation and identification of a genus, *Chromobacterium*, not previously found on processed poultry. Appl Microbiol. 29: 864.
Cox, N. A., and Bailey, J. S. 1985. Performance of the DMS Rapid E system with stock cultures and food isolates. Proceedings, 84th Annual Meeting, Amer. Soc. Microbiol. (In press).
Cox, N. A., and Blankenship, L. C. 1975. Comparison of rinse sampling methods for detection of salmonellae on eviscerated broiler carcasses. J. Food Sci. 40: 1333.
Cox, N. A., Mercuri, A. J., Juven, B. J., and Thomson, J. E. 1975. Enterobacteriaceae at various stages of poultry chilling. J. Food Sci. 40: 44.
Cox, N. A., Mercuri, A. J., Tanner, D. A., Carson, M. O., Thomson, J. E., and Bailey, J. S. 1978. Effectiveness of sampling methods for *Salmonella* detection on processed broilers. J. Food Prot. 41: 341.
Cox, N. A., Mercuri, A. J., Carson, M. O., and Tanner, D. A. 1979. Comparative study of Micro-ID, Minitek and conventional methods with *Enterobacteriaceae* freshly isolated from foods. J. Food Prot. 42: 735.
Cox, N. A., Bailey, J. S., and Thomson, J. E. 1983. Comparison of preenrichment to direct enrichment and the effect of pyruvate in media for recovery of salmonellae in feed. Poultry Sci. 62: 947.
da Silva, G. A. N. 1967. Incidences and characteristics of *Staphylococcus aureus* in turkey products processing plants. PhD. Thesis, Iowa State University, Ames, IA.
De Boer, E., Hartog, B. J., and Oosterom, J. 1982. Occurrences of *Yersinia enterocolitica* in poultry products. J. Food Prot. 45: 322.
Dische, F. E., and Elek, S. D. 1957. Experimental food poisoning by *Clostridium welchii*. Lancet. 2: 71.
Dixon, J. M. S., and Pooley, F. E. 1961. Salmonellae in a poultry processing plant. Mon. Bull. Minist. Health Pub. Health Lab. Serv. (G. Brit) 21: 138.
Dougherty, T. J. 1974. *Salmonella* contamination in a commercial poultry (broiler) processing operation. Poultry Sci. 53: 814.
Dowell, V. R., Jr. 1982. Salmonellosis associated with the ingestion of meats in the United States. *In* "Control of the Microbial Contamination of Foods and Feeds in International Trade; Microbial Standards and Specifications," edited by Kurata, H., and Hesseltine, C. W., Saikon Publishing Co., Ltd., Tokyo. pp. 95.
Doyle, L. P. 1944. A vibrio associated with swine dysentery, Am. J. Vet. Res. 5: 3.
Doyle, L. P. 1948. The etiology of swine dysentery. Am. J. Vet. Res. 9: 50.
Doyle, M. P., Hugdahl, M. B., Chang, M. T., and Berry, J. T. 1982. Serological relatedness of mouse-virulent *Yersinia enterocolitica*. Infect. Immun. 37: 1234.
Duitschaever, C. L. 1977. Incidence of salmonellae in retailed raw cut-up chicken. J. Food Prot. 40: 191.
Edwards, P. R., and Ewing, W. H. 1972. "Identification of *Enterobacteriaceae*," 3rd ed. Burgess Publication Co., Minneapolis, MN.
Eiden, J. J., and Dalton, H. P. 1980. An animal reservoir for *Campylobacter fetus* ss. *jejuni*. Presented at 29th Antimicrobial Agents and Chemotherapy, Am. Soc. for Microbiology, New Orleans, LA, September (Abstract 694).
Erwin, L. E. 1955. Examination of prepared poultry feeds for the presence of *Salmonella* and other enteric organisms. Poultry Sci. 34: 215.
Feely, J. C., Lee, W. H., and Morris, G. K. 1976. *Yersinia enterocolitica*. *In* Compendium of

methods for the microbiological examination of foods. Amer. Public Health Assoc. Speck, M. L. (ed.). pp.351.

Felsenfeld, O., Young, V. M., and Yoshimura, T. 1950. A survey of salmonella organisms in market meat, eggs, and milk. J. Am. Vet. Med. Assn. 116: 17.

Gangarosa, E. J. 1978. What have we learned from 15 years of *Salmonella* surveilance? *In* "National Salmonellosis Seminar." Washington, D. C.

Gartner, A. 1888. Pathogene and saparophytische Bacterien in ihrem Verthaltniss zum wasser, insonderlich zum Trinkwaseer. Cor. Bl. d. Allg. Arztl. XVII, 233.

Gemski, P., Lazere, J. R., and Casey, T. 1980. Plasmid associated with pathogenicity and calcium dependency of *Yersinia enterocolitica*. Infect. Immun. 27: 682.

Genigeorgis, C., and Sadler, W. W. 1966. Characterization of strains of *Staphylococcus aureus* isolated from livers of commercially slaughtered poultry. Poultry Sci. 45: 973.

Gibbs, P. A., and Patterson, J. T. 1978. Biochemical characteristics and enterotoxigenicity of *Staphylococcus aureus* strains isolated from poultry. J. Appl. Bact. 44: 57.

Gibbs, P. A., Patterson, J. T., and Thompson, J. K. 1978a. Characterization of poultry isolates of *Staphylococcus aureus* by a new set of poultry phages. J. Appl. Bact. 44: 387.

Gibbs, P. A., Patterson, J. T., and Thompson, J. K. 1978b. The distribution of *Staphylococcus aureus* in a poultry processing plant. J. Appl. Bact. 44: 401.

Gill, C. O., and Harris, L. M. 1984. Hamburgers and broiler chickens as potential sources of human *Campylobacter enteritis*. J. Food Prot. 47: 96.

Glezen, W. P., Hines, M. P., Kerbaugh, M., Green, M. E., and Koomer, J. 1966. *Salmonella* in two poultry processing plants. J. Amer. Vet. Med. Assoc. plants. J. Amer. Vet. Med. Assoc. 148: 550.

Gordon, R. F., and Tucker, J. F. 1965. The epizootiology of *Salmonella menstron* infection of fowls and the effect of feeding poultry food artificially infested with *Salmonella*. Br. Poultry Sci. 6: 251.

Goren, E., and de Jong, W. A. 1980. *Campylobacter fetus* subspecies *jejuni* in chickens. Tijdschr. Diergeneesk 105: 724.

Grant, I. H., Richardson, N. J., and Bokkenheuser, V. D. 1980. Broiler chickens as a potential source of *Campylobacter* infections in humans. J. Clin. Microbiol. 11: 508.

Guthertz, L. S., Fauin, J. T., Spicer, D., and Fowler, J. L. 1976. Microbiology of fresh comminuted turkey meat. J. Milk Food Technol. 39: 823.

Hacking, M. A., and Sileo, L. 1974. *Yersinia enterocolitica* and *Yersinia pseudotuberculosis* from wildlife in Ontario. J. Wildlife Dis. 10: 452.

Hacking, W. C., Mitchell, W. R., and Carlson, H. C. 1978. *Salmonella* investigation in an Ontario feed mill. Can. J. Comp. Med. 42: 400.

Harvey, J., Patterson, J. T., and Gibbs, P. A. 1982. Enterotoxigeniciy of *Staphylococcus aureus* strains isolated from poultry: raw poultry carcasses as a potential food poisoning hazard. J. Appl. Bact. 52: 251.

Hauschild, A. H. W., Nilo, L., and Dorward, W. J. 1968. *Clostridium perfringens* type A infection of legated intestinal loops in lambs. Appl. Microbiol. 16: 1235.

Hobbs, B. C. 1961. Public health significance of *Salmonella* carriers in livestock and birds. J. Appl Bact. 24: 340.

Hobbs, B. C. 1971. *In* "Poultry Disease and World Economy." Gordon, R. F., and Freeman, B. M. (eds.), Edinburgh, British Poultry Sci. 65.

Hobbs, B. C., Smith, M. E., Oakley, C. L., Warrack, G. H., and Cruickshank, J. C. 1953. *Clostridium welchii* food poisoning. J. Hyg. 51: 75.

Hafsad, M. S. 1956. Hepatitis in chickens—a report of progress in veterinary medicine research. Vet. Med. Res. Inst. Iowa State College, Ames, IA.

Hurvell, B., Danielsson-Tham, M. L., and Olsson, E. 1980. Zoonotic aspects of *Yersinia enterocolitica* with special reference to its ability to grow at low temperature. *In* "Psy-

chrotropic Microorganisms in Spoilage and Pathogenicity." Roberts, T. A., Hobbs, G., Christian, J. B. H., and Skovgaard, N. (eds.) Academic Press, p.393.
Inove, M., and Kurose, M. 1975. Isolation of *Yersinia enterocolitica* from cow's intestinal contents and beef meat. Jap. J. Vet. Sci. 37: 91.
Isa, J. M., Boycott, B. R., and Broughton, E. 1963. A survey of *Salmonella* contamination in animal feed and feed constituents. Can. Vet. J. 11: 41.
Johnston, R. W. 1982. *Salmonella* in meat and poultry in the United States and in imports to the United States. *In* "Control of the Microbial Contamination of Foods and Feeds in International Trade: Microbial Standards and Specifications. Kurata, H., and Hesseltine, C. W. (eds.), Saikon Publishing Co. Ltd., Tokyo, pp.149.
Jones, F. S., and Little, R. B. 1931a. The etiology of infectious diarrhea (winter scours) in cattle. J. Exp. Med. 53: 835.
Jones, F. S., and Little, R. B. 1931b. Vibrionic enteritis in calves. J. Exp. Med. 53: 845.
Jones, F. S., Orcutt, M., and Little, R. B. 1931. Vibrios (*Vibrio jejuni* N. SP.) associated with intestinal disorders of cows and calves. J. Exp. Med. 53: 863.
Karmali, M. A., and Fleming, P. C. 1979. *Campylobacter* enteritis. Can. Med. Assoc. J. 120: 1525.
Karmali, M. A., and Skirrow, M. B. 1984. Taxonomy of the genus *Campylobacter*. *In* Butzler, Jean-Paul (ed.) "*Campylobacter* Infection in Man and Animals." CRC Press, Boca Raton, FL.
Kiggings, E. M., and Plastridge, W. N. 1956. Effects of gaseous environment on growth and catalase content of *Vibrio fetus* cultures of bovine origin. J. Bact. 72: 397.
Kinde, H., Genigeorgis, C. A., and Pappaioanou, M. 1984. Prevalence of *Campylobacter jejuni* in chicken wings. Appl. & Environ. Microb. 45: 1116.
Kraft, A. A. 1971. Microbiology of poultry products. J. Milk Food Technol. 34: 23.
Ladiges, W. C., and Foster, J. F. 1974. Incidence of *salmonella* in beef and chicken. J. Milk Food Technol. 37: 213.
Laird, W. J., and Cavanaugh, D. C. 1980. Correlation of autoagglutination and virulence in yersiniae. J. Clin. Microbiol. 11: 430.
Lee, W. H., McGrath, P. P., Carter, P. H., and Eide, E. L. 1977. The ability of some *Yersinia enterocolitica* strains to invade HeLa cells. Can. J. Microbiol. 23: 1714.
Leistner, L., Hechelmann, H., Kashiwazaki, M., and Albertz, R. 1975. Nachwies von *Yersinia enterocolitica* in Faeces und Fleisch von Schweinen, Rinden und Geflugel. Fleischwirtschaft 11: 1599.
Leuchtefeld, N. W., and Wang, W-LL 1981. Culture survey of *Campylobacter fetus* subsp *jejuni* at a turkey processing plant. J. Clin. Microbiol. 13: 266.
Lillard, H. S. 1971. Occurrance of *C. perfringens* in broiler processing and further processing operations. J. Food Sci. 36: 1008.
Lillard, H. S. 1973. Contamination of blood system and edible parts of poultry with *Clostridium perfringens* during water scalding. J. Food Sci. 38: 151.
Lovett, J., Hunt, J., and Park, C. E. 1981. Incidence of *Campylobacter fetus* ss *jejuni* in southern Ohio fresh whole (eviscerated) market chickens. Presented at 81st Ann. Meeting, Am. Soc. for Microbiology, Dallas, TX, March (Abstract P11).
MacKenzie, M. A., and Bains, B. S. 1976. Dissemination of *Salmonella* serotypes from raw feed ingredients to chicken carcasses. Poultry Sci. 55: 957.
Magwood, S. E., Rigby, C., and Fung, R. H. J. 1967. *Salmonella* contamination of the product and environment of selected Canadian chicken processing plants. Can. J. Comp. Med. 31: 88.
May, K. N. 1974. Changes in microbial numbers during final washing and chilling of commercially slaughtered broilers. Poultry Sci. 53: 1282.
McBride, G. B., Skura, B. J., Yada, R. Y., and Bowmer, E. J. 1980. Relationship between

incidence of *Salmonella* contamination among pre-scalded eviscerated and post-chilled chicken in a poultry processing plant. J. Food Prot. 43: 538.

McClung, L. S. 1945. Human food poisoning due to growth of *Clostridium perfringens* in freshly cooked chicken: preliminary note. J. Bact. 50: 229.

McFadyean, J., and Stockman, S. 1913. Report of the departmental committee appointed by the Board of Agriculture and Fisheries to enquire into epizootic abortion. His Majesty's Stationary Office, London.

Mead, G. C., and Impey, C. S. 1970. The distribution of clostridia in poultry processing plants. Br. Poultry Sci. 11: 407.

Mollaret, H. H., and Thal, E. 1974. *Yersinia*. In "Bergey's Manual of Determinative Bacteriology," 8th Ed. Buchanan, R. E., and Gibbons, N. E. (eds.) The Williams and Wilkins Col, Baltimore, MD. pp.330.

Morehouse, L. G., and Wedman, E. E. 1961. *Salmonella* and other disease-producing organisms in animal by-products—A survey. J. Am. Vet. Med. Assoc. 139: 989.

Morris, G. K., and Ayres, J. C. 1960. Incidence of *salmonella* on commercially processed poultry. Poultry Sci. 39: 1131.

Morris, G. K., and Wells, J. G. 1970. *Salmonella* contamination in a poultry processing plant. Appl. Microbiol. 19: 795.

Morris, G. K., McMurray, B. L., Galton, M. M., and Wells, J. G. 1969. A study of the dissemination of *Salmonella* in commercial broiler chicken operation. Am. J. Vet. Res. 30: 1413.

Mulder, R. W. A. W., Notermans, S., and Kampelmacher, E. H. 1977. Inactivation of *Salmonella* on a chilled and deep frozen broiler carcasses by irradiation. J. Appl. Bact. 42: 179.

Murphy, C. D. 1969. Detection of low level salmonellae contamination of feed and potential infectivity for poultry. M.S. Thesis, Texas A&M Univ., College Station, TX.

Nilehn, B. 1969. Electron microscopic studies on flageallation in different strains of *Yersinia enterocolitica*, Acta Path. Microbiol. Scand 77: 527.

Norberg, P. 1981. Enteropathogenic bacteria in frozen chicken. Appl. Envir. Microbiol. 42: 32.

Notermans, S., Schothorst, M. van, Leusden, F. M. van, and Kampelmacher, E. H. 1975. Quantitative studies for the presence of salmonellae in deep frozen broiler chickens. Tijdschr. Diergeneesk 100: 648.

Notermans, S., Dufrenne, J., and van Leeuwen, W. J. 1982. Contamination of broiler chickens by *Staphylococcus aureus* during processing: incidence and origin. J. Appl. Bact. 52: 275.

Otsuki, Koichi, Tsubokyra, Misao, and Itagak, Keizabaro, Itagaki, 1973. Isolation of *Yersinia enterocolitica* from monkeys and deers. Jap. J. Vet. Sci. 35: 447.

Pai, C. H., and Mors, V. 1978. Production of enterotoxin by *Yersinia enterocolitica*. Infect. and Immun. 19: 908.

Parks, C. E., and Stankiewicz, Z. K. 1981. Incidence of *Campylobacter jejuni* in fresh eviscerated whole market chickens. Can. J. Microbiol. 27: 841.

Patterson, J. T. 1969. Bacterial contamination of processed poultry. Br. Poultry Sci. 10: 89.

Peckman, M. C. 1958. Avian vibrionic hepatitis. Avian Dis. 2: 348.

Prescott, J. F., and Bruin-Mosch, C. W. 1981. Carriage of *Campylobacter jejuni* in healthy and diarrheic animals. Am. J. Vet. Res. 1: 164.

Rakovsky, J., Pauckova, V., and Aldova 1973. Human *Yersinia enterocolitica* infections in Czechoslovakia. Contributions to Microbiology and Immunology (*Yersinia, Pasteurella, Francisella*) 2: 93.

Ribeiro, C. D. 1978. *Campylobacter enteritis*. Lancet. 2: 270.

Sadler, W. W., and Corstvet, R. E. 1965. Second survey of market poultry for *Salmonella* infection. Appl. Microbiol. 13: 348.
Sadler, W. W., Yamamoto, R., Adler, H. E., and Stewart, G. F. 1961. Survey of market poultry for *Salmonella* infection. Appl. Microbiol. 9: 72.
Salmon, D. E., and Smith, T. 1885. Report on swine plague. U.S. Bureau of Animal Industries, 2nd Annual report, U.S. Government Printing Office. Washington, D.C.
Salzer, R. H., Kraft, A. A., and Ayres, J. C. 1964. Bacteria associated with giblets of commercially processed turkeys. Poultry Sci. 43: 934.
Schiemann, D. A. 1980. Isolation of toxigenic *Yersinia enterocolitica* from retail pork products. J. Food Prot. 43: 360.
Schiemann, D. A., Devenish, J. A., and Toma, S. 1981. Characteristics of virulence in human isolates of *Yersinia enterocolitica*. Infect. Immun. 32: 400.
Schleifer, J. H., Juven, B. J., Beard, C. W., and Cox, N. A. 1984. The suspectibility of chicks to *Salmonella montivideo* in artificially contaminated poultry feed. Avian Dis. 28: 497.
Sebald, M., and Veron, M. 1963. Teneur en bases de l'ADN et classification des vibrions. Ann. Inst. Pasteur. 105: 897.
Shanker, S., Rosenfield, J. A., Davey, G. R., and Sorrell, T. C. 1982. *Campylobacter jejuni*: Incidence in processed broilers and biotype distribution in human and broiler isolates. Appl. Environ. Microbiol. 43: 1219.
Shayegani, M., Morse, O., Deforge, I., Root, T., Malberg-Parons, L, and Maupins, P. S. 1983. Microbiology of a major foodborne outbreak of gastroenteritis caused by *Yersinia enterocolitica* serotype 0:8. J. Clin. Microbiol. 17: 35.
Simmons, N. A., and Gibbs, F. J. 1979. *Campylobacter* spp in oven-ready poultry. J. Infect. 1: 159.
Smibert, R. M. 1978. The genus *Campylobacter*. Ann Rev. Microbiol. 32: 673.
Smith, T. 1918. Spirilla associated with disease of the fetal membranes in cattle. J. Exp. Med. 28: 701.
Smith, T., and Taylor, M. S. 1919. Morphology and biology of *Vibrio fetus*. J. Exp. Med. 30: 299.
Smith, M. V., and Muldoon, P. J. 1974. *Campylobacter fetus* subspecies *jejuni* (*Vibrio fetus*) from commercially processed poultry. App. Microbiol. 27: 995.
Stern, N. J. 1982. *Yersinia enterocolitica*: Recovery from foods and virulence characterization. Food Technol. 36: 84.
Stern, N. J., and Oblinger, J. L. 1980. Recovery of *Yersinia enterocolitica* from surfaces of inoculated hearts and livers. J. Food Prot. 43: 706.
Sun, M. 1984. New study adds to antibiotics debate. Science 226: 818.
Surkiewicz, B. F., Johnston, R. W., Moran, A. B., and Krumm, G. W. 1969. A bacteriological survey of chicken eviscerating plants. Food Technol. 23: 80.
Svedhem, A., and Kaijser, B. 1980. *Campylobacter fetus* subspecies *jejuni*: A common cause of diarrhea in Sweden. J. Infect. Dis. 142: 353.
Swaminathan, B., Harmon, M. C., and Mehlman, I. J., 1982. *Yersinia enterocolitica*, A Review. J. Appl. Bact. 52: 151.
Thompson, J. K., Gibbs, P. A., and Patterson, J. T. 1980. *Staphylococcus aureus* in commercial laying flocks: Incidence and characteristics of strains isolated from chicks, pullets, and hens in an integrated commercial enterprise. Br. Poultry Sci. 21: 315.
Toma, S. 1973. Survey of the incidence of *Yersinia enterocolitica* in the Province of Ontario. Can. J. Pub. Health. 64: 477.
Toma, S., and Diedrick, V. R. 1975. Isolation of *Yersinia enterocolitica* from swine. J. Clin. Microbiol. 2: 478.
Vandepitte, G., Wauters, G., and Isebaert, A. 1973. Epidemiology of *Yersinia enterocolitica*

infections in Belgium. Contributions to Microbiology and Immunology. (*Yersinia, Pasteurella,* and *Francisella*) 2: 111.

Vinzent, R., Dumas, J., and Picard, N. 1947. Septicemine grave au cours de la grossesse, due a un vibrion. Avortmentconsecutif. Bull. Acad. Nat. Med. 131: 90.

Walker, H. W. 1975. Foodborne illness from *Clostridium perfringens*. Crit. Rev. Food Sci. Nutr. 7: 71.

Walker, H. W., and Ayres, J. C. 1956. Incidence and kinds of microorganisms associated with commercially dressed poultry. Appl. Microbiol. 4: 345.

Walker, H. W., and Ayres, J. C. 1959. Microorganisms associated with commercial processed turkeys. Poultry Sci. 38: 1351.

Wempe, J. M., Genigeogis, C. A., Farver, T. B., and Yusufu, H. I. 1983. Appl. Environ. Microbiol. 45: 355.

Wilder, A. N., and MacCready, R. A. 1966. Isolation of *Salmonella* from poultry, poultry products and poultry processing plants in Massachusetts. New Engl. J. Med. 274: 1453.

Wilson, E., Paffenbarger, R. S., Foter, M. J., and Lewis, K. H. 1961. Prevalence of salmonellae in meat and poultry products. J. Infect. Dis. 109: 166.

Woodburn, M. 1964. Incidence of salmonellae in dressed broiler-fryer chickens. Appl Microbiol. 12: 492.

Wooley, R. E., Shotts, E. B., Jr., and McConnell, J. W. 1980. Isolation of *Yersinia enterocolitica* from selected animal species. Am. J. Vet. Res. 41: 1667.

Zen-Yoji, H., Sakai, S., Maruyama, T., and Yanagawa, Y., 1974. Isolation of *Yersinia enterocolitica* and *Yersinia pseudotuberculosis* from swine, cattle and rats at an abattoir. Japan. J. Microbiol. 18: 103.

Zink, D. L., Freely, J. C., Wells, J. G., Vanderzant, C., Vickery, J. C., Roof, W. D., and O'Donavan, G. A. 1980. Plasmid-medicated tissue invasiveness in *Yersinia enterocolitica*. Nature 283: 224.

Zottola, E. A., and Busta, F. F. 1971. Microbiological quality of further-processed turkey products. J. Food Sci. 36: 1001.

CHAPTER 11

Further-Processed Poultry Meat Products

G. W. Froning

*Department of Food Science and Technology
University of Nebraska
Lincoln, Nebraska 68583-0919*

I. INTRODUCTION

Further processing of poultry products has grown substantially in recent years. The turkey industry now markets more than 50% of its finished product in the further-processed form. Products such as turkey bologna, salami, turkey roasts, breaded steaks, turkey ham, etc., are now easily found on the supermarket shelves. Chicken is also now being marketed in a variety of forms including breaded chicken patties, nuggets, chicken sticks, chicken frankfurters, and breaded precooked pieces. The development of these new, further-processed poultry items has been nurtured by the demand for convenience foods and the competitively low price of poultry meat. Also, the excellent nutritional properties of poultry meat have acted as a stimulus for its increased consumption.

As poultry meat is fabricated into further-processed poultry meat products, new problems are presented from the microbiological standpoint. The cutting, handling, mixing, formulation, deboning, and other processing operations of poultry meat increase the possibility of microbial contamination. Changes in the nature and numbers of bacteria are potential problems that may surface as a result of further processing. Considerable progress has been made in this important area in recent years.

TABLE I

Mean Number of Bacteria per Square Centimeter on Poultry Parts and Equipment during Cutting in Commercial Processing Plants[a]

Plant no.	No. of replications	No. before cutting off		No. after cutting off		No. after packaging		No. on conveyor	
		Mean[b]	Range of rep. means	Mean[b]	Range of rep. means	Mean[b]	Range of rep. means	Mean[b]	Range of rep. means
1	5	191	61–590	2,400	360–6,600	9,400	680–31,000	95,000	24,000–166,000
2	1	2,700	—	2,400	—	10,000	—	—	No conveyor[c]
3	1	8,100	—	26,400	—	40,000	—	167,000	—
4	1	440	—	1,800	—	5,100	—	2,400	—
5	1	1,100	—	5,200	—	7,300	—	73,000	—
6	6	306	160–710	390	160–790	375	330–580	234	104–610
All plants		2,100		6,400		12,000		56,600	

[a] Reprinted with permission from May (1962). Copyright © by Institute of Food Technologists.
[b] Each mean represents 30–180 individual observations.
[c] Parts were dropped directly into shipping boxes.

II. CUTTING

Microorganisms are largely found on the surface of freshly slaughtered poultry. Cutting carcasses into their component parts increases the possibility of increased microbial growth. Today, more and more poultry is being cut up in centralized processing plants rather than the retail store. Centralized cutting and packaging of poultry has led to better control of bacterial contamination through improved quality assurance and handling practices.

May (1962) reported changes in total bacteria count on the surface of chicken parts and work surfaces during cutting and packaging of chickens in commercial plant and retail stores. Initial counts on the uncut surface of chicken broilers were low in both plants and stores: 2100 and 1500 per square centimeter, respectively. Counts increased approximately sixfold in plants (Table I) and eightfold in stores during cutting and packaging. Contamination was largely attributed to working surfaces, with some coming from handling by the workers. May's work indicated considerable differences in bacterial counts between plants, which is probably related to general quality assurance programs in the plants. Continuous spray washing of the conveyor belt with chlorinated water appeared to virtually eliminate bacterial buildup during cutting under plant conditions. It is also generally felt that hand-washing facilities adjacent to operators in plants will reduce bacterial contamination during cutting and packaging.

III. DEBONING

A. Conventional Deboning

Brant and Guion (1972) obtained swab and direct microbiological samples of equipment, clothing, turkey skin, and turkey meat from four commercial turkey deboning plants. Cutting boards and conveyor belts were found to have high total aerobic mesophilic counts in all plants, with one plant running consistently high. Skin counts were higher generally on the back of the carcass. With reference to meat samples, the scrap meat had somewhat higher aerobic mesophilic counts. Coliform counts varied among plants, with cutting boards and steel tables carry-

TABLE II
Effect of Tissue Preparation Systems on Bacterial Concentrations of Further-Processed Turkey Meat Samples[a]

Sample location	Mesophilic[b]	Psychrotrophic[b]	Coliform[c]	Salmonellae[d]	S. aureus[d]	Clostridium perfringens[c]	Shigellae[d]
Breast							
incoming carcass skin surface	2.04g	2.26f	0.6 (4/8)	3.1 (3/4)	13.1 (1/4)	ND (0/8)	ND (0/4)
precut surface	2.66e,f	2.75e,f	0.3 (3/8)	1.1 (1/4)	9.2 (2/4)	ND (0/8)	ND (0/4)
precut tissue	2.08f,g	1.94f	ND (0/8)	13.5 (2/4)	22.3 (1/4)	ND (0/8)	ND (0/4)
hand deboned tissue	3.46e	3.68e	8.9 (3/8)	383.7 (3/4)	26.4 (2/4)	6.2 (1/8)	2.2 (1/4)
shaped tissue postwrapping	3.00e	3.20e	132.0 (4/8)	2.1 (1/4)	ND (0/4)	ND (0/8)	ND (0/4)
Thigh							
precut surface	2.84e	2.89e	1.5 (5/8)	3.9 (4/4)	25.0 (2/4)	0.3 (1/8)	ND (0/4)
precut tissue	1.63f	1.57g	ND (0/8)	17.9 (2/4)	130.6 (1/4)	ND (0/8)	ND (0/4)
hand deboned tissue	1.93f	2.09f	ND (0/8)	308.6 (2/4)	ND (0/4)	ND (0/8)	ND (0/4)
MDTM[h]							
predeboner	3.47f	3.52f	53.0 (8/8)	19.4 (3/4)	61.3 (3/4)	ND (0/8)	ND (0/4)
postdeboner	4.34e	4.37e	276.0 (8/8)	50.4 (3/4)	690.3 (4/4)	7.6 (1/8)	4.1 (2/4)

[a] Reprinted from Denton and Gardner (1981) with permission. Copyright © by Institute of Food Technologists.
[b] Bacterial concentrations expressed as logarithm of number of bacteria per square centimeter of surface area or per gram of tissue. Each figure is a geometric mean of eight samples.
[c] Geometric mean of bacteria per square centimeter or per gram of tissue (number of positive samples/total samples). ND, none detected.
[d] Geometric mean of most probable number of bacteria per 1000 cm² or 100 g of tissue (number of positive samples/total samples). ND, none detected.
[e,f,g] Breast, thigh, or MDTM sample means in the same column not followed by the same letter differ significantly at $p < .5$.
[h] MDTM, mechanically deboned turkey meat.

ing a generally higher level of coliform counts. There was no consistent trend with respect to coliforms isolated on the skin and meat from various plants. Brant and Guion (1972) further concluded that staphylococci and salmonellae were infrequently present, and when present they occurred in low numbers.

Other work (Table II) reported by Denton and Gardner (1981) indicated that total mesophilic bacterial numbers from turkey breast samples increased significantly as a result of carcass handling and hand deboning. Coliforms, *Clostridium perfringens*, and *Shigella* were generally low in both breast and thigh meat as a result of hand deboning. Salmonella numbers increased in muscle tissue during the hand deboning operation. No real increases were noted in coagulase-positive *Staphlococcus aureus* numbers from hand deboned breast tissue, with none detected from thigh muscle.

From these studies, it can be concluded that hand deboning operations do increase bacterial numbers. If good quality control practices are used, however, growth of microorganisms can be minimized.

B. Hot Deboning

Studies on the effect of hot deboning on bacteriological quality of further-processed poultry products is limited. Lillard *et al.* (1984) hot deboned uneviscerated broiler carcasses and observed significantly higher total aerobic microorganisms and *Escherichia coli* than that observed from fully processed chilled broilers. When uneviscerated carcasses were spray-washed immediately after defeathering, hot-deboned carcasses were similar to fully processed chilled broilers with respect to bacterial quality. Several workers have reported microbiological data related to hot deboning of red meats. These studies on red meats should provide further clues as to the effect of hot deboning on microbiology of poultry meat products.

Lin *et al.* (1979) reported that pork sausage made from prerigor meat exhibited higher total aerobic mesophiles and lipolytic bacterial counts than that observed from postrigor muscle. Davidson *et al.* (1968) also found that hot-processed pork sausage had higher mesophilic and psychrotrophic bacterial counts as compared to postrigor tissue.

Lee *et al.* (1982) studied the bacterial flora of hot-deboned and conventionally processed beef. Both hot-deboned and conventionally processed beef were observed to have similar bacterial flora, consisting mainly of *Streptococcus* and *Lactobacillus* species after 14 days storage. Their research indicated that hot-deboned beef was as safe as conventionally processed beef under the conditions of their study.

Berry and Kotula (1982) observed slightly higher aerobic plate counts and coliform counts from hot-deboned beef strip loins as compared to the cold-deboned treatment. They concluded, however, that under the conditions of their study hot-deboning of beef does not produce any major microbial problems.

Many of these studies indicate that microbial growth may be increased by hot deboning methods. Although these differences appear to be minimal, handling of hot deboned products may require some special precautions.

C. Mechanical Deboning

The nature of the mechanical deboning process may be conducive to microbial growth in the resultant meat product. Mechanical deboning may cause considerable cellular destruction and thereby provide a suitable medium for bacterial growth.

Ostovar et al. (1971) observed that the total microbial and fecal coliform counts were relatively high in mechanically deboned poultry meat. Delaying the mechanical deboning process (holding poultry parts 5 days at 3–5°C) increased the microbial population (Table III), which

TABLE III

Changes in Total Bacterial Counts and Fecal Coliforms of Conventional and Delayed Deboned Meat (Whole Fowl) during Storage at 3 and $-15°C$[a]

Storage (days)	Conventional process		Delayed process	
	Total counts/g	MPN[b] fecal coliform/g	Total counts/g	MPN[b] fecal coliform/g
At 3°C				
0	1.49×10^5	460	3.30×10^5	460
3	1.64×10^5	460	5.15×10^6	460
6	6.35×10^6	240	6.30×10^7	460
12	2.24×10^8	240	3.15×10^8	460
At $-15°C$				
90	1.81×10^4	<10	6.32×10^4	<10
180	4.19×10^3	<10	3.58×10^4	<10
270	3.63×10^3	<10	2.89×10^4	<10

[a] Reprinted with permission from Ostovar et al. (1971). Copyright © by Institute of Food Technologists.
[b] MPN, most probable number.

was reported to be due to psychrotrophic microorganisms. The known proteolytic ability of these organisms can greatly contribute to increased spoilage. For this reason, they recommend deboning carcasses soon after they are processed. Their study further indicated that only 6 out of 54 samples were contaminated with salmonellae, while 4 showed the presence of *Clostridium perfringens*, and none were contaminated with *S. aureus*. *Pseudomonas, Achromobacter,* and *Flavobacterium* were the predominant psychrotrophic genera isolated in this investigation.

Maxcy *et al.* (1973) reported that hand deboned ground poultry and mechanically deboned poultry were comparable with counts ranging from 100,000 to 1,000,000 per gram. The microflora was noted to be diverse, with *Bacillus* sp. accounting for the greatest percentage. Total microbial load, the nature of the microflora, and proteolytic activity of the contaminants indicated that the challenge of microbial spoilage was similar to red meat products, with no apparent unique microbial quality problems.

Some studies have emphasized various approaches to reducing bacterial loads in mechanically deboned poultry products. Mast and MacNeil (1975) pasteurized mechanically deboned poultry meat at various temperatures between 59 and 71°C for 1–6 min. Using data from total aerobic plate counts, a thermal destruction curve was constructed showing the time–temperature relationships to achieve equal degrees of pasteurization. Pasteurization extended the shelf life of mechanically deboned poultry meat (MDPM) and reduced bacterial loads. Shelf life of MDPM, as measured by bacterial numbers, was improved as either temperature or time of pasteurization was increased. The most effective treatment in prolonging shelf life was a pasteurization temperature of 71°C for 6 min which produced a cooked appearance and which reduced functional properties of the resultant meat product. Shelf life was substantially improved in MDPM which had been heat-treated at 65°C for 4 min.

Raccach *et al.* (1979) observed that the shelf life of MDPM was extended 2 days by inoculating with resting cells of starter cultures of *Pediococcus cerevisiae* and *Lactobacillus plantarium*. The microbial population of the treated sample was 10% of that noted in the control group at the onset of "off-odor" in the control treatment. These lactic acid cultures did not decrease the pH of the meat. Fluorescent psychrotrophic colonies were not detected in the treated samples.

Mechanically deboned poultry meat is similar microbiologically to ground meat products. If this product is handled and processed properly, bacterial numbers can be effectively controlled. Improved engineering and handling systems have greatly improved the bacterial quality of MDPM.

D. Ground Poultry Meat

Maxcy et al. (1973) observed that bacterial counts increased rapidly in ground chicken and turkey when stored at 5°C. After 4 days of storage, microbiological numbers were in the range commonly associated with spoilage. Initial total plates counts in ground poultry were 100,000–1,000,000 per gram. Coliform counts ranged from 10 to 1000 per gram. The microflora was diverse and similar to that observed in ground red meat products.

Guthertz et al. (1977) evaluated the microbial quality of frozen ground turkey purchased from the retail store. Various species of primarily mesophilic, aerobic gram-positive and gram-negative organisms were isolated and identified. *Salmonella* spp. were isolated from 38% of the samples, while *Clostridium perfringens* were detected in 20%. Under the conditions of their study, they concluded that ground turkey meat carries a dense and diverse microbial load.

Raccach et al. (1979) observed that inoculating ground poultry meat with starter cultures of lactic acid microorganisms reduced bacterial loads and improved shelf life about 2 days over that noted from the control treatment. They further indicated that lactic acid starter cultures inhibited the growth of *Salmonella typhimurium* and *Staphylococcus aureus*.

IV. PRECOOKED POULTRY PRODUCTS

A. Roasts and Rolls

Roasts or rolls made from poultry meat have continued to be popular further-processed items. They have become particularly adaptable to the institutional market, because of portion control, convenience, and less waste. Since these products are precooked and "ready-to-eat," they present special problems from a public safety standpoint. They may be consumed without reheating and must be free of pathogenic bacteria.

Wilkinson et al. (1965) studied the relationship of the internal end point cooking temperature on the kill of pathogenic bacteria in turkey rolls cooked in an oven. When inoculated with bacteria prior to cooking, none of the *Salmonella typhimurium, S. thompson, S. enteritidus,* or *Staphylococcus aureus* survived an end point temperature of 66°C (150°F) or more. In two rolls, *Streptococcus faecalis* survived an end point temperature of 66°C, but none survived 71°C (160°F) or higher for the times involved. Wilkinson *et al.* concluded that there is no danger of "food poisoning" due to the pathogens studied when unfrozen turkey rolls were cooked at an internal temperature of 71°C in an oven set at 107°C (225°F). These workers indicated that postcooking contamination may be a problem, and care should be taken to minimize recontamination.

Bryan *et al.* (1968) investigated the effectiveness of commercial water-cooking procedures in eliminating viable salmonellae from encased turkey rolls. Salmonellae were not recovered from rolls cooked in this manner, although these organisms were isolated from rolls prior to cooking. The thermal processing procedures used by two plants of 73.9°C (165°F) for $5\frac{1}{2}$ hr or 85°C (185°F) for $4\frac{1}{2}$ hr were sufficient to kill salmonellae and to greatly reduce total aerobic and indicator organism counts. Loads of more than 1 million salmonellae per square centimeter were reduced to nondetectable levels (less than 0.3 cells per gram and per square centimeter) when internal temperatures of at least 65.6°C (150°F) were reached or when temperatures of 62.8°C (145°F) or higher were applied for 4 hr.

Lillard (1971) observed that *Clostridium perfringens* was rarely isolated by direct plating from cooked chicken rolls or ingredients, indicating either low counts, absence of the organism, or failure to grow on direct plating due to cell-damaging treatments (hot and cold temperatures). Because of this, enrichment techniques were incorporated; the organism was most frequently isolated from samples of flour and batter, and only 2.6% of 118 samples of cooked chicken product were positive.

Zottula and Busta (1971) studied microbiological quality of a number of further-processed turkey products, including roasts, rolls, turkey in gravy, and turkey with stuffing and gravy. A wide range of total numbers of microorganisms was found to be present. In the 35 raw products examined, all were found to contain coliform types, 19 contained *Escherichia coli*, 25 contained *Staphylococcus aureus*, 7 contained *Clostridium perfringens*, and only 3 samples of salmonellae were detected. The incidence of these types was lower in cooked products. Neither salmonellae nor *Escherichia coli* were found in cooked samples. In a survey of the cooked samples, coliform types were noted in 16 of 38, while 6 of 38 contained *Clostridium perfringens*, and only 1 of the 38 was positive for *Staphylococcus aureus*.

Denton and Gardner (1981) investigated three different cooking methods (oven roasting, immersion water, and smokehouse cooking processes) for comminuted turkey rolls. Results showed that the three cooking procedures were equally effective in reducing bacterial numbers. Coliform, salmonellae, *Staphylococcus aureus*, *Clostridium perfringens*, and shigellae numbers were observed to be negative after cooking. The work of Denton and Gardner indicated that recontamination could be a possibility if cooked products are repackaged after cooking. They recommend that personnel wear disposable gloves in an effort to reduce contamination. In addition, work surfaces should be periodically cleaned with soap and water and sanitized with iodine solution.

B. Pot Pies

Several workers have investigated the bacteriological quality of frozen pot pies. Canole-Parola and Ordal (1957) examined five brands of pot pies and observed considerable variation in bacterial contamination. Of the 40 unbaked pies sampled, 20 had total counts above 100,000 per gram, and 18 had coliform counts over 10 per gram. Enterococci were present in all samples tested, and coagulase-positive staphylococci were noted in 37 of 40 pies. Five *Salmonella* cultures were detected from the unbaked pies. After baking, it was indicated that baking times and temperatures recommended by the manufacturers were not sufficient to kill all pathogens. The researchers recommended improved sanitation practices and modification of cooking instructions to the consumer.

Hucker and David (1957) noted that alternate freezing and thawing of commercial frozen chicken pot pies did not increase the total flora unless the conditions of the thawed phase initiated growth. When held at 21 and 32°C, chicken pies showed the first increase in total flora after about 10 hr of storage. Hucker and David further observed that approximately 50% of the commercial frozen chicken pies contained coliform types.

Huber *et al.* (1958) compared bacteriological quality of commercial chicken pot pies and pies from plants made according to military specifications (Table IV). The extremely low counts and the absence of enterococci and coagulase-positive staphylococci were attributed to the military specifications. Military specifications required that vegetables be cooked, the temperature of the sauce be at least 160°F when added, and the crust be baked. Samples from commercial plants 1 and 2 were observed to have low bacterial counts, although enterococci and staphylococci were present.

Fanelli and Ayres (1959) observed that numbers of organisms in chicken pot pies were reduced by blast freezing. Freezing also reduced the incidence of fecal streptococci, staphylococci, coliforms, yeasts, and molds.

Peterson and Gunderson (1960) noted that definite off-flavors and

11. Further-Processed Poultry Meat Products

TABLE IV

Comparison of Bacteriological Findings on Chicken Pot Pie from Three Commercial Plants and Two Plants under Military Supervision[a]

Plant source	S.P.C. per gram	Coliform plate count per gram	Presence of Enterococci[b]	Staphylococci[b]
Commercial[c]				
plant 1	8,500	<10	+	+
	9,000	<10	+	+
	6,300	<10	+	−
	6,800	<10	+	−
	11,000	<10	+	−
plant 2	11,500	<10	+	−
	13,500	<10	+	+
	25,500	<10	+	+
	25,500	<10	+	+
	54,000	<10	+	+
plant 3	36,000,000	200	+	+
	11,700,000	110	+	+
	28,000,000	900	+	+
	38,000,000	1,000	+	+
	69,000,000	2,400	+	+
Military[d]				
plant 4	10	<10	−	−
	<10	<10	−	−
	<10	<10	−	−
	<10	<10	−	−
	<10	<10	−	−
plant 5	10	<10	−	−
	40	<10	−	−
	<10	<10	−	−
	10	<10	−	−
	25	<10	−	−

[a] Reprinted with permission from Huber et al. (1958). Copyright © by the Institute of Food Technologists.
[b] Enterococci and coagulase-positive staphylococci present in 0.1 gram.
[c] Commercial pot pies = unbaked crusts.
[d] Military pot pies = baked crusts.

aroma began in chicken pot pies on the third day of storage at 41°F. The number of bacteria at this time was about 10,000 per gram. They further isolated potent extra- and endocellular enzymes from *Pseudomonas fluorescens* which apparently initiated unfavorable flavor changes in chicken pies.

Incidence of *Clostridium perfringens* vegetative cells and spores in chicken pot pies and other frozen foods was investigated by Trakulchang and Kraft (1977). Survival of *C. perfringens* in a bacteriological medium was high, whereas vegetative cells were virtually eliminated. Thawing at low temperatures for 2 days resulted in further reduction of vegetative cells but not spores. About 50% of the frozen samples were positive for vegetative cells, and about 15% demonstrated spores of *C. perfringens*.

C. Creamed and Gravy Products

Creamed and gravy products are particularly subject to microbiological spoilage, because of the desirable media for bacterial growth. Peterson and Gunderson (1960) studied the role of psychrotrophic bacteria in spoilage of gravy-type products. They indicated that enzymes from bacteria which grow below 41°F were responsible for flavor damage in these products. Their work further showed that psychrotrophic bacteria (*Pseudomonas fluorescens*) produces extra- and endocellular enzymes that destroy the colloidal nature of gravies.

Felstenhausen et al. (1963) observed a change in the flavor of a water extract from chicken gravy when inoculated with various microorganisms (*Escherichia coli*, *Pseudomonas aeruginosa*, *P. fluorescens*, *P. geniculata*, or *Proteus vulgaris*). Detectable flavor differences were noted with inoculations of as little as 100 bacterial cells per gram when evaluated after 4 hr. Freezing reduced the number of organisms in chicken gravy to 50–60% of the original number of viable cells found after 8 weeks of storage.

Surkiewicz et al. (1973) reported that poultry and gravy products produced under good manufacturing practices contained no salmonellae, coliforms, or *Staphylococcus aureus*. Aerobic counts were also quite low. It was indicated that good sanitary practices were necessary to prevent an increase in coliforms during the day's production.

D. Battered Products

Battered poultry products are now being widely marketed. Several fast-food firms regularly offer battered fillet sandwiches, nuggets, and other forms of battered and breaded products. Microbiological safety of these products is important, particularly as it relates to precooking and final preparation at a fast-food restaurant or within the home.

Lillard (1971) investigated the incidence of *C. perfringens* using enrichment techniques in breaded broiler parts, flour, batter, and spice. The organism was frequently isolated from samples of flour and batter. Only 2.6% of 118 cooked parts were positive for *C. perfringens*.

Craven and Lillard (1974) studied the effect of reheating precooked chicken in a microwave oven on survival of *C. perfringens*. Their research indicated that microwave heating of inoculated chicken stimulated germination and outgrowth of spores when the internal temperature was 49 or 84°C. They concluded that heating precooked chicken by microwave to internal temperatures of 84°C is not adequate to eliminate the possibility of *C. perfringens* food poisoning.

V. CURED POULTRY PRODUCTS

Ayres *et al.* (1980) reviewed the role of nitrite in cured meat products. Nitrite inhibits anaerobic bacteria, including *C. botulinum*. The major role of nitrite in cured meats has been to prevent growth of germinating spores.

Sofos *et al.* (1979) studied *C. botulinum* growth and toxin production in a mechanically deboned chicken meat frankfurter product during 27°C temperature abuse. Low nitrite concentrations (10 and 40 ppm) did not influence *C. botulinum* growth and toxin production. The addition of sorbic acid (0.2%) at these nitrite levels resulted in a significant extension of the time needed for toxin to develop. Nitrite levels of 156 ppm doubled the time for botulinum toxin production. The delay in toxin production was increased fivefold when 156 ppm nitrite and 0.2% sorbic acid were combined.

In another study, Sofos *et al.* (1980) investigated the influence of pH on the effectiveness of sodium nitrite and/or sorbic acid to control *C. botulinum* growth during elevated temperature abuse (27°C) of chicken frankfurters. The effect of sorbic acid (0.2%), alone or in combination with nitrite (40, 156 ppm), was pH dependent when inhibiting spore germination, growth, and toxin production. The effect was not observed at pH values above 6.20, and it increased with decreasing pH.

More recently, Bushway *et al.* (1982) examined the effects of nitrite, sorbate, and combinations of these two ingredients plus salt (sodium chloride) on the number of aerobic microorganisms in chicken meat patties. Nitrite concentrations of 400 and 2500 ppm were effective in preventing bacterial growth in chicken white meat patties, while 2500 ppm was required to prevent growth in dark meat. The use of sorbate (0.26%) in combination with nitrite (40 ppm) was effective in reducing bacterial growth in white meat patties. It was postulated that the higher concentration of iron in dark chicken muscle may bind nitric oxide produced by nitrite, thereby reducing the nitric oxide available to interact with aerobic microorganisms.

Smoked chickens and turkeys have been a popular item in recent years. Wisniewski and Maurer (1979) reported that treatments of dry curing plus smoking and conventional brining plus roasting of turkeys resulted in low bacterial counts. Oblinger et al. (1978) observed that smoked broiler halves had a mixed microbial flora with visible mold growth when stored at 25°C. At 5 and 15°C, yeast molds were the predominate microbial flora.

VI. STUFFED PRODUCTS

Dawson et al. (1979) reviewed previous work on prestuffed poultry products. It was indicated that an internal temperature of 71°C must be reached in the stuffing, and roasting should not be interrupted before this temperature was reached. The U.S. Public Health Service recommends that meat or poultry not be held in the temperature zone between 10 and 49°C for more than 3 hr (U.S. Public Health Service, 1962). This temperature range is particularly critical in stuffed products which possess an excellent media for harboring pathogenic organisms.

For optimum safety it is generally recommended that stuffing not be cooked inside the bird. In order to reach the necessary internal temperature in the center of the stuffing, the bird itself is often overcooked and less palatable than when the stuffing is cooked separately.

REFERENCES

Ayres, J. C., Mundt, J. O., and Sandine, W. E. 1980. Microbiology of Foods. W. H. Freeman and Company, San Francisco.

Berry, B. W., and Kotula, A. W. 1982. Effects of electrical stimulation, temperature of boning and storage time on bacterial counts and shelf-life characteristics of beef cuts. J. of Food Sci. 47: 852.

Brant, A. W., and Guion, C. W. 1972. Microbiology of commercial turkey deboning. Poultry Sci. 51: 423.

Bryan, F. L., Ayres, J. C., and Kraft, A. A. 1968. Destruction of Salmonella and indicator organisms during thermal processing of turkey rolls. Poultry Sci. 47: 1966.

Bushway, A. A., Ficker, N., and Jen, C. W. 1982. Effect of nitrite and sorbate on total number of aerobic microorganisms in chicken white and dark meat patties. J. of Food Sci. 47: 858.

Canole-Parola, E., and Ordal, Z. J. 1957. A survey of the bacteriological quality of frozen pies. Food Technol. 11: 578.

Craven, S. E., and Lillard, H. S. 1974. Effect of microwave heating of precooked chicken on Clostridium perfringens. J. of Food Sci. 39: 211.

Davidson, W. D., Clipfel, R. L., Meade, R. S., and Hanson, L. E. 1968. Postmortem processing treatment on selected characteristics of ham and fresh pork sausage. Food Technol. 22: 772.

Dawson, L. E., Chipley, J. R., Cunningham, F. E., and Kraft, A. A. 1979. Incidence and control of microorganisms on poultry products. North Central Regional Research Publication No. 260.

Denton, J. H., and Gardner, F. A. 1981. Effect of further processing systems on selected microbial attributes of turkey meat products. J. of Food Sci. 47: 214.

Fanelli, M. J., and Ayres, J. C. 1959. Methods of detection and effect of freezing on the microflora of chicken pies. Food Technol. 13: 294.

Felstenhausen, V. C., Strong, H. S., and Torrie, J. H. 1963. The influence of selected bacteria upon the flavor of a precooked frozen poultry product. Food Technol. 17: 146.

Guthertz, L. A., Fruin, J. T., Okoluk, R. L., and Fowler, J. L. 1977. Microbial quality of frozen comminuted turkey meat. J. Food Sci. 42: 1344.

Huber, D. A., Zaborowski, H., and Rayman, M. M. 1958. Studies on the microbiological quality of precooked frozen meals. Food Technol. 12: 190.

Hucker, G. J., and David, E. R. 1957. The effect of alternate freezing and thawing on the total flora of frozen chicken pies. Food Technol. 11: 354.

Lee, C. Y., Fung, D. Y. C., and Kastner, C. L. 1982. Computer-assisted identification of bacteria on hot-boned and conventionally processed beef. J. of Food Sci. 47: 363.

Lillard, H. S. 1971. Occurrence of Clostridium perfringens in broiler processing and further processing operations. J. of Food Sci. 36: 1008.

Lillard, H. S., Hamm, D., and Thomson, J. E. 1984. Effect of spray-washing uneviscerated carcasses on microbiological quality of hot-boned poultry meat harvested after reduced processing. Poultry Sci. 63: 661.

Lin, H. S., Topel, D. G., and Walker, H. W. 1979. Influence of prerigor and postrigor muscle on the bacteriological and quality characteristics of pork sausage. J. of Food Sci. 44: 1055.

Mast, M. G., and MacNeil, J. H. 1975. Heat pasteurization of mechanically deboned poultry meat. Poultry Sci. 54: 1024.

Maxcy, R. B., Froning, G. W., and Hartung, T. E. 1973. Microbial quality of ground poultry meat. Poultry Sci. 52: 486.

May, K. N. 1962. Bacterial contamination during cutting and packaging chickens in processing plants and retail stores. Food Technol. 16: 89.

Oblinger, J. L., Koo, L. C., Koborger, J. A., and Janky, D. M. 1978. Changes in the microbial flora of smoked chicken during storage. Poultry Sci. 57: 123.

Ostovar, K., MacNeil, J. H., and O'Donnell, K. 1971. Poultry product quality. 5. Microbiological evaluation of mechanically deboned poultry meat. J. of Food Sci. 36: 1005.

Peterson, A. C., and Gunderson, M. F. 1960. Role of psychrophilic bacteria frozen food spoilage. Food Technol. 14: 413.

Raccach, M., Baker, R. C., Regeinstein, J. M., and Mulnix, E. J. 1979. Potential application of microbial antagonism to extended storage stability of a flesh food. J. of Food Sci. 44: 43.

Sofos, J. N., Busta, F. F., Bhothipaksa, K., and Allen, C. E. 1979. Sodium nitrite and sorbic acid effects on Clostridium botulinum toxin formation in chicken frankfurter-type emulsions. J. of Food Sci. 44: 668.

Sofos, J. N., Busta, F. F., and Allen, C. E. 1980. Influence of pH on Clostridium botulinum control by sodium nitrite and sorbic acid in chicken emulsion. J. of Food Sci. 45: 7.

Surkeiwicz, B. F., Harris, M. E., and Johnston, R. W. 1973. Bacteriology survey of frozen meat and gravy produced at establishments under federal inspection. Applied Micro. 26: 574.

Trakulchang, S. P., and Kraft, A. A. 1977. Survival of Clostridium perfringens in refrigerated and frozen meat and poultry items. J. of Food Sci. 42: 519.

U.S. Public Health Service. 1962. Food Service Sanitation Manual. U.S. Dept. of Health Education and Welfare, Washington, D.C. p. 44.

Wilkinson, R. J., Mallmann, W. L., Dawson, L. E., Irmiter, T. F., and Davidson, J. A. 1965. Effective heat processing for the destruction of pathogenic bacteria in turkey rolls. Poultry Sci. 44: 131.

Wisniewski, G. D., and Maurer, A. J. 1979. A comparison of five cure procedures for smoked turkeys. J. of Food Sci. 44: 130.

Zottula, E. A., and Busta, F. F. 1971. Microbiological quality of further-processed turkey products. J. of Food Sci. 36: 1001.

CHAPTER 12

Microbiology of Frozen Poultry Products

E. A. Sauter

Department of Animal Sciences
University of Idaho
Moscow, Idaho 83843

I. INTRODUCTION AND HISTORY OF FROZEN POULTRY PRODUCTS

Development of mechanical refrigeration during the latter part of the nineteenth century made possible the frozen poultry products industry. However, much work and experimentation were required to develop the quality products we have today.

The change from frozen, New York–dressed birds to the frozen, ready-to-cook poultry of today is little short of revolutionary. Probably, ready-to-cook turkeys are the most commonly recognized frozen poultry meat product; however, frozen fryers in various forms are available in most supermarkets. In addition, there is an ever-increasing variety of frozen, further-processed poultry items largely taken for granted by the U.S. consumer. Control of microorganisms has been an essential part of the large research and development program which has made the flavorful, wholesome, and nutritious frozen poultry products possible.

II. MICROBIOLOGY OF FROZEN POULTRY

Substantial quantities of the various poultry meat products produced are frozen as ready-to-cook products by the processor. Turkeys are probably the best example of which approximately 90% reach the consumer frozen. Consumers often purchase fryers at supermarket specials and freeze them in home freezers for consumption at a later time. Between these extremes, various quantities of different poultry products are frozen by processors and others who have access to cold storage facilities to reduce the perishability of the products. All of these procedures affect the microbiological properties of the product. While freezing may reduce numbers of bacteria, it cannot be counted upon to eliminate microbial populations. Hooshyar et al. (1982) reported that both *Salmonella infantis* and *Staphylococcus aureus* survived freezing for 28 days at $-18°C$ on frozen smoked broilers. Ayres et al. (1950) reported isolation and identification of the bacterial genera *Achromobacter, Aerobacter, Alcaligenes, Bacillus, Eberthella, Escherichia, Flavobacterium, Micrococcus, Proteus, Pseudomonas, Salmonella, Sarcina,* and *Streptococcus,* as well as the mold genera, *Cryptococcus, Oospora, Penicillium,* and *Rhodotorula* from eviscerated, cut-up poultry. A similar variety of microorganisms could be expected on frozen poultry. Elliot and Straka (1964) reported that shelf life of frozen chicken after thawing was comparable to unfrozen chicken when held at 2°C, based on total counts of psychrophiles and odor tests, and that time in frozen storage did not affect shelf life after thawing. Similar results were reported by Spencer et al. (1961), who compared the shelf life at 1.1°C of broilers frozen 2–12 weeks and thawed, with unfrozen fresh broilers. Several workers including Ayres et al. (1950), Fromm (1959), and Kinsley and Mountney (1966) compared various sampling procedures for determining the microbial load on poultry carcasses. Use of a sterile template to swab a specific area was usually considered the method of choice for total aerobic counts when speed of manipulation and variability as well as precision are taken into account. However, Mallmann et al. (1958) reported that a rinse method more accurately reflected number of surface bacteria, and Cox et al. (1976) found that blending excised skin tissue resulted in higher counts for both Enterobacteriaceae and total plate counts than swabbing. Coliform numbers are sometimes used as indicators of bacterial contamination; however, Wilkerson et al. (1961) reported that enterococci gave a better indication of initial bacterial load on frozen poultry carcasses, since they survive freezing better than coliforms.

A very large percentage of whole, ready-to-cook turkeys are merchandised as frozen turkeys. Kraft et al. (1963) compared microbial survival on turkeys during liquid immersion and air blast freezing and reported a 99% destruction of the surface organism by either method. Some *Salmonella* were recovered after 1 month of frozen storage. Hartung (1965) reported that numbers of viable bacteria on the surface of frozen turkeys declined as the time in frozen storage increased. Broken bags and storage temperature did not affect survival rate after 6 months. Pickett and Miller (1967) compared liquid nitrogen with conventional freezing of turkeys; both methods resulted in an 80% reduction of microorganisms. Reddy et al. (1978) reported about a two–log cycle reduction in microbial counts from evisceration to frozen storage.

Numerous studies concerned with the isolation of pathogenic microorganisms from frozen poultry have been conducted. Surkiewicz et al. (1969) was able to isolate salmonellae in low numbers from poultry carcasses. Sauter et al. (1978) isolated *Salmonella typhimurium* from 10 replicates of commercially frozen fryers. Cox et al. (1980), after comparing lactose preenrichment and direct enrichment, concluded that direct enrichment was adequate for detection of low levels of salmonellae on deep-chilled and frozen broilers and required 24 hr less time than preenrichment methods. Enkiri and Alford (1971) reported that there was a tendency for low-incidence strains of salmonella to be more susceptible to destruction during frozen storage, but no samples were free of salmonella after 10 weeks of frozen storage. Norberg (1981) tested 82 samples of six brands of frozen chicken and found only one sample with salmonella; however, 22 and 24% of the samples contained *Campylobacter fetus* and *Yersinia enterocolitica*, respectively.

Frozen ground turkey meat was evaluated by Guthertz et al. (1977). Results from 50 samples indicated average standard plate count of $9.4 \times 10^7/g$. Psychrotrophic counts averaged $2.4 \times 10^7/g$. *Salmonella* spp. were isolated from 38% of samples, *Staphylococcus aureus* from 42%, *S. epidermidis* from 40%, and *Micrococcus* sp. from 54% of samples. Other microorganisms isolated from a large percentage of the samples were *Corynebacteria* (54%), *Streptococcus liquefaciens* (78%), *S. faecalis* (56%), *Citrobacter freundii* (64%), *Enterobacter cloacae* (66%), *Escherichia coli* (80%), *Klebsiella pneumoniae* (66%), and *Clostridium perfringens* from 20% of the samples, as well as numerous other genera. A total of 36 species of 15 genera are listed, several of which may have public health significance. Lillard (1977) reported, in a study of commercially deboned chicken, that freezing (4–6 weeks) significantly reduced survival of both vegetative cells and spores of *C. perfringens*, but there were no differences in aerobic plate counts before and after freezing. She concluded that after frozen storage *C. perfringens* should present no special hazard when deboned chicken meat was used in food products. In contrast, Trakulchang and Kraft (1977), using inoculated samples, reported that after 42 days of storage at $-29°C$, *C. perfringens* spores had a high rate of survival, although the rate for vegetative cells was low. Cotterill et al. (1977) reported low levels of salmonella in frozen, commercially ground turkey meat, similar to levels reported by Ostovar et al. (1971) for mechanically deboned fowl and turkey meat.

Other frozen, further-processed turkey products all fabricated from deboned turkey were studied by Zottola and Busta (1971), who sampled 50 different items from eight different processors. Of 35 raw products sampled, all contained coliforms, 25 contained *S. aureus*, 19 had *E. coli*, 7 contained *C. perfringens*, while only 3 samples contained *Salmonella*. Thirty-eight cooked products were tested. No *E. coli* or *Salmonella* were found, but 16 contained other coliforms, 6 contained *Clostridium perfringens*, and 1 contained *S. aureus*. Strong and Canada (1964) studied survival of *C. perfringens* spores in frozen chicken gravy. When the organism was growing in the gravy, 4.29% survived after 90 days, and 3.69% after 180 days at $-17.7°C$. Wang et al. (1976) reported that all samples of frozen fried chicken obtained from retail stores were negative for *Salmonella*. The log range of mesophilic, psychrophilic, and salt-tolerant (Staph. 110 media) bacteria were 2.90 to 4.78, 2.74 to 4.66, and 2.84 to 4.54 per gram, respectively. Canola-Parola and Ordal (1957), after examination of precooked frozen chicken and turkey pies from five different producers, reported that recommended times and temperatures for some brands were not adequate. Plate counts for uncooked chicken pies ranged from 3.7×10^3 to 3.9×10^6 per gram. The range for turkey pies was from 1.4×10^4 to 3.3×10^6 per gram. For pies baked under simulated home conditions, the counts ranged from 9.1×10^2 to 1.8×10^6 per gram for chicken and from 1.0×10^3 to 4.4×10^4 per gram. Some salmonella-paracolon survived baking; 92% of pies had coagulase-positive *Staphylococcus aureus*, and enterococci were present in all pies. Fanelli and Ayres (1959) found that numbers of anaerobes in chicken pies were not reduced by freezing, although total counts, coliforms, streptococci, and staphylococci, as well as yeasts and molds, declined with freezing.

III. THAWING OF POULTRY PRODUCTS

Thawing at refrigerator temperatures (2–4°C) is generally recommended for frozen poultry. The Poulty and Egg National Board (1965) recommendation for frozen broiler/fryer chicken was to thaw it in the refrigerator for 12-24 hr or to leave the chicken in its original wrapping and to place it in cool running water from $\frac{1}{2}$ to 2 hr. Slow thawing is important for frozen turkeys; 1–2 days should be allowed to complete defrosting prior to cooking. When thawing the numerous, frozen, further-processed poultry products, one should follow the instructions on the label as to defrosting time and procedures. Some of these items may be cooked directly from the frozen state. Iacano et al. (1955) compared roasting of frozen and thawed turkeys and reported that more time was required for the frozen product to pass through the temperature range at which bacteria could multiply. However, the time was not long enough to permit rabid buildup of microorganisms. Cornforth et al. (1982) compared six methods for roasting turkeys from the frozen state and reported that when turkeys were cooked to a temperature of 71.1°C, all methods essentially eliminated vegetative bacterial cells.

Hartung et al. (1966) studied thawing temperatures of turkeys in relation to further processing and reported that thawing in cold water at 10°C or at 7°C in a cold room did not affect the microbial load of the finished product. However, there was a significant increase in bacterial load of the raw product, with a trend toward larger numbers of bacteria in cooked products when thawing was done at 25°C. Olson et al. (1981) reported that five rapid freeze–rapid thaw cycles resulted in a 99% reduction of *Salmonella typhimurium* cells in 0.1% buffered peptone, with 75% of the surviving cells being injured. The same treatment resulted in a 90% reduction of *S. typhimurium* on chicken wings.

IV. MICROBIAL BUILDUP AFTER THAWING

Increase in microbial numbers after thawing depends primarily on the holding temperature. Elliott and Straka (1964) reported that the shelf life at 2°C for thawed, minced poultry meat was about the same as for similar, unfrozen meat. Maxey et al. (1973) reported that at 5°C there was a 2-day lag phase after thawing of frozen ground poultry meat before rapid bacterial buildup began. Gram-negative bacteria other than coliforms produced most of the microbial buildup at 5°C. Sauter et al. (1978) compared the microbial buildup at 3.3°C on fresh and thawed frozen fryers. Total aerobic plate count increased from 2.2×10^4 to 7.5×10^7 for fresh fryers from 1 day through spoilage after 8 days. Similar data for thawed frozen fryers were 2.0×10^4 to 4.7×10^7 per square cm.

Recently, U.S. Department of Agriculture researchers (Stern et al., 1984) analyzed for *Campylobacter jejuni* approximately 800 fresh and frozen meat and poultry samples collected at the point of slaughter. They presented strong evidence that freezing or frozen storage was deleterious to *C. jejuni* survival. They isolated the organism from 12.1% of the fresh samples and 2.3% of the frozen samples.

REFERENCES

Ayres, J. C., Ogilvy, W. S., and Stewart, G. F. 1950. Post-mortem changes in stored meats. I. Microorganisms associated with development of slime on eviscerated cut-up poultry. Food Technol 4: 199.
Canola-Parola, E., and Ordal, Z. J. 1957. A survey of the bacteriological quality of frozen poultry pies. Food Technol. 11: 578.
Cornforth, D. P., Brennand, C. P., Brown, R. J., and Godfrey, D. 1982. Evaluation of various methods for roasting frozen turkeys. J. Food Sci. 47: 1108.
Cotterill, O. J., Glauert, H. P., and Russell, W. D. 1977. Microbial counts and thermal resistance of *Salmonella oranienburg* in ground turkey meat. Poultry Sci. 56: 1889.
Cox, N. A., Mercuri, A. J., Thomson, J. E., and Chew, V. 1976. Swab and excised tissue sampling for total and *Enterobacteriaceae* counts of fresh and surface frozen broiler skin. Poultry Sci. 55: 2405.
Cox, N. A., Bailey, J. S., Thomson, J. E., and Carson, M. O. 1980. Lactose pre-enrichment versus direct enrichment for recovering Salmonella from deep-chilled broilers and frozen meat products. Poultry Sci. 59: 2431.
Elliott, R. P., and Straka, R. P. 1964. Rate of microbial deterioration of chicken meat at 2°C after freezing and thawing. Poultry Sci. 43: 81.
Enkiri, N., and Alford, J. A. 1971. Relationship of frequency of isolation of *Salmonella* to their resistance to drying and freezing. Appl. Microbiol. 21: 381.
Fanelli, M. J., and Ayres, J. C. 1959. Methods of detection and effect of freezing on the microflora of chicken pies. Food Technol 13: 299.
Fromm, D. 1959. An evaluation of techniques commonly used to quantitatively determine the bacterial population on chicken carcasses. Poultry Sci. 38: 887.
Guthertz, L. S., Fruin, J. T., Okoluk, R. L., and Fowler, J. L. 1977. Microbial quality of frozen comminuted turkey meat. J. Food Sci. 42: 1344.
Hartung, T. E., 1965. The influence of temperature and bag breakage on the qtality of frozen turkey carcasses. Poultry Sci. 44: 459.
Hartung, T. E., Olson, W. E., and Moreng, R. E. 1966. Quality characteristics of turkey products as influenced by antefabrication and fabrication environment of the carcasses. Poultry Sci. 45: 612.
Hooshyar, P., Oblinger, J. L., and Janky, D. M. 1982. Survival of *Salmonella infantis* and *Staphylococcus aureus* on smoked broiler halves. Poultry Sci. 61: 79.
Iacano, P. O., Ball, C. O., and Stier, E. F. 1955. A correlation between heat penetration and the rate of bacterial growth during roasting of turkeys. Food Technol. 10: 159.
Kinsley, R. M., Jr., and Mountney, G. J. 1966. A comparison of methods used for microbiological examination of poultry carcasses. Poultry Sci. 45: 1211.
Kraft, A. A., Ayres, J. C., Weiss, K. F., Marion, W. W., Balloun, S. L., and Forsyth, R. H. 1963. Effect of method of freezing on survival of microorganisms on turkeys. Poultry Sci. 42: 128.
Lillard, H. S. 1977. Effect of freezing on incidence and levels of *Clostridium perfringens* in mechanically deboned chicken meat. Poultry Sci. 56: 2052.
Mallmann, W. L., Dawson, L. E., Sultzer, B., and Wright, H. 1958. Studies on microbiological methods for predicting shelf-life of dressed poultry. Food Technol. 12: 122.
Maxey, R. B., Froning, G. W., and Hartung, T. E. 1973. Microbial quality of ground poultry meat. Poultry Sci. 52: 486.
Norberg, P. 1981. Enteropathogenic bacteria in frozen chicken. Appl Environ. Microbiol. 42: 32.

Olson, V. M., Swaminathan, B., and Stadelman, W. J. 1981. Reduction in numbers of *Salmonella typhimurium* on poultry parts by repeated freeze-thaw treatments. J. Food Sci. 46: 1323.

Ostovar, K., MacNeil, J. H., and O'Donnell, K. 1971. Poultry product quality. 5. Microbial evaluation of mechanically deboned meat. J. Food Sci 36: 1005.

Pickett, L. D., and Miller, B. F. 1967. The effect of liquid nitrogen freezing on taste tenderness and keeping qualities of dressed turkey. Poultry Sci. 46: 1148.

Poultry and Egg National Board. 1965. Broiler-fryer the all purpose chicken. Leaflet, Chicago, Ill.

Reddy, K. V., Kraft, A. A., Hasiak, R. J., Marion, W. W., and Hotchkiss, D. K. 1978. Effect of spin chilling and freezing on bacteria on commercially processed turkeys. J. Food Sci. 43: 334.

Sauter, E. A., Petersen, C. F., and Parkinson, J. F. 1978. Microfloral comparison of fresh and thawed frozen fryers. Poultry Sci. 57: 422.

Spencer, J. V., Sauter, E. A., and Stadelman, W. J. 1961. Effect of freezing thawing and storing broilers on spoilage, flavor and bone darkening. Poultry Sci. 40: 128.

Stern, N. J., Green, S. S., Thaker, N., Krout, D. J., and Chiu, J. 1984. Recovery of *Campylobacter jejuni* from Fresh and Frozen Meat and Poultry Collected at Slaughter. J. Food Protection. 47(5): 372.

Strong, D. H., and Canada, J. C. 1964. Survival of *C. perfringens* in frozen chicken gravy. J. Food Sci. 29: 479.

Surkiewicz, B. F., Johnston, R. W., Moran, A. B., and Krumm, G. W. 1969. A bacteriological survey of chicken eviscerating plants. Food Technol. 23: 1066.

Trakulchang, S. P., and Kraft, A. A. 1977. Survival of *Clostridium perfringens* in refrigerated meat and poultry items. J. Food Sci. 42: 518.

Wang, P. L., Day, E. J., and Chen, T. C. 1976. Microbiological quality of frozen fried chicken products obtained from a retail store. Poultry Sci. 55: 1290.

Wilkerson, W. B., Ayres, J. C., and Kraft, A. A. 1961. Occurrence of enterococci and coliform organisms on fresh and stored poultry. Food Technol. 15: 286.

Zottola, E. A., and Busta, F. F. 1971. Microbiological quality of further-processed turkey products. J. Food Sci. 36: 1001.

CHAPTER **13**

Microbiology of Cooked and Canned Poultry Products

W. J. Stadelman

Department of Food Science
Purdue University
West Lafayette, Indiana 47907

I. INTRODUCTION

Thermally processed foods are packaged in containers that offer a complete barrier to gas and microbial transfer. The common packages are metal cans, glass, and retortable flexible pouches. Since poultry products are low-acid foods, government regulations covering thermal processing are covered by 21 CFR Part 113 (Anon., 1979).

A few definitions from the federal code are as follows:

I. Low acid foods are all products with a pH of more than 4.6 and high acid foods are those of pH 4.6 or less.
II. Commercial sterility:
   A. Commercial sterility of thermally processed foods means the conditions achieved:
      1. By the application of heat which renders the food free of:
         a. Microorganisms capable of reproduction in the food under normal non-refrigerated conditions of storage and distribution; and
         b. Viable microorganisms (including spores) of public health significance; or
      2. By the control of water activity and the application of heat, which renders the food free of microorganisms capable of reproducing in the food under normal nonrefrigerated conditions of storage and distribution.
   B. Commercial sterility of equipment and containers used for aseptic processing and packaging of food means the condition achieved by application of heat, chemical sterilant(s), or other appropriate treatment that renders the equipment and containers free of viable microorganisms having public health significance, as well as microorganisms of non-health significance, capable of reproducing in the food under normal nonrefrigerated conditions of storage and distribution.

III. Scheduled process means the process selected by the processor as adequate under the conditions of manufacture for a given product to achieve commercial sterility. This process may be in excess of that necessary to ensure destruction of microorganisms of public health significance, and shall be at least equivalent to the process established by a competent processing authority to achieve commercial sterility.
IV. Operating process means the process selected by the processor that equals or exceeds the minimum requirements set forth in scheduled process.
V. Critical factor means any property, characteristic, condition, aspect, or other parameter variation of which may affect the scheduled process and the attainment of commercial sterility. Equipment operation critical factor measurements shall be made and recorded at intervals not to exceed 15 minutes but frequently enough to insure adherence to the scheduled process. Scheduled processes for low-acid foods shall be established by qualified persons having expert knowledge of thermal processing requirements for low-acid foods in hermetically sealed containers and having adequate facilities for making such determinations. The type, range, and combination of variations encountered in commercial production shall be adequately provided for in establishing the scheduled process. Critical factors, e.g., minimum headspace, consistency, maximum fill-in or drained weight, a_w, etc., that may affect the scheduled process, shall be specified in the scheduled process. Acceptable scientific methods of establishing heat sterilization processes shall include, when necessary, but shall not be limited to, microbial thermal death time data, process calculations based on product heat penetration data, and inoculated packs. Calculation shall be performed according to procedures recognized by competent processing authorities. If incubation tests are necessary for process confirmation, they shall include containers from test trials and from actual commercial production runs during the period of instituting the process. The incubation tests for confirmation of the scheduled processes should include the containers from the test trials and a number of containers from each of four or more actual commercial production runs. The number of containers from actual commercial production runs should be determined on the basis of recognized scientific methods to be of a size sufficient to ensure the adequacy of the process. Complete records covering all aspects of the establishment of the process and associated incubation tests shall be prepared and shall be permanently retained by the person or organization making the determination. It is essential that a scheduled process be established for each thermally processed product in each size container.

Thermally processed poultry meat products would include soups, chicken, turkey, or game bird muscle with bones, boneless intact muscle products, and boneless comminuted items.

Baker et al. (1983) prepared chicken burgers, chicken loaves, and roasted chicken and inoculated each with known concentrations of *Salmonella typhimurium* and *Staphylococcus aureus*. The products were cooked in a microwave oven and by conventional methods (electric frying pan and oven) to an end point temperature ranging from 64 to 84°C. They found that *S. aureus* was recovered from the different products, regardless of method used for cooking. The one exception was roasted chicken cooked in the electric oven. Except for chicken burger cooked in the microwave oven, *S. typhimurium* was destroyed.

Chicken soups are generally liquid, free-flowing products, so that both convection and conduction will aid in heating the canned material during thermal processing. A rotary retort greatly speeds the rate of heating, due to continual mixing action of fluids in the rotated cans. Package sizes are usually small, so that processing schedules call for relatively short times in the rotary retorts. If stationary retorts are used, the time for heating must be extended, since the only mixing in the cans will be due to convection currents.

Examples of chicken, turkey, or game bird parts with bones would include canned whole birds, which is usually limited to chickens and game birds, or canned parts. Since large pieces of meat are involved, the retort time is lengthy, as all heating is by conduction. In the packing of such products it is essential that headspace in the cans be kept to a minimum. An underfilled can could result in inadequate heating of the solid contents surrounded by gas instead of liquid.

Comminuted meat products could be party franks, meat spreads, or pâtés. If there is liquid in the package a rotary retort could be used to reduce processing times. With a highly viscous material such as spreads, package rotation will not significantly affect processing times.

Most retorts are operated with 15 psi pressure to achieve a temperature of about 121°C. Using this temperature, heat penetration rates through the various products must be determined so as to establish the scheduled process. In establishing the scheduled process the time to achieve pressure and the times at pressure and venting procedures must be included. Also, the cooling schedule should be maintained so as to assure safe, wholesome products.

II. EQUIPMENT

Equipment needed for production of thermally processed poultry meat products would include all of the facilities for processing and further processing. In addition, package fillers, sealers, and conveyors would be needed. The heart of the system would be the retort. A recording thermometer on each retort is essential so that temperature records for each lot in a batch system, or for the time of day in a continuous retort, are maintained. These temperature records should be carefully monitored to assure that critical times and temperatures are achieved for each package.

Retorts come in many sizes and degrees of sophistication. The pressure cooker used in many households is a retort vessel. Such a pressurized vessel of a larger size could be used in commercial application. For speed in heating, commercial retorts are usually connected to a pressurized steam line. No attempt will be made here to discuss the various types of retort vessels. Desrosier and Desrosier (1977) offer such information. For details on the canning of meat products, Krokha et al. (1983) is a valuable reference. These researchers cover both sterilized and pasteurized canned meat products which would apply equally for poultry meat.

In establishing the operating process there are several factors that must be carefully considered and maintained constant by quality assurance personnel. There must be a separate operating process, primarily in duration of the process, for each package size. Beyond this, in order to maintain validity in the operation process, the viscosity, degree of agitation, and the composition of each product must be maintained on a uniform basis. With respect to composition, one of the most frequent problem areas is in particle size. The possibilities of using flexible films for heat processing and distribution of cooked meats, including poultry, is reviewed by Terlizzi et al. (1984). They indicate that such packaging is possible, but Nelson (1984) is less optimistic in that he believes that a major technological advance is necessary before flexible sterile packaging of food products with particulate materials will be widely utilized. A shift to using flexible film packaging for thermal processing or of aseptic packaging will require an entirely different set of equipment for packaging and sterilization.

III. MICROORGANISMS AND CANNED POULTRY PRODUCTS

The organism of greatest concern is *Clostridium botulinum*. The operation process is largely controlled by times and temperatures necessary for spore inactivation of the clostridia spores. Since this level of heat is well above the requirements for destruction of the vegetative cells of other foodborne pathogens, *Salmonella* spp., *Campylobacter jejuni*, and *Staphylococcus* spp. are of no significance until after the thermally processed package is opened. At this time the canned poultry products are a highly perishable food and should be kept for relatively short times in a refrigerated environment. For more information the reader is referred to Fields (1979).

REFERENCES

Anonymous. 1979. 21. Code of Federal Regulations, Part 113. Thermally processed low-acid foods packaged in hermetically sealed containers. Office of the Federal Register, National Archives and Records Administration,Washington, DC.

Baker, R. C., Poon, W., and Vadehra, D. V. 1983. Destruction of *Salmonella typhimurium* and *Staphylococcus aureus* in poultry products cooked in conventional and microwave ovens. Poultry Sci. 62: 805.

Desrosier, N. W. and Desrosier, J. N. 1977. "The Technology of Food Preservation," 4th ed. AVI Publishing Company, Westport, CT.

Fields, M. L. 1979. "Fundamentals of Food Microbiology." AVI Publishing Company, Westport, CT.

Krokha, Y. A., Oreshkin, E. P., and Ustinova, A. V. 1983. "Konservitonannye Myasoprodukty" [Preserved Meat Products.] 216 pp. Legkaya i Pishchevaya Promyshlennost, Moscow, USSR.

Nelson, P. E. 1984. Outlook for asceptic bag-in-box packaging of products for remanufacture. Food Technol. 38(3): 72.

Terlizzi, F. M., Perdue, R. R., and Young, L. L. 1984. Processing and distributing cooked meats in flexible films. Food Technol. 38(3): 67.

CHAPTER 14

Microbiology of Salted and Smoked Poultry Products

D. M. Janky*
J. L. Oblinger†
J. A. Koburger†

*Poultry Science Department
†Department of Food Science and Human Nutrition
University of Florida
Gainesville, Florida 32611

I. INTRODUCTION

Salting and/or smoking have been used to preserve meats since the beginning of recorded history. However, since the advent and widespread use of refrigeration in the more developed countries, the major function of salting and/or smoking for preservation has been replaced by that of a flavoring function. In addition, salt solutions are often injected into meat products or used as a marination medium to achieve improved juiciness and tenderness (Hale et al., 1977). The relatively long cooking time and low humidity involved in the smoking process tends to reduce product yield by drying. Brining, the infusion of tissue with sodium chloride, tends to improve water-holding capacity and is often used in conjunction with smoking to assist in preserving yield.
In sufficient quantities salting and smoking have preservation functions. Smoking preserves meat by the deposition of bacteriostatic and bacteriocidal compounds such as formaldehyde and phenols on product surfaces (Draudt, 1963). In addition, smoking reduces the water activity of the product by drying the surface. Depending on concentration, salt may also reduce the water activity of these products, thus providing a selective bacteriostatic effect (Scott, 1957). In order to preserve the mild

flavor of poultry, however, levels of salt and/or smoke applied are not sufficient to allow for unrefrigerated storage (Mast, 1978) or for extended storage at refrigeration temperatures (Oblinger et al., 1978).

II. SALTED AND SMOKED PRODUCTS

Almost any type of poultry meat product or carcasses can be brined and/or smoked (e.g., quail, turkey, broilers). Microbiologically, however, only good quality products should be used, since salting and/or smoking at today's levels will not mask an excessive population of organisms or development of off-flavors. Yields should also be considered for poultry product types. Smoked or salted meat from older birds tends to be tougher and to yield less due to increased fat loss during processing.

Most smoked products undergo a curing (brining) step prior to the actual smoking process. The curing solution may contain salt, sugar, nitrates, and nitrites. The nitrites are added to produce the cured color (pink) of smoked products and may contribute to the development of cured meat flavor (Hale et al., 1977), as well as providing a bacteriostatic effect on *Clostridium botulinum* (Perigo and Roberts, 1968). The potential for growth of *C. botulinum* on smoked poultry products cannot be ignored, as indicated by the problems experienced by the smoked fish industry (Pace and Krumbiegel, 1973). Both salt and sugar contribute to flavor development, inprove water-holding capacity and cooked yield, and lower water activity to produce a limited bacteriostatic effect. The curing solution may be pumped deep into the tissues or used as a soaking marinade, or both, depending upon the size of the bird.

Raw poultry generally serves as host to a broad spectrum of microorganisms. The most prominent spoilage organisms (*Pseudomonas, Acienetobacter*) are aerobic and psychrotrophic. Several pathogenic species are also associated with raw poultry, especially *Salmonella* and *Campylobacter*. In a curing and smoking operation, the necessary increased handling of the raw product may introduce *Staphylococcus aureus*. Aerobic Plate Counts for raw products are usually in the range of log 3 to log 4 colony-forming units per milliliter of rinse solution (100 ml of rinse per bird) or per square centimeter of skin (Oblinger et al., 1976, 1978). Coliform counts are generally lower at log 2, while yeast and mold counts are usually about log 1 (Table I). Time, temperature, and competition will dictate the predominant flora during storage.

III. SALTING

Brine soaking or curing generally has little effect on the initial microbial load of raw carcasses (Table I); however, in some instances, brine soaking has been shown to reduce both total aerobe and coliform counts by a factor of log 1 (Janky et al., 1976). This is especially true in the case of fecal coliforms (Janky et al., 1978) and is probably dependent on salt concentration and temperature.

Brine solution (5% NaCl) without added spices and other ingredients is relatively free of microorganisms until carcass addition occurs (Janky et al. 1976, 1978). Within minutes after carcass addition both total aerobic and coliform counts tend to increase to levels similar to those observed on the carcasses (Table II). These counts then decrease as brining time approaches 8 hr (Janky et al., 1976). The initial microbial load of cure solutions may be somewhat higher than those values if seasoning and spices are added. These materials would contribute their own microbial load to the solution.

There are limited data available on the microbiological effects of dry salt massaging or dry salt packing as currently practiced in some roll formation and in kosher processing, respectively. Nonhalophilic microorganisms are often adversely affected by these proccesses. The predominant organisms at spoilage of salted poultry belonged to the *Moraxella* group (Juven and Gertshovski, 1976), whereas pseudomonads were responsible for spoilage of untreated meat. This shift in spoilage organisms was attributed to the increased salt tolerance of the *Moraxella*. Dry salt "cured" hams have a relatively long shelf life at room temperatures; however, poultry products are seldom salted and dried to a level which would provide a major extension of shelf life.

TABLE I

Microbial Levels of Raw Brined and Unbrined Broilers[a]

Sample	Total aerobes (CFU/ml rinse)[b]	Coliforms (CFU/ml rinse)[b]	Yeast and molds (CFU/ml rinse)[b]
Unbrined broiler carcasses	4.6×10^3	6.9×10^2	8.3×10^1
Brined[c] broiler carcasses	4.8×10^3	1.6×10^2	1.0×10^2

[a] From Oblinger et al. (1978).
[b] CFU, colony-forming units.
[c] 5% NaCl, 2°C, 16 hr.

TABLE II

Coliform and Total Counts of Brining Solution before the Addition of Carcasses and 5 min, 8 hr, and 16 hr after the Addition of Carcasses[a]

Time of analysis	Total aerobes (CFU/ml)[b]	Coliforms (MPN)[c]
0	<10	0
5 min	1.4×10^4	4.6×10^1
8 hr	2.8×10^2	1.4×10^1
16 hr	6.0×10^2	1.6×10^1

[a] From Janky et al. (1976).
[b] CFU, colony-forming units.
[c] MPN, most probable number.

IV. SMOKING

In the United States, the cooking process is nearly always accomplished as an integral part of the smoking process. This has been termed "hot-smoke" processing as opposed to "cold-smoke" processing. In this latter process, smoke is applied at relatively low temperatures (<40°C), then the final product may be cooked as is commonly done with poultry. The major difference between the two methods is that the yield obtained with hot-smoke processing generally is lower than cold-smoke processing. Both processes should produce similiar microbiological profiles if the profiles are made after the cooking step is completed. Increasing surface temperature during smoking has been observed to improve phenol and formaldehyde deposition, thus increasing bacteriostatic and bacteriocidal effects (Heiszler et al., 1972). In addition, surface moisture and smokehouse humidity may affect deposition of phenolic compounds (Koburger et al., 1979). As one would expect, microbial counts on freshly smoked, cooked products (Table III) are generally the more heat-resistant gram-positive species and in one study have been identified as predominantly *Staphylococcus epidermidis* (Hooshyar et al., 1982). Other organisms isolated by these authors were species of *Micrococcus* and *Bacillus*. No gram-negative organisms were detected. Similar results have been reported for smoked fish (Lee and Pfeifer, 1973; Koburger et al., 1977) and would be expected in light of the nature of the basic processing technique.

TABLE III

Microbial Populations of Smoked, Cooked Broilers

Total aerobes[a] (CFU/ml rinse)	Coliforms[b] (CFU/ml rinse)	Yeast and molds[a] (CFU/ml rinse)
<10	0	<10

[a] From Hooshyar et al. (1982).
[b] From Oblinger et al. (1976).

V. STORAGE

As occurs with most fully cooked poultry products, microbial safety and shelf life of smoked cooked products depends to a large extent on postprocessing contamination during packaging and storage temperature. Handling of the product has been shown to decrease shelf life of smoked turkeys held at 2.2°C (Hale et al., 1977).

Even when good manufacturing processes are employed, smoked broiler carcass halves had only 3 weeks' shelf life at 5°C and considerably shorter shelf life at higher temperatures (Oblinger et al., 1978) (Table IV). Growth of molds has been shown to be a problem with both smoked fish (Waters, 1961) and smoked poultry (Oblinger et al., 1978) and can be controlled with the use of antimycotic agents. Oblinger et al. (1978)

TABLE IV

Total Aerobic, Coliform, Yeast, and Mold Counts for Smoked Broiler Carcass Halves Stored at 25, 15, and 5°C[a]

Storage temp. (°C)	Storage time (days)	Total aerobes (CFU/ml rinse)	Coliforms (CFU/ml rinse)	Yeast and molds (CFU/ml rinse)
25	0	<10	0	0
	3	4.9×10^6	0	3.8×10^3
	6	3.7×10^7	0	2.0×10^6
	9	—	—	Overgrowth
15	4	8.2×10^2	0	1.8×10^2
	8	2.9×10^5	0	2.2×10^4
	12	—	—	Overgrowth
5	7	1.4×10^2	0	3.0×10^1
	14	3.1×10^2	0	1.4×10^2
	21	1.8×10^3	0	8.1×10^2
	28	1.0×10^6	0	5.6×10^4

[a] From Oblinger et al. (1978).

TABLE V

Microbial Quality of Raw and Smoked Broilers[a]

		Months of storage ($-18°C$)					
Parameter	Raw	0	1	4	6	9	12
Aerobic plate count (20°C)	6.2×10^3	<10	<10	4×10^1	4×10^1	2×10^2	4×10^1
Total coliforms	3.5×10^1	<10	<10	<10	<10	<10	<10
Fecal coliforms	2.7×10^1	<10	<10	<10	<10	<10	<10
E. coli	2.7×10^1	<10	<10	<10	<10	<10	<10
S. aureus	4.3×10^0	<10	<10	<10	<10	<10	<10

[a] From Koburger et al. (1981).

suggested that the sawdust used for smoke generation might be the mold source.

Depending on the degree of postprocessing contamination, the shelf life of poultry products can be markedly affected. Gardner et al. (1980) observed a shelf life for smoked broilers of at least 28 days at 7°C storage temperature. Such shelf lives at refrigeration temperatures are probably the maximum limits and indicate that the curing and smoking of poultry has very little effect on microbial growth.

TABLE VI

Counts (Colony-Forming Units/ml) of High and Low Inoculation Levels of *Salmonella infantis* and *Staphylococcus aureus* on Smoked Broiler Halves[a]

Inoculation levels	Organisms	Days of storage at 5°C				
		0	5	10	15	25
High	S. infantis	2.2×10^6	1.1×10^6	7.4×10^5	3.7×10^5	1.5×10^5
	Staph. aureus	2.0×10^6	1.4×10^6	1.2×10^6	7.0×10^5	7.3×10^5
Low	S. infantis	9.9×10^2	2.1×10^2	2.3×10^2	1.7×10^2	4.0×10^1
	Staph. aureus	3.6×10^2	1.8×10^2	1.1×10^2	3.2×10^1	2.8×10^1

		Days of storage at 18°C				
		0	7	14	21	28
High	S. infantis	2.2×10^6	6.7×10^5	4.6×10^5	1.6×10^5	2.7×10^5
	Staph. aureus	2.0×10^6	1.2×10^6	1.1×10^6	1.0×10^6	1.2×10^6
Low	S. infantis	9.9×10^2	2.2×10^2	1.8×10^2	2.1×10^2	3.3×10^1
	Staph. aureus	3.6×10^2	1.4×10^2	9.9×10^1	1.1×10^2	8.7×10^1

[a] From Hooshyar et al. (1982).

Koburger et al. (1981) observed little change in microbial loads of smoked broilers over 12 months of storage at −18°C (Table V). These authors also found only slight changes in organoleptic and physical attributes of the smoked broilers during the 12 months of frozen storage. They concluded that smoked poultry could be held frozen for at least 1 year without encountering serious quality changes.

Safety of smoked and/or salted poultry is basically a function of storage conditions, end point cooking temperatures, and postcooking contamination. Temperatures of the salted raw product should be maintained below 4°C, since *Staphylococcus aureus* is salt tolerant and has the potential to produce a heat-stable toxin at higher temperatures. After proper cooking, postprocessing contamination with pathogens must be avoided, since neither refrigerated nor frozen storage decreases pathogens to nonharmful levels (Hooshyar et al., 1982) (Table VI).

VI. CONCLUSIONS

The nature of poultry husbandry, as well as modern processing techniques, has not resulted in a need for "hard" preservation methods as with beef, pork, or fish. Most further processing of poultry occurs because of the desire for specialty products. In addition, the sophistication of our palates today rejects most traditional preserved foods as being "too strong" regarding flavor, texture, or odor.

The use of salting and smoking of poultry, therefore, is limited to such procedures as the kosher salting of poultry to remove blood or the mild salting and smoking of turkeys at holiday time. Mild salting generally has little effect on the microflora present. Kosher salting, however, significantly reduces the population of organisms and brings about a shift in the nature of the spoilage population. Little is gained in extension of shelf life by salt treatment unless the product undergoes an additional drying step.

Smoking of poultry, on the other hand, markedly reduces the microbial level and with proper handling and refrigeration can bring about a significant extension of shelf life of up to 21 days at 5°C. Those organisms surviving as spoilage organisms on smoked poultry and capable of growth are generally gram-positive bacteria or molds. It is this visible mold growth in many cases which limits the shelf life of refrigerated smoked poultry, however, when frozen, it has an acceptable storage life of at least a year.

REFERENCES

Draudt, H. N. 1963. The meat smoking process: A review. Food Technol. 17: 1557.
Gardner, F. A., Denton, J. G., and Mellor, D. B. 1980. The effects of curing and cooking methods on the yield of smoked chicken broilers. Poultry Sci. 59: 1612. (Abstract).
Hale, K. K., Jr., Cohn, D. D,. and Stubblefield, J. D. 1977. Effects of finishing diet and cure procedures on quality of smoked poultry. Poultry Sci. 56: 211.
Heiszler, M. G., Kraft, A. A., Rey, C. R., and Rust, R. E. 1972. Effect of time and temperature of smoking in microorganisms on frankfurters. J. Food Sci. 37: 845.
Hooshyar, P., Oblinger, J. L., and Janky, D. M. 1982. Survival of *Salmonella infantis* and *Staphylococcus aureus* on smoked broiler halves. Poultry Sci. 61: 79.
Janky, D. M., Koburger, J. A., Oblinger, J. L., and Riley, P. K. 1976. Effect of salt brining and cooking procedure on tenderness and microbiology of smoked cornish game hens. Poultry Sci. 55: 761.
Janky, D. M., Arafa, A. S., Oblinger, J. L., Koburger, J. A., and Fletcher, D. D. 1978. Sensory, physical, and microbiological comparisons of brine-chilled, water chilled, and hot packaged (no chill) broilers. Poultry Sci. 57: 417.
Juven, B. J., and Gertshovski, R. 1976. The effect of salting on the microbiology of poultry meat. J. Milk Food Technol. 39: 13.
Koburger, J. A., Oblinger, J. L., and Janky, D. M. 1977. Microbiology of the smoked mullet process. Proc. Second. Ann. Trop. Subtrop. Subtrop. Fish Technol. Conf. of the Amer. 2: 165.
Koburger, J. A., Janky, D. M., and Oblinger, L. J. 1979. Some observations on the brining of mullet. Proc. Fourth Ann. Trop. Subtrop. Fish. Technol. Conf. of the Amer. 4: 128.
Koburger, J. A., Janky, D. M., and Oblinger, J. L. 1981. Quality changes during frozen storage of smoked broilers. Poultry Sci. 60: 2463.
Lee, J. S., and Pfeifer, D. K. 1973. Aerobic microbial flora of smoked salmon. J. Milk Food Technol. 36: 143.
Mast, M. G. 1978. Curing and smoking poultry products. World's Poultry Sci. J. 34: 107.
Oblinger, J. L., Janky, D. M., and Koburger, J. A. 1976. The effect of water soaking, brining, and cooking procedure on tenderness of broilers. Poultry Sci. 55: 1494.
Oblinger, J. L., Koo, L. C., Koburger, J. A., and Janky, D. M. 1978. Changes in the microbial flora of smoked chicken during storage. Poultry Sci. 57: 123.
Pace, P. J., and Krumbiegel, E. R. 1973. *Clostridium botulinum* and smoked fish production: 1963–1972. J. Milk Food Technol. 36: 42.
Perigo, J. A., and Roberts, T. A. 1968. Inhibition of clostridia by nitrite. J. Food Technol. 3: 91.
Scott, W. J. 1957. Water relations to food spoilage microorganisms. Advances Food Res. 7: 83.
Waters, M. E. 1961. Inhibition of mold on smoked mullet. Comm. Fish. Rev. 23(4): 8.

Index

A

Acetobacter, 13
Achromobacter, 195
Acinetobacter, 262
Additives, 24, 25
Aflatoxins, 17
Antibiotics, 24, 277
Antimetabolites, 24
Ascomycetes, 16

B

Bacillus, 11, 33, 74, 195, 334, 351
  B. cereus, 34, 65, 69, 74
  B. subtilis, 33
    detection, 52
    sampling, 83, 84
Bacteria, 6
  description, 7
  growth, 17
  live poultry, 193, 195
Basidiomycetes, 16
Bioassay systems, 68
Bioluminescence, 57
Botulism, 33

C

Campylobacter, 34, 293–294
  Campylobacter fetus ssp. jejuni, 34, 65
    detection, 45, 48, 60

Canned poultry, 23, 341
Carbon dioxide (CO_2), 228, 229, 281
Chemiluminescence, 57
Chickens, 219
Chillers, 2, 200, 203
Chlorination, 36, 202, 278
Citrobacter, 14, 293
Classes, 219
Clostridia, 11, 32–33, 56, 195, 294, 299
  C. botulinum, 11, 23, 33, 64, 261, 283
    detection, 51, 54, 69
  C. perfringens, 11, 29, 32, 64, 65, 200, 283, 299
    detection, 48, 62, 69
Coatings, 286, 328
Coliforms, 33, 90
  detection, 56
  Escherichia coli, 91
Colony-forming units (CFU), 20
Colorimetry, 48
Condemned poultry, 217
Contamination, 1
  Bacillus cereus, 34
  Campylobacter fetus ssp. jejuni, 35
  Clostridium perfringens, 33
  coliforms, 91
  decontamination, 2
  detection, 56, 79
  processing, 193
  Salmonella, 30
  scalding, 196
Coryneforms, 12, 34, 195

Cross-contamination, 1, 2, 193, 203
Cured poultry, 329, 349
Cutting, 319

D

Deboning, 319, 321
Dehydration, 20
Deuteromycetes, 17
Direct microscopic count, 44, 49–52
Disease, 30
  poultry, 30, 34
  sheep, 34
Disinfectants, 285
Dosimetry, 239
Drying, 20
Ducks, 220
Dye reduction, 60

E

Edwardsiella, 15
Endotoxins, 59, 64–65, 89
Enterobacter, 14, 33
Enterobacteriaceae, 93
Enterococci, 92
  enterocolitis, 37
Enterotoxins, 10, 64, 66
Equipment, 2, 343
  contamination, 2
  Campylobacter fetus ssp. *jejuni*, 36
  Clostridium perfringens, 33
  chemical treatment, 24
Erwinia, 14
Escherichia, 13, 14, 65, 195
  E. coli, 14, 64, 65, 195, 198
    contamination, 91–92
    detection, 48, 58, 59, 62, 70
    dye reduction, 60
    enumeration, 53
    source, 91
Evisceration, 31, 198
Exotoxins, 65–66, 89

F

Fermentation, 5, 8–9
Food acids, 280

Food poisoning, 10
  causes, 11, 33–34, 64
  outbreaks, 11, 29, 31, 33, 65, 89
  symptoms, 10
  toxins, 10–11
Freeze-drying, 22
Freezing, 21
Frozen products, 333
  thawing, 336
Fungi, 15–17, 95
  Ascomycetes, 16
  Basidiomycetes, 16
  Deuteromycetes, 17
  mold, 6, 15, 95
  yeast, 6, 15
Fungi imperfecti, 17

G

Gamma rays, 238, 251
Gas–liquid chromotography, 69
Geese, 221
Giblets, 2, 204
Grading Service, 218
Gram Stain, 7, 21–22, 65
Gray (Gy), 239
Growth Curve, 18
Guineas, 221

H

Hafnia, 15
Halobacterium, 13
Halogens, 25

I

Indicator organisms, 87
  coliforms, 90
  enterobacteriaceae, 93
  enterococci, 92
  fungi, 95
  total count, 88
  viruses, 96
Injured microorganisms, 76
Ionizing radiation, 23

K

Klebsiella, 14, 293

Index

L

Lactobacillus, 9, 195
Leuconostoc, 9
Limulus lysate assay, 59

M

Membrane filtration, 52
Microbacterium, 12, 34
Microbial load, 20
  CFU, 20
Microbial toxins, 64
Microcalorimetry, 56
Micrococcus, 10
Microorganisms, 5, 20–25
  incidence, 29, 30
  sources, 1, 194
Microwaves, 24
Mold, 6, 15, 95
Moraxella, 262
Morganella, 15
Most probable number, 47
Mycotoxins, 67

N

Nonionizing radiation, 23
  microwaves, 24
  ultraviolet radiation, 23

O

Ozone, 279

P

Packaging, 223
  films, 227
  function, 224
  materials, 226
  microbial control, 228
    CO_2, 228–230
  package material, 230
Pasteurization, 22, 275
Pathogens, 2, 64, 293
  Bacillus, 11
  Campylobacter, 294
  Clostridium botulinum, 64
  Clostridium perfringens, 20, 64, 299
  detection, 87
  Escherichia coli, 64
  Salmonella, 14, 20, 64, 301
  Staphylococcus aureus, 20, 31–32, 64, 305
  Yersinia, 15, 37, 307
Pediococcus, 9
Penicillium, 16
Phosphates, 265, 286
Photobacterium, 13
Phycomycetes, 16
Picking, 198
Pigeons, 221
Plate count, 44
Precooked products, 324
  battered, 328
  pot pies, 326
  roasts, 324
  rolls, 324
Preservation, 20, 275
  antibiotics, 277
  coatings, 286
  controlled atmosphere, 281
  disinfectants, 285
  drying, 20
  food acids, 280
  freeze drying, 22
  high temperature, 22, 275
  low temperature, 21
  pasteurization, 22, 275–276
  polyphosphates, 286
  radiation, 23
  salts, 282
  sanitizers, 277
Processing, 2, 31, 193
  chilling, 2, 200–204
  contamination, 1, 31, 193
  cross-contamination, 193, 203
  cutting, 319
  deboning, 319
  evisceration, 198
  giblets, 204
  picking, 198
  scalding, 195
Products, 317
  canned, 341
  cooked, 341
  cured, 329
  frozen, 333
  precooked, 324
  salted, 347, 349

Products (*Cont.*)
  smoked, 347, 350
  stuffed, 330
Propionibacterium, 12, 293
Proteus, 15
Protista, 6
Providencia, 15
Pseudomonas, 13, 33, 197

Q

Quality, 29
  measurement, 60
  indicator microorganisms, 87
  radiometry, 56
  USDA, 214

R

Radiation, 23, 235
  dosimetry, 239
  food irradiation, 247
  gamma ray, 238, 251
  high-dose, 259
    toxicity, 263
  medium-dose, 251
    nutritional aspects, 256
    toxicity, 256
  preservation, 250
  protection, 240
  radappertization, 245
  radicidation, 245, 253
  radurization, 245, 253
  regulation, 249
  shielding, 242
  sources, 236
    Cesium-137, 237, 251
    Cobalt-60, 237, 251
  X-rays, 235
Radiometry, 56
Rapid methods, 71, 75

S

Salmonella, 14, 29, 30, 64, 69, 301
  chill water, 201
  detection, 50, 56–58, 70, 73–74
  giblets, 205
  incidence, 30, 89
  processing equipment, 31
  radiation, 254
  recovery, 78
  salmonellosis, 14, 65
  sampling, 80, 81, 84
  scalding, 195
Salts, 265, 282, 349
Sampling techniques, 79
  atmospheric sampling, 86
  rinse method, 81
  RODAC plate counting, 82
  stomacher, 84
  swab/swab rinse, 79
Sanitizers, 277
  chlorination, 36, 202, 278
  ozone, 279
  quaternary ammonium compounds, 279
Scalding, 195
Serratia, 14
Shielding, 242
Shigella, 14
Shigella dysenteriae, 14
Smoked products, 347, 350
Spectrophotometry, 48
Spiral plate count, 46
Spores, 7, 21, 22, 25
Staphylococcus, 10, 31, 70, 195, 305
  detection, 51, 56, 73
  dye reduction, 60–61
  food poisoning, 10, 65
  incidence, 32
  outbreaks, 29, 31
  S. aureus, 10, 31, 64, 77, 280, 283, 305
    chill water, 201
    enumeration, 53
    giblets, 205
    recovery, 83
  toxins, 10, 32, 58
Streptococcus, 9, 92, 195

T

Thawing, 336
Total counts, 2, 88
Toxins, 10, 11, 17, 64, 89
  aflatoxins, 17
  detection, 51, 59, 65
  endotoxins, 59, 64–66, 89
  enterotoxins, 10, 51, 64, 66
  exotoxins, 65–67, 89
  mycotoxins, 67
  producers, 10, 11, 17, 64–66

Index

U

Ultraviolet radiation, 23
USDA, 213, 214
  classes, 219
  Food Safety and Inspection Service (FSIS), 215
    condemned, 217
    grading service, 218

V

Viruses, 6, 22, 96
  detection, 58
  indicator organism, 96

X

X-Rays, 235

Y

Yeast, 6, 15
  Candida, 16
  detection, 49, 55
  Saccaromyces, 16
Yersinia, 15, 37, 307
  sources, 37
  Yersinia enterocolitica, 37, 65, 70
    detection, 51, 58